MODERN CERAMIC ENGINEERING

MANUFACTURING ENGINEERING
AND MATERIALS PROCESSING

A Series of Reference Books and Textbooks

SERIES EDITORS

Geoffrey Boothroyd

Department of Mechanical Engineering
University of Massachusetts
Amherst, Massachusetts

George E. Dieter

Dean, College of Engineering
University of Maryland
College Park, Maryland

MODERN CERAMIC ENGINEERING :

Properties, Processing, and Use in Design

David W. Richerson
Garrett Turbine Engine Company
Phoenix, Arizona

Keywords

1 - Ceramics
2. Design
3. Mechanical
       engineering
4. Mechanical properties
5. Physical properties
6. Production

MARCEL DEKKER, INC.   New York and Basel

16.9.87

Library of Congress Cataloging in Publication Data

Richerson, David W., [date]
    Modern ceramic engineering.

    (Manufacturing engineering and materials process-
ing ; 8)
    Includes index.
    1. Ceramics. I. Title. II. Series.
TP807.R53   1982   666   82-9004
ISBN 0-8247-1843-7   AACR2

MARCEL DEKKER, INC.
270 Madison Avenue, New York, New York 10016

Current printing (last digit):
10 9 8 7 6 5 4

PRINTED IN THE UNITED STATES OF AMERICA

Ceramic materials have become increasingly important in modern industrial and consumer technology, yet most engineers and technologists receive little or no training in ceramics and are unprepared to take advantage of their unique properties. This book was prepared to provide engineers, students, teachers, and technicians with an introduction to the structure, properties, processing, design concepts, and applications of advanced ceramics. Emphasis is on developing an understanding of why ceramics are different from metals and organics and then applying this understanding to optimum material selection. Many specific applications of advanced ceramics are discussed in this context. Some include heat engine components, armor, permanent magnets, phosphors, igniters, capacitor dielectrics, thermal barrier coatings, and oxygen sensors.

This book was initially begun to fill a gap as a textbook for engineering students in fields other than ceramics. However, the book evolved during preparation into a final form equally suitable as a textbook and a reference book for individuals in disciplines other than ceramics and in ceramics.

The book is divided into three parts, each of which augments the others, but which can also stand alone. The first part is entitled "Structures and Properties" and describes the source of physical, thermal, mechanical, electrical, magnetic, and optical properties in ceramics as compared to metals and organic materials. Specific ceramic materials with optimum properties for various types of applications are discussed together with potential limitations.

The second part is entitled "Processing of Ceramics" and reviews the fabrication processes used for manufacturing ceramic components. It discusses the specific steps in each process where property-limiting material flaws are likely to occur, and provides the reader with approaches for detecting these flaws and working with the ceramic fabricator to eliminate them.

The third part is entitled "Design with Ceramics" and applies the property, fabrication, and inspection principles learned in the first two sections to the selection and design of ceramic components for advanced engineering applications.

The need for a new text or reference book with a broad treatment of ceramic materials and technology was identified through preparation and teaching of a three-semester-hour ceramics course at Arizona State University and a series of half-day and three-day ceramic courses for the American Society for Metals (ASM).

The course at Arizona State University was begun in 1975 with the objective of providing students of engineering and technology and experienced engineers of nonceramic disciplines with a better understanding of ceramic materials and their unique properties and processing and design requirements. Different textbooks were tried for this course each time it was taught, including Introduction to Ceramics by Kingery, Bowen, and Uhlmann; Physical Ceramics for Engineers by Van Vlack; and Ceramics for High Performance Applications edited by Burke, Lenoe, and Katz. All three were excellent books, but none turned out to be appropriate for the objectives of the course. Introduction to Ceramics was too detailed and analytical and did not cover processing, the effects of processing on properties, machining, nondestructive inspection, failure analysis, and ceramic design approaches. Physical Ceramics for Engineers was more engineering oriented, but also did not cover the scope required by the course. Ceramics for High Performance Applications was directed largely toward gas turbine and advanced heat engine applications and was therefore also too narrow in scope. As a result, the bulk of the course at ASU was presented from notes derived from numerous sources and personal experience.

The two three-day courses taught for ASM were entitled "High-Strength Ceramics for Engineering Applications" and "Impact of Ceramics on Modern Engineering." The half-day sessions were part of a three-day course entitled "High-Temperature Structural Materials." These courses were presented to engineers and technicians from a variety of disciplines. As was the case with the ASU course, no suitable text or reference book was available. The notes were further expanded to prepare for these courses, especially in the areas of the relationships of structure and processing to properties, design approaches, failure analysis, and discussion of specific applications.

The ASU and ASM courses clearly identified the need for a new text or reference book that would provide a broad treatment of ceramic materials and technology that could be understood and implemented by engineers, technicians, students, and teachers. The resulting book was two years in writing and is based on the course notes plus considerable additional literature research.

Many organizations and individuals have supported this effort, and I would like to take this opportunity to thank as many as possible: the Arizona State University Division of Technology and the American Society for Metals (especially Nick Jessen) for sponsoring and supporting the ceramic courses; the Defense Advanced Research Projects Agency, the Air Force Materials Laboratory and the Air Force Aero-Propulsion Laboratory for sponsoring

the programs which produced many of the examples and photographs included within this book; Garrett Turbine Engine Company for providing technical data and photographs and for aiding in manuscript preparation; Greg
Brigham for preparing most of the line illustrations; Floyd Brown for his
encouragement and help in manuscript preparation; Dr. Nelson W. Hope
and Denise B. Birnbaum for help in obtaining references and data sources;
Dr. Robert Shane for his patience, encouragement, and technical and edi-
torial suggestions; and particularly to Christy F. Johnson, Judith A. Mar-
tindale, Angie F. Peters, and Michael Anne Richerson for the extensive
task of word processing in preparing initial drafts and the final manuscript.

David W. Richerson

Introduction

The objective of this book is to provide an increased understanding of ceramic technology and its practical application. The approach is non-mathematical and concentrates on the basic material and property concepts needed to provide the reader with a working knowledge of the more important ceramic materials and design considerations.

The book is divided into three parts:

Part I: Structures and Properties
Part II: Processing of Ceramics
Part III: Design with Ceramics

Part I explores the physical, thermal, and mechanical properties of ceramics and their relationship to atomic bonding, crystal structure, and microstructure. Comparison with metals and organic materials is emphasized and concepts of ceramic design begin to evolve.

Part II studies the ceramic fabrication processes, describing in detail each step from raw material selection to shape forming to quality control. Each process step is discussed in terms of its relationship to the properties and acceptability of the final ceramic component. Understanding of the way a component is fabricated can often help an engineer resolve an application problem.

Part III applies the information covered in the first two parts to the design of ceramics. Emphasis is on the differences in design approach required for ceramics as compared to metals and plastics. The importance and techniques of fracture analysis are also described in Part III. The final chapter discusses material selection for specific applications.

Ceramics are encountered in virtually every facet of industry and everyday life. An understanding of what a ceramic is and what it can do significantly broadens the scope and effectiveness of an engineer, technician, or instructor.

The following exercise will help illustrate the variety of engineering applications of ceramic materials. Prior to reading the rest of the book the reader is requested to write down his or her best estimate (based on previous experience) of which ceramic material or materials would be optimum for each of the following categories and which special properties of each ceramic makes it the best choice.

1. Sandblast nozzle
2. Insulating refractory for furnace lining
3. Seal
4. Pottery
5. High-temperature heat exchanger
6. Armor
7. Permanent magnet
8. Ceramic quench block to draw heat away from a heat source as fast as possible
9. Thermal protection material such as that required for a reentry vehicle such as the Space Shuttle
10. Transparent material requiring uniform properties in all directions (isotropic)
11. Material having high thermal expansion in one direction and low thermal expansion in the other direction
12. Material that can be formed to shape in the green state and does not change dimension during the densification or firing operation
13. Grinding media for a ball mill
14. Low-cost fibers to be mixed with organics into a composite for structural applications
15. Coating that can be applied by molten-particle-spray techniques which significantly reduces the temperature of the substrate during subsequent high-temperature exposure
16. Low-cost mold material for slip casting
17. Material to separate and protect thermocouple wires
18. Material capable of surviving extreme thermal shock
19. Substrate for electrical devices
20. Kiln furniture for a high-purity diffusion furnace for doping diodes and other electrical devices
21. Ceramic material that increases the charge storage capability of a capacitor by a factor of 1600
22. Very low density material used for insulation in high-temperature furnace construction
23. Radome
24. Gas turbine stator
25. High-temperature cement that can be applied on-site
26. Material used in cutoff and grinding wheels

Although this list represents only a small sampling of the many types of ceramics and their diverse applications, it illustrates how widely used ceramics are and how important it is that engineers be familiar with the terminology and with the types and properties of ceramic materials. To illustrate this point further, the reader is requested to answer in writing another question: What are the applications of ceramic materials within a typical home? A list of 10 should be easy; a list of 20 is more challenging, but achievable.

Contents

I
STRUCTURE AND PROPERTIES

Individuals who have not previously studied ceramics typically ask: "What is a ceramic?" or "What is the difference between a ceramic and metal?" Most people have the concept that a ceramic is brittle, has a high melting temperature, is a poor conductor of heat and electricity, and is nonmagnetic, and that a metal is ductile, is a good conductor of heat and electricity, and can be magnetic. These stereotyped viewpoints are not necessarily true for either ceramics or metals. In fact, there is no clear-cut boundary that separates ceramics into one class and metals into another. Rather, there are intermediate compounds which have some aspects typical of ceramics and some typical of metals.

The nature of a material is largely controlled by the atoms present and their bond mechanism. Chapter 1 discusses the types of atomic bonding, the resulting structures, and the atomic elements that are most likely to combine to form ceramic, metallic, and organic materials. Chapter 2 discusses the physical, electrical, magnetic, and thermal properties in terms of the atomic elements present, the bond mechanism, and the crystal structure. Chapter 3 does the same for the mechanical properties. Chapter 4 discusses time-dependent and environment-dependent properties. Chapters 2, 3, and 4 all consider the application limitations imposed by these properties for important ceramic materials.

Atomic Bonding and Crystal Structure

The properties of a material are controlled largely by the type of bonding between atoms, which in turn is controlled by the electron configuration of the atoms. Initially, we review briefly the principles of atomic structure that are needed later in the chapter for an understanding of bonding and crystal structure.

1.1 ELECTRONIC CONFIGURATION OF ATOMS

An atom can be visualized in a simplified manner as a positively charged nucleus surrounded by negatively charged electrons. The energy of the electrons varies such that specific electrons are located in specific shells around the nucleus. These are called quantum shells. Each shell is referred to by a principal quantum number n, where n = 1, 2, 3 .... The total number of electrons in a shell is $2n^2$. Thus the lowest-energy quantum shell (n = 1) has only 2 electrons and succeedingly higher energy shells have 8 (n = 2), 18 (n = 3), 32 (n = 4), and so on, electrons, respectively.

Although electrons within a quantum shell have similar energy, no two are identical. To distinguish between these electrons, shells are divided into subshells called orbitals, which describe the probability of where pairs of electrons will be within the shell with respect to the nucleus. The first quantum shell has only two electrons, both in the s orbital, with a spherical probability distribution around the nucleus at a radius of approximately 0.5 Å. These two electrons have identical energy, but opposite magnetic behavior or spin.

The second shell has eight electrons, two in s orbitals and six in p orbitals. All have higher energy than the two electrons in the first shell and are in orbitals farther from the nucleus. (For instance, the s orbitals of the second shell of lithium have a spherical probability distribution at about 3 Å radius.) The p orbitals are not spherical, but have dumbbell-shaped probability distributions along the orthogonal axes, as shown in Fig. 1.1.

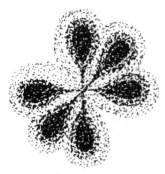

Figure 1.1 Electron probability distributions for p orbitals. The highest probability electron positions are along the orthogonal axes. Two electrons, each with opposite spin, are associated with each axis, resulting in a total of six p electrons if all the p orbitals in the shell are filled.

These p electrons have slightly higher energy than that of s electrons of the same shell and are in pairs with opposite spins along each axis when the shell is full.

The third quantum shell has d orbitals in addition to s and p orbitals. A full d orbital contains 10 electrons. The fourth and fifth shells contain f orbitals in addition to s, p, and d orbitals. A full f orbital contains 14 electrons.

A simple notation is used to show the electron configurations within shells, to show the relative energy of the electrons, and thus to show the order in which the electrons can be added or removed from an atom during bonding. This notation can best be illustrated by a few examples.

EXAMPLE 1.1 Oxygen has eight electrons and has the electron notation $1s^2 2s^2 2p^4$. The 1 and 2 preceding the s and p designate the quantum shell, the s and p designate the subshell within each quantum shell, and the superscripts designate the total number of electrons in each subshell. For oxygen the 1s and 2s subshells are both full, but the 2p subshell is two electrons short of being full.

EXAMPLE 1.2 As the atomic number and the number of electrons increase, the energy difference between electrons and between shells decreases and overlap between quantum groups occurs. For example, the 4s subshell of iron fills before the 3d subshell is full. This is shown in the electron notation by listing the order of fill of energy levels in sequence from the left of the notation to the right:

$$Fe = 1s^2 2s^2 2p^6 3s^2 3p^6 3d^6 4s^2$$

IA	IIA	IIIB	IVB	VB	VIB	VIIB	VIII	VIII	VIII	IB	IIB	IIIA	IVA	VA	VIA	VIIA	O
1 H $1s$																1 H $1s$	2 He $1s^2$
3 Li $2s$	4 Be $2s^2$											5 B $2s^2 2p$	6 C $2s^2 2p^2$	7 N $2s^2 2p^3$	8 O $2s^2 2p^4$	9 F $2s^2 2p^5$	10 Ne $2s^2 2p^6$
11 Na $2p^6 3s$	12 Mg $2p^6 3s^2$											13 Al $3s^2 3p^1$	14 Si $3s^2 3p^2$	15 P $3s^2 3p^3$	16 S $3s^2 3p^4$	17 Cl $3s^2 3p^5$	18 Ar $3s^2 3p^6$
19 K $3p^6 4s$	20 Ca $3p^6 4s^2$	21 Sc $3d4s^2$	22 Ti $3d^2 4s^2$	23 V $3d^3 4s^2$	24 Cr $3d^4 4s^2$	25 Mn $3d^5 4s^2$	26 Fe $3d^6 4s^2$	27 Co $3d^7 4s^2$	28 Ni $3d^8 4s^2$	29 Cu $3d^{10} 4s$	30 Zn $3d^{10} 4s^2$	31 Ga $4s^2 4p$	32 Ge $4s^2 4p^2$	33 As $4s^2 4p^3$	34 Se $4s^2 4p^4$	35 Br $4s^2 4p^5$	36 Kr $4s^2 4p^6$
37 Rb $4p^6 5s$	38 Sr $4p^6 5s^2$	39 Y $4d5s^2$	40 Zr $4d^2 5s^2$	41 Nb $4d^4 5s$	42 Mo $4d^5 5s$	43 Tc $4d^6 5s$	44 Ru $4d^7 5s$	45 Rh $4d^8 5s$	46 Pd $4d^{10}$	47 Ag $4d^{10} 5s$	48 Cd $4d^{10} 5s^2$	49 In $5s^2 5p$	50 Sn $5s^2 5p^2$	51 Sb $5s^2 5p^3$	52 Te $5s^2 5p^4$	53 I $5s^2 5p^5$	54 Xe $5s^2 5p^6$
55 Cs $5p^6 6s$	56 Ba $5p^6 6s^2$	57–71 La $5p^6 5d6s^2$	72 Hf $5d^2 6s^2$	73 Ta $5d^3 6s^2$	74 W $5d^4 6s^2$	75 Re $5d^5 6s^2$	76 Os $5d^6 6s^2$	77 Ir $5d^9$	78 Pt $5d^4 6s$	79 Au $5d^{10} 6s$	80 Hg $5d^{10} 6s^2$	81 Tl $6s^2 6p$	82 Pb $6s^2 6p^2$	83 Bi $6s^2 6p^3$	84 Po $6s^2 6p^4$	85 At $6s^2 6p^5$	86 Rn $6s^2 6p^6$
87 Fr $6p^6 7s$	88 Ra $6p^2 7s^2$	89 Ac $6d7s^2$															

58 Ce $4f^2 6s^2$	59 Pr $4f^3 6s^2$	60 Nd $4f^4 6s^2$	61 Pm $4f^5 6s^2$	62 Sm $4f^6 6s^2$	63 Eu $4f^7 6s^2$	64 Gd $4f^7 5ds^2$	65 Tb $4f^8 5d6s^2$	66 Dy $4f^{10} 6s^2$	67 Ho $4f^{11} 6s^2$	68 Er $4f^{12} 6s^2$	69 Tm $4f^{13} 6s^2$	70 Yb $4f^{14} 6s^2$	71 Lu $4f^{14} 5d6s^2$
90 Th $6d^2 7s^2$	91 Pa $5f^2 6d7s^2$	92 U $5f^3 6d7s^2$	93 Np $5f^5 7s^2$	94 Pu $5f^6 7s^2$	95 Am $5f^7 7s^2$	96 Cm $5f^7 6d7s^2$	97 Bk $5f^8 6d7s^2$	98 Cf $5f^9 6d7s^2$	99 Es	100 Fm	101 Md	102 No	103 Lw

Figure 1.2 Periodic table, including abbreviated electron configuration notation for each element.

EXAMPLE 1.3 Electronic notation helps a person visualize which electrons are available for bonding and to estimate the type of bond that is likely to result. Unfilled shells contribute to bonding. Electron notation is often abbreviated to include only the unfilled and outer shells. The iron electron notation is thus abbreviated to $3d^64s^2$, which tells the reader that all the subshells up to and including 3s are filled. Yttrium is abbreviated from $1s^22s^22p^63d^{10}4s^24p^64d^15s^2$ to $4d^15s^2$ or even more simply, $4d5s^2$. Figure 1.2 lists the abbreviated electron configurations of the elements arranged according to the periodic table.

1.2 BONDING

The unfilled outermost electron shells are involved in bonding. The elements He, Ne, Ar, Kr, Xe, and Rn have full outer electron shells and thus are very stable and do not easily form bonds with other elements. Elements with unfilled electron shells are not as stable and interact with other atoms in a controlled fashion such that electrons are shared or exchanged between these atoms to achieve stable full outer shells.

The three primary interatomic bonds are referred to as metallic, ionic, and covalent. These provide the bond mechanism for nearly all the solid ceramic and metallic materials discussed in later chapters. Other secondary mechanisms referred to as van der Waals bonds also occur, but are discussed only briefly.

Metallic Bonding

As the name implies, metallic bonding is the predominant bond mechanism for metals. It is also referred to as electronic bonding, from the fact that the valence electrons (electrons from unfilled shells) are freely shared by all the atoms in the structure. Mutual electrostatic repulsion of the negative charges of the electrons keeps their distribution statistically uniform throughout the structure. At any given time, each atom has enough electrons grouped around it to satisfy its need for a full outer shell. It is the mutual attraction of all the nuclei in the structure for this same cloud of shared electrons that results in the metallic bond.

Because the valence electrons in a metal distribute themselves uniformly and because all the atoms in a pure metal are of the same size, close-packed structures result. Such close-packed structures contain many slip planes along which movement can occur during mechanical loading, producing the ductility that we are so accustomed to for metals. Pure metals typically have very high ductility and can undergo 40 to 60% elongation prior to rupturing. Highly alloyed metals such as the superalloys also have close-packed structures, but the different-size alloying atoms disrupt movement along slip planes and decrease the ductility. Superalloys typically have 5 to 20% elongation.

The free movement of electrons through the structure of a metal results in high electrical conductivity under the influence of an electrical field and high thermal conductivity when exposed to a heat source. These properties are discussed in more detail in Chap. 2.

Metallic bonding occurs for elements to the left and in the interior of the periodic table (see Fig. 1.2). Alkali metals such as sodium (Na) and potassium (K) are bonded by outer s electrons and have low bond energy. These metals have low strength, low melting temperatures, and are not overly stable. Transition metals such as chromium (Cr), iron (Fe), and tungsten (W) are bonded by inner electrons and have much higher bond strengths. Transition metals thus have higher strength, higher melting temperatures, and are more stable.

Ionic Bonding

Ionic bonding occurs when one atom gives up one or more electrons and another atom or atoms accept these electrons such that electrical neutrality is maintained and each atom achieves a stable, filled electron shell. This is best illustrated by a few examples.

EXAMPLE 1.4 Sodium chloride (NaCl) is largely ionically bonded. The Na atom has the electronic structure $1s^2 2s^2 2p^6 3s$. If the Na atom could get rid of the 3s electron, it would have the stable neon (Ne) structure. The chlorine atom has the electronic structure $1s^2 2s^2 2p^6 3s^2 3p^5$. If the Cl atom could obtain one more electron, it would have the stable argon (Ar) structure. During bonding, one electron from the Na is transferred to the Cl, producing a sodium ion (Na^+) with a net positive charge and a chlorine ion (Cl^-) with an equal negative charge, resulting in a more stable electronic structure for each. These opposite charges provide a Coulombic attraction which is the source of ionic bonding. To maintain overall electrical neutrality, one Na atom is required for each Cl atom and the formula becomes NaCl.

EXAMPLE 1.5 The aluminum (Al) atom has the electronic structure $1s^2 2s^2 2p^6 3s^2 3p^1$. To achieve a stable Ne structure three electrons would have to be given up, producing an ion with a net positive charge of 3 (Al^{3+}). The oxygen (O) atom has the electronic structure $1s^2 2s^2 2p^4$ and needs two electrons to achieve the stable Ne structure. Bonding occurs when two Al atoms transfer three electrons each to provide the six electrons required by three O atoms to produce the electrically neutral Al_2O_3 compound.

The crystal structure of an ionically bonded material is determined by the number of atoms of each element required for electrical neutrality and the optimum packing based on the relative sizes of the ions. The size of the ions is usually stated in terms of ionic radius. The ionic radius can vary slightly depending on the number and type of oppositely charged ions surrounding an ion. Table 1.1 shows ionic radii for ions with coordination numbers of 4 and 6 [1,2]. The <u>coordination number</u> is defined as the number of anions (negative ions) surrounding the cation (positive ion).

The relative size of the ions determines the coordination number. Figure 1.3 shows stable and unstable configurations. For cation-to-anion ratios between 0.155 and 0.225 a coordination number of 3 is most probable, with the anions at the corners of a triangle around the cation. For ratios in the range 0.225 to 0.414 a coordination number of 4 is most probable, with the anions at the corners of a tetrahedron around the cation. Similarly, ratios of 0.414 to 0.732 and greater than 0.732 are most likely to result in coordination numbers of 6 and 8, respectively.

EXAMPLE 1.6 What is the most likely coordination number for a structure made up of Mg^{2+} and O^{2-}? Si^{4+} and O^{2-}? Cr^{3+} and O^{2-}?
From Table 1.1,

$$\frac{Mg^{2+}}{O^{2-}} = \frac{0.72}{1.40} = 0.51 \qquad \text{coordination number} = 6$$

$$\frac{Si^{4+}}{O^{2-}} = \frac{0.40}{1.40} = 0.29 \qquad \text{coordination number} = 4$$

$$\frac{Cr^{3+}}{O^{2-}} = \frac{0.62}{1.40} = 0.44 \qquad \text{coordination number} = 6$$

As predicted, Mg^{2+} has a coordination number of 6 in MgO. The O^{2-} ions also have a coordination number of 6 and are arranged in a cubic close-packed structure with the Mg^{2+} ions filling the octahedral interstitial positions. This structure is called the rock salt structure (after NaCl). Other ionic structures are listed in Table 1.2.

NOT STABLE   STABLE   STABLE

Figure 1.3 Stable and unstable configurations which determine atomic coordination number within a structure. (Adapted from Ref. 2.)

Table 1.1 Ionic Radii for 6 and 4 Coordination (4 Coordination in Parentheses)

Ag+ 1.15 (1.02)	Al3+ 0.53 (0.39)	As5+ 0.50 (0.34)	Au+ 1.37 --	B3+ 0.23 (0.12)	Ba2+ 1.36 --	Be2+ 0.35 (0.27)	Bi5+ 0.74 --	Br- 1.96 --	C4+ 0.16 (0.15)	Ca2+ 1.00 --	Cd2+ 0.95 (0.84)	Ce4+ 0.80 --
Cl- 1.81 --	Co2+ 0.74 --	Co3+ 0.61 --	Cr2+ 0.73 --	Cr3+ 0.62 --	Cr4+ 0.55 (0.44)	Cs+ 1.70 --	Cu+ 0.96 --	Cu2+ 0.73 (0.63)	Dy3+ 0.91 --	Er3+ 0.88 --	Eu3+ 0.95 --	F- 1.33 (1.31)
Fe2+ 0.77 (0.63)	Fe3+ 0.65 (0.49)	Ga3+ 0.62 (0.47)	Gd3+ 0.94 --	Ge4+ 0.54 (0.40)	Hf4+ 0.71 (0.96)	Hg2+ 1.02 --	Ho3+ 0.89 --	I- 2.20 --	In3+ 0.79 --	K+ 1.38 --	La3+ 1.06 --	Li+ 0.74 (0.59)
Mg2+ 0.72 (0.49)	Mn2+ 0.67 --	Mn4+ 0.54 --	Mo3+ 0.67 --	Mo4+ 0.65 --	Na+ 1.02 (0.99)	Nb5+ 0.64 (0.32)	Nd3+ 1.00 --	Ni2+ 0.69 --	O2- 1.40 (1.38)	P5+ 0.35 (0.33)	Pb2+ 1.18 (0.94)	Pb4+ 0.78 --
Rb+ 1.49 --	S2- 1.84 --	S6+ 0.30 (0.12)	Sb5+ 0.61 --	Sc3+ 0.73 --	Se2- 1.98 --	Se6+ 0.42 (0.29)	Si4+ 0.40 (0.26)	Sm2+ 0.96 --	Sn2+ 0.93 --	Sn4+ 0.69 --	Sr2+ 1.16 --	Ta5+ 0.64 --
Te2- 2.21 --	Te6+ 0.56 --	Th4+ 1.00 --	Ti2+ 0.86 --	Ti4+ 0.61 --	Tl+ 1.50 --	Tl3+ 0.88 --	U4+ 0.97 --	U5+ 0.76 --	V2+ 0.79 --	V5+ 0.54 (0.36)	W4+ 0.65 --	W6+ 0.58 (0.41)
Y3+ 0.89 --	Yb3+ 0.86 --	Zn2+ 0.75 (0.60)	Zr4+ 0.72 --									

Source: Compiled from Refs. 1 and 2.

Table 1.2 Ionic Crystal Structures

Name of structure	Packing of anions	Coordination of anions	Coordination of cations	Examples
Rock salt	Cubic close–packed	6	6	$NaCl$, MgO, CaO, LiF, CoO, NiO
Zinc blende	Cubic close–packed	4	4	ZnS, BeO, SiC
Perovskite	Cubic close–packed	6	12, 6	$BaTiO_3$, $CoTiO_3$, $SrZrO_3$
Spinel	Cubic close–packed	4	4, 6	$FeAl_2O_4$, $MgAl_2O_4$, $ZnAl_2O_4$
Inverse spinel	Cubic close–packed	4	4 (6, 4)[a]	$FeMgFeO_4$, $MgTiMgO_4$
CsCl	Simple cubic	8	8	$CsCl$, $CsBr$, CsI
Fluorite	Simple cubic	4	8	CaF_2, ThO_2, CeO_2, UO_2, ZrO_2, HfO_2
Antifluorite	Cubic close–packed	8	4	Li_2O, Na_2O, K_2O, Rb_2O
Rutile	Distorted cubic close–packed	3	6	TiO_2, GeO_2, SnO_2, PbO_2, VO_2
Wurtzite	Hexagonal close–packed	4	4	ZnS, ZnO, SiC
Nickel arsenide	Hexagonal close–packed	6	6	$NiAs$, FeS, $CoSe$
Corundum	Hexagonal close–packed	4	6	Al_2O_3, Fe_2O_3, Cr_2O_3, V_2O_3
Ilmenite	Hexagonal close–packed	4	6, 6	$FeTiO_3$, $CoTiO_3$, $NiTiO_3$
Olivine	Hexagonal close–packed	4	6, 4	Mg_2SiO_4, Fe_2SiO_4

[a] First Fe in tetrahedral coordination, second Fe in octahedral coordination.

Source: Adapted from Ref. 2.

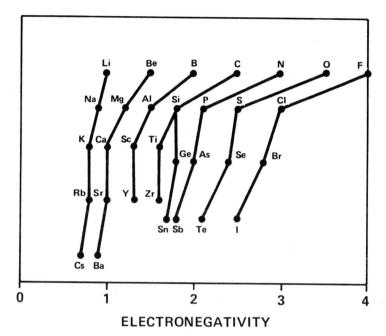

ELECTRONEGATIVITY

Figure 1.4 Pauling's electronegativity scale. Elements to the left have a low affinity for electrons and those to the right have a high affinity. (Reprinted with slight modifications from Linus Pauling, The Nature of the Chemical Bond, 3rd ed., © 1960 by Cornell University. Used by permission of the publisher, Cornell University Press.)

Most of the ionic structures are close packed. Bonding is associated with the s electron shells (which have a spherical probability distribution) and would be nondirectional if purely ionic. However, there is a tendency for increased electron concentration between atom centers, which provides a degree of nonionic character. The degree of ionic character of a compound can be estimated using the electronegativity scale (Fig. 1.4) derived by Pauling [3]. Electronegativity is a measure of an atom's ability to attract electrons and is roughly proportional to the sum of the energy to add an electron (electron affinity) and to remove an electron (ionization potential). The larger the electronegativity difference between atoms in a compound, the larger the degree of ionic character. The semiempirical curve derived by Pauling is shown in Fig. 1.5.

EXAMPLE 1.7 What is the degree of ionic character of MgO? of SiO_2? of SiC?
From Fig. 1.4 and Fig. 1.5,

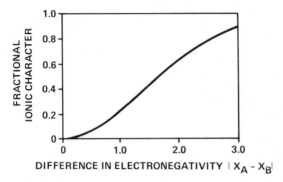

Figure 1.5 Semiempirical curve derived by Pauling for using the electro-negativity difference between two elements to estimate the degree of ionic character. (Reprinted with slight modifications from Linus Pauling, The Nature of the Chemical Bond, 3rd ed., © 1960 by Cornell University. Used by permission of the publisher, Cornell University Press.)

$$E_{Mg} - E_O = 2.3 \qquad \text{fraction ionic MgO} \simeq 0.75$$

$$E_{Si} - E_O = 1.7 \qquad \text{fraction ionic SiO}_2 \simeq 0.5$$

$$E_{Si} - E_C = 0.3 \qquad \text{fraction ionic SiC} < 0.1$$

The monovalent ions in groups IA (Li, Na, K, etc.) and VIIA (F, Cl, Br, etc.) produce compounds that are highly ionic, but that have relatively low strength, low melting temperatures, and low hardness. Ionic compounds with more highly charged ions such as Mg^{2+}, Al^{3+}, and Zr^{4+} have stronger bonds and thus have higher strength, higher melting temperatures, and higher hardness. Specific properties for specific materials are discussed in Chaps. 2 through 4.

In summary, the following are characteristic of ionic bonding and the resulting ceramic materials:

1. Electron donor plus electron acceptor to achieve electrical neutrality
2. Structure determined by atom (ion) size and charge with a tendency to achieve as close packing as sizes will permit
3. Bonding nondirectional
4. Transparent to visible wavelengths of light
5. Absorb infrared wavelengths
6. Low electrical conductivity at low temperature
7. Ionic conductivity at high temperature
8. Metal ions with group VII anions are strongly ionic (NaCl, LiF)

9. Compounds with higher-atomic-weight elements of group VI (S, Se, Te) are increasingly less ionic in nature

10. Strength of ionic bonds increases as charge increases; many oxides composed of multiple-charged ions are hard and melt at high temperatures (Al_2O_3, ZrO_2, Y_2O_3)

Covalent Bonding

Covalent bonding occurs when two or more atoms share electrons such that each achieves a stable, filled electron shell. Unlike metallic and ionic bonds, covalent bonds are directional. Each covalent bond consists of a pair of electrons shared between two protons such that the probability distribution for each electron resembles a dumbbell. This produces the directionality of the bond. The bonding of carbon atoms to produce diamond is a good example.

Carbon has an atomic number of 6 and an electronic structure of $1s^2 2s^2 2p^2$ and thus has four valence electrons available for bonding. Each 2s and 2p electron shares an orbital with an equivalent electron from another carbon atom, resulting in a structure in which each carbon atom is covalently bonded to four other carbon atoms in a tetrahedral orientation. This is shown schematically in Fig. 1.6a for one tetrahedral structural unit. The central carbon atom has its initial six electrons plus one shared electron from each of the adjacent four carbon atoms, resulting in a total of 10 electrons. This is equivalent to the filled outer shell of a neon atom and is a very stable condition. Each of the four outer carbon atoms of the tetrahedron are bonded directionally to three additional carbon atoms to produce a periodic tetrahedral structure with all the atoms in the structure (except the final outer layer at the surface of the crystal) sharing four electrons to achieve the stable electronic structure of neon.

(a) DIAMOND STRUCTURE UNIT   (b) METHANE MOLECULE

Figure 1.6 Schematic example of covalently bonded materials. (a) Diamond with periodic three-dimensional structure. (b) Methane with single-molecule structure. Shaded regions show directional electron probability distributions for pairs of electrons.

The continuous periodic covalent bonding of carbon atoms in diamond results in high hardness, high melting temperature, and low electrical conductivity at low temperature. Silicon carbide has similar covalent bonding and thus high hardness, high melting temperature, and low electrical conductivity at low temperature.[*]

Covalently bonded ceramics typically are hard, strong, and have high melting temperatures. However, these are not inherent traits of covalent bonding. For instance, most organic materials have covalent bonds but do not have high hardness or high melting temperatures. The deciding factor is the strength of the bond and the nature of the structure. For instance, methane (CH_4) forms a tetrahedral structural unit like diamond, but the valence electrons of both the carbon atom and the four hydrogen atoms are satisfied within a single tetrahedron and no periodic structure results. Methane is a gas under normal ambient conditions. A methane molecule is shown schematically in Fig. 1.6b. Organic bonding and structures are discussed in more detail later in the chapter.

The directional bonding of covalent materials results in structures that are not close packed. This has a pronounced effect on the properties, in particular density and thermal expansion. Close-packed materials such as the metals and ionic-bonded ceramics have relatively high thermal expansion coefficients. The thermal expansion of each atom is cumulated through each close-packed adjacent atom throughout the structure to yield a large thermal expansion of the whole mass. Covalently bonded ceramics typically have a much lower thermal expansion because some of the thermal growth of the individual atoms is absorbed by the open space in the structure.

Covalent bonding occurs between atoms of similar electronegativity which are not close in electronic structure to the inert gas configuration. (Refer to the electronegativity scale in Fig. 1.4.) Atoms such as C, N, Si, Ge, and Te are of intermediate electronegativity and form highly covalent structures. Atoms with a greater difference in electronegativity form compounds having a less covalent bond nature. Figures 1.4 and 1.5 can be used to estimate the relative covalent bond nature. However, it should be noted that the curve in Fig. 1.5 is empirical and can be used only as an approximation, especially in intermediate cases.

EXAMPLE 1.8 What is the approximate degree of covalent character of diamond? of Si_3N_4? of SiO_2?
From Fig. 1.4,

$$E_C - E_C = 0$$

$$E_{Si} - E_N = 1.2$$

---

[*]Silicon carbide doped with appropriate impurities has significantly increased electrical conductivity and is an important semiconductor material.

$E_{Si} - E_O = 1.7$

From Fig. 1.5,

Fraction covalent $C = 1 -$ fraction ionic $C = 1 - 0 = 1.0$

Fraction covalent $Si_3N_4 = 1 -$ fraction ionic $Si_3N_4 = 1.0 - 0.3 = 0.7$

Fraction covalent $SiO_2 = 1 -$ fraction ionic $SiO_2 = 1.0 - 0.5 = 0.5$

In summary, the following are characteristics of covalent bonding and the resulting ceramic materials:

1. Electron sharing to fill outer electron shells and achieve electrical neutrality
2. Formed by atoms having similar electronegativity
3. Bonding highly directional
4. Structures not close packed, but typically three-dimensional frameworks containing cavities and channels
5. Compounds typically have high strength, hardness, and melting temperature
6. Structures often have low thermal expansion

Ionic and Covalent Bond Combinations

Many ceramic materials have a combination of ionic and covalent bonding. An example is gypsum ($CaSO_4$), from which plaster is manufactured. The sulfur is covalently bonded to the oxygen to produce $SO_4{}^{2-}$, which is two electrons short of having full outer electron shells for each of the five atoms. The calcium donates its two valence electrons and is thus bonded ionically to the $SO_4{}^{2-}$:

$$
\text{Ca} + \cdot \overset{\cdot\cdot}{\underset{\cdot\cdot}{\text{O}}} : \overset{\cdot\cdot}{\underset{\cdot\cdot}{\text{S}}} : \overset{\text{O}}{\underset{\cdot\cdot}{\text{O}}} \cdot \longrightarrow \text{Ca}^{2+} + : \overset{\cdot\cdot}{\underset{\cdot\cdot}{\text{O}}} : \overset{:\text{O}:}{\underset{:\text{O}:}{\text{S}}} : \overset{\cdot\cdot}{\underset{\cdot\cdot}{\text{O}}} :
$$

A similar type of combined bonding results in the many silicate compositions which are so important to ceramic technology. These silicate structures are based on the $SiO_4{}^{4-}$ tetrahedron and the various ways these tetrahedra can be linked. As illustrated in Table 1.3, the tetrahedra are linked by sharing of corners, resulting in framework, chain, ring, and layer structures. This is analogous to polymerization in organic materials, but alternating Si and O bonds are most commonly involved in the ceramics compared to the C—C bonds in organics.

Table 1.3 Bonding of Silicate Structures

Bonding of tetrahedra	Structure classification	Schematic	Examples
Independent tetrahedra	Orthosilicates	SiO_4^{4-}	Zircon ($ZrSiO_4$), mullite ($Al_6Si_2O_{13}$), forsterite (Mg_2SiO_4), kyanite (Al_2SiO_5)
Two tetrahedra with one corner shared	Pyrosilicates	$Si_2O_7^{6-}$	Ackermanite ($Ca_2MgSi_2O_7$)
Two corners shared to form ring or chain structures	Metasilicates	$Si_6O_{18}^{12-}$ $(SiO_3)_n^{2n-}$	Spodumene [$LiAl(SiO_3)_2$], wollastonite ($CaSiO_3$), beryl ($Be_3Al_2Si_6O_{18}$), asbestos [$Mg_3Si_2O_5(OH)_4$]
Three corners shared	Layer silicates	See Ref. 2, pp. 75–79	Kaolinite clay [$Al_2(Si_2O_5)(OH)_4$], mica (muscovite var.) [$KAl_2(OH)_2(AlSi_3O_{10})$], talc [$Mg_3(Si_2O_5)_2(OH)_2$]
Four corners shared	Framework silicates	See Ref. 2, pp. 71–74, for SiO_2 and derivatives	Quartz, cristobalite, tridymite (SiO_2), feldspars (albite var.) ($NaAlSiO_3$), zeolites (mordenite var.) [$(Ca,Na_2)Al_2Si_9O_{22} \cdot 6H_2O$]

Van der Waals Bonds

Ionic and covalent bonds are referred to as primary bonds and account for the atomic bonding of most ceramic materials. However, other weaker secondary bond mechanisms also occur which have major effects on the properties of some ceramic materials. These secondary bonds are grouped together under the name van der Waals forces.

One type of van der Waals force is referred to as dispersion. In all molecules (including monatomic molecules) there is a fluctuating electrical dipole that varies with the instantaneous positions of the electrons. Interaction of these fluctuating dipoles between molecules leads to very weak forces of attraction, which can result in bonding when other forces are absent. The dispersion effect is the mechanism of condensation of the noble gases at very low temperature.

Another type of van der Waals force is molecular polarization, in which an electrical dipole forms in asymmetrical molecules. Hydrogen fluoride can be used as a simple example. When the single shared electron is orbiting the fluorine nucleus, a dipole moment results, where the hydrogen side of the molecule has a net positive charge and the fluorine side has a net negative charge. Because the fluorine nucleus is larger than the hydrogen nucleus, the shared electron spends more time around the fluorine nucleus. Consequently, the center of positive charge and the center of negative charge do not coincide and a weak electric dipole results which can contribute to weak bonding of one molecule to another.

A third type of van der Waals force is the hydrogen bridge or hydrogen bonding. It is a special case of molecular polarization in which the small hydrogen nucleus is attracted to the unshared electrons in a neighboring molecule. The most common case is water (H_2O), although this type of bonding is also found with other hydrogen-containing molecules, such as ammonia (NH_3).

Van der Waals forces are very important in layer structures, such as the clays, micas, graphite, and hexagonal boron nitride. All of these ceramic materials have strong primary bonding within the layers, but depend on van der Waals-type bonds to hold the layers together. Highly anisotropic properties result. All of these layer structures have easy slip between layers. In the clay minerals this property makes possible plasticity with the addition of water and was the basis of the early use of clay for pottery. In fact, it was the basis of almost all ceramic fabrication technology prior to the twentieth century and is still an important factor in the fabrication of pottery, porcelain, whiteware, brick, and many other items.

The easy slip between layers in graphite and hexagonal boron nitride has also resulted in many applications of these materials. Both can be easily machined with conventional cutting tools and provide low-friction, self-lubricating surfaces for a wide variety of seals. Both are also used as solid lubricants and as boundary layer surface coatings.

The weak bonds between layers of mica and the resulting easy slip has recently led to a new application for these materials. Small synthetic mica crystals are dispersed in glass to form a nonporous composite having excellent electrical resistance properties. The presence of the mica permits machining of the composite to close tolerances with no chipping or breakage, using conventional low-cost machine tools.

Although van der Waals forces are weak, they are adequate to cause adsorption of molecules at the surface of a particle. For particles of colloid dimensions (100 Å to 3 μm), adsorbed ions provide enough charge at the surface of a particle to attract particles of opposite charge and to repel particles of like charge. This has a major effect on the rheology (flow characteristics of particles suspended in a fluid) of particle suspensions used for slip casting and mixes used for extrusion, injection molding, and other plastic-forming techniques (see Chap. 6).

The discussions in this chapter of electronic structure, bonding, and crystal structure have been brief and simplified. More detailed discussions are available in Refs. 1 through 7.

1.3 POLYMORPHIC FORMS AND TRANSFORMATIONS

As described in the sections on bonding, the stable crystal structure for a composition is dependent on the following:

1. Balance of electrical charge
2. Densest packing of atoms consistent with atom size, number of bonds per atom, and bond direction
3. Minimization of the electrostatic repulsion forces

As the temperature of or the pressure on a material changes, interatomic distance and the level of atomic vibration change such that the initial structure may not be the most stable structure under the new conditions. Materials having the same chemical composition but a different crystal structure are called polymorphs and the change from one structure to another is referred to as a polymorphic transformation.

Polymorphism is common in ceramic materials and in many cases has a strong impact on the useful limits of application of the material. For instance, the stable form of zirconium oxide (ZrO_2) at room temperature is monoclinic, which transforms to a tetragonal form at about 1100°C. This transformation is accompanied by a large volume change that results in internal stresses in the ZrO_2 body large enough to cause fracture or substantial weakening. In attempts to avoid this problem, it was discovered that appropriate additions of MgO, CaO, or Y_2O_3 to ZrO_2 produced a cubic form that did not undergo a transformation and was thus useful over a broader temperature range.

Before selecting a material for an application, it is necessary for an engineer to verify that the material does not have an unacceptable transformation. A good first step is to check the phase equilibrium diagram for the composition. Even if more than one polymorph is present within the intended temperature range of the application, the material may be acceptable. The important criterion is that no large or abrupt volume changes occur. This can be determined by looking at the thermal expansion curve for the material.

Many ceramic materials exist in different polymorphic forms. Among these materials are SiO_2, SiC, C, Si_3N_4, BN, TiO_2, ZnS, $CaTiO_3$, Al_2SiO_5, FeS_2, and As_3O_5. The properties of most of these are discussed in later chapters.

Two types of polymorphic transformations occur. The first, displacive transformation, involves distortion of the structure, such as a change in bond angles, but does not include breaking of bonds. It typically occurs rapidly at a well-defined temperature and is reversible. The martensite transformation in metals is a displacive transformation. So also are the cubic-tetragonal $BaTiO_3$ and tetragonal-monoclinic ZrO_2 transformations.

Displacive transformations are common in the silicate ceramics. In general, the high-temperature form has higher symmetry, larger specific volume, larger heat capacity, and is always the more open structure. The low-temperature form typically has a collapsed structure achieved by rotating the bond angle of alternating rows of SiO_4 tetrahedra in opposite directions.

The second type of transformation is the reconstructive transformation. Bonds are broken and a new structure formed. Much greater energy is required for this type of transformation than for a displacive transformation. The rate of reconstructive transformation is sluggish, so the high-temperature structure can usually be retained at low temperature by rapid cooling through the transformation temperature.

The activation energy for a reconstructive transformation is so high that transformation frequently will not occur unless aided by external factors. For example, the presence of a liquid phase can allow the unstable form to dissolve, followed by precipitation of the new stable form. Mechanical energy can be another means of overcoming the high activation energy.

Silica (SiO_2) is a good example for illustrating transformations. Both displacive and reconstructive transformations occur in SiO_2 and play an important role in silicate technology. Figure 1.7 shows the temperature-initiated transformations for SiO_2. The stable polymorph of SiO_2 at room temperature is quartz. However, tridymite and cristobalite are also commonly found at room temperature in ceramic components as metastable forms because the reconstructive transformations in SiO_2 are very sluggish and do not normally occur. Quartz, tridymite, and cristobalite all have displacive transformations in which the high-temperature structures distort

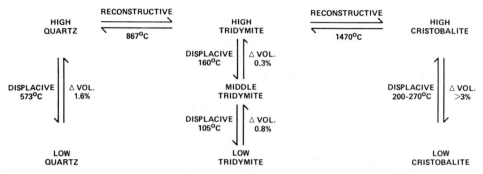

Figure 1.7 Transformations and volume changes for SiO_2 polymorphs.
(Adapted from Ref. 2.)

by changes in bond angle between SiO_4 tetrahedra to form the low-temperature structures. These displacive transformations are rapid and cannot be
restrained from occurring.

It is important to note the size of the volume changes associated with
displacive transformations in SiO_2. These limit the applications, especially
of cristobalite and quartz. Ceramic bodies containing moderate to large
amounts of quartz or cristobalite either fracture during thermal cycling
through the transformation temperature or are weakened. In the fabrication
of silica brick for high-temperature applications, a small amount of $CaCO_3$
or CaO is added to act as a flux at the firing temperature to dissolve the
quartz and precipitate the SiO_2 as tridymite. The tridymite has a much
lower shrinkage during transformation and is thus less likely to result in
fracture or weakening of the refractory brick.

1.4 NONCRYSTALLINE STRUCTURES

The structures described so far have all had units of atomic arrangement
that were repeated uniformly throughout the solid. For example, in a silicate the crystal structure is made up of an ordered repetition of SiO_4 tetrahedra. Each atom of a given type has the same neighboring atoms at the
same bond angles and the same interatomic distances. This type of structure, in which both short-range and long-range order occur, is called a
crystalline structure.

Structures that have short-range order but no long-range periodicity
are referred to as noncrystalline. Figure 1.8 illustrates the difference
between a crystalline and a noncrystalline material. Noncrystalline solids
such as glass, gels, and vapor-deposited coatings have many applications
and are very important to a broad range of engineering disciplines.

Glasses

Glasses are the most widely used noncrystalline ceramic. A glass is
formed when a molten ceramic composition is cooled so rapidly that the
atoms do not have time to arrange themselves in a periodic structure. At
temperatures below the solidification temperature, glasses are not stable
thermodynamically and the atoms would rearrange into a crystalline structure if they had the mobility. Over long periods of time glass can crystallize, as evidenced by the presence of cristobalite in some volcanic glass
(obsidian). The crystallization can be speeded up by increasing the temperature to a level at which atomic mobility is increased. Most engineers
who have used fused silica in high-temperature applications have encountered this. At use temperatures well below the melting temperature of
1713°C, cristobalite crystals form in the fused silica (slowly at 1200°C and
relatively rapidly at 1400°C). Fused silica has a very low, nearly
linear thermal expansion curve and is one of the best thermal-shock-
resistant ceramic materials for applications where rapid thermal cycling
occurs. Crystallization to cristobalite is undesirable because the large
volume change of the polymorphic transformation from high cristobalite to

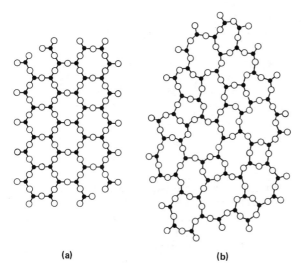

(a)   (b)

Figure 1.8 (a) Crystalline material, characterized by both short-range
and long-range periodicity of the atomic structure. (b) Noncrystalline
material, characterized by short-range order but no long-range periodicity.
(From Ref. 2.)

low cristobalite in the range 200 to 270°C results in cracking of the fused silica part.

Some of the lesser known but important engineering applications of glasses are discussed later, particularly in Chaps. 2 and 13. Further information about the structure and thermodynamics of glass can be found in Refs. 8 and 9.

Although a wide variety of properties can be achieved with glasses, the following are general characteristics:

1. Short-range atomic order but no long-range order
2. Structure isotropic, so the properties are uniform in all directions
3. Typically transparent to optical wavelengths, but can be formulated to absorb or transmit a wide variety of wavelengths
4. Typically good electrical and thermal insulators
5. Soften before melting, so they can be formed by blowing into intricate hollow shapes

Gels

Gels are noncrystalline solids that are formed by chemical reaction rather than melting. Silica gel, which is highly useful as a bonding agent in the ceramic and metal industries, is produced by a reaction of ethyl silicate with water in the presence of a catalyst. $Si(OH)_4$ results, which is then dehydrated to form SiO_2. A silica gel can also be formed by the reaction of sodium silicate with acid.

Another noncrystalline inorganic gel, $Al(H_2PO_4)_3$, can be produced by reacting aluminum oxide (Al_2O_3) with phosphoric acid (H_3PO_4). Like the silica gels, this aluminum phosphate gel is produced at room temperature and is an excellent inorganic cement. The technology and important applications of ceramic cements are discussed in Chap. 7.

Vapor Deposition

An important class of noncrystalline materials is produced by rapid condensation of a vapor on a cold substrate or by reaction from a gas at a hot substrate. The vapor can be produced by sputtering, electron-beam evaporation, or thermal evaporation. Vapor contacting a cold substrate solidifies so rapidly that the atoms do not have time to rearrange into a crystalline structure.

Condensation from a vapor has been used to produce noncrystalline coatings of materials that are difficult or impossible to produce as noncrystalline solids by other approaches. These coatings are usually nonporous, very fine grained, and have unique properties.

1.5 MOLECULAR STRUCTURES

So far we have discussed the bonding and structures of metals and ceramics, but have ignored organic materials. Organic materials are extremely important in modern engineering and their general characteristics should be understood just as well as metals and ceramics.

The majority of organic materials are made up of distinct molecules. The atoms of each molecule are held together strongly by covalent bonds with the outer electron shells filled. Because all the shells are filled, the individual molecules are stable and do not have a drive to bond with other molecules (as mentioned earlier for methane).

Organic molecular structures are usually formed from the nonmetallic elements and hydrogen. The most common are the hydrocarbons which consist primarily of carbon and hydrogen, but may also have halogens (especially Cl and F), hydroxide (OH), acetate ($C_2H_3O_2$), or other groups replacing one or more of the hydrogens. Other molecular structures include ammonia, which is made up of N and H, and the silicones, which contain Si in the place of carbon.

Hydrocarbons

The hydrocarbons and modified hydrocarbons are perhaps the most frequently encountered engineering organic materials. Some of the simple compositions and molecular structures are illustrated in Fig. 1.9. The straight lines between the atoms represent individual covalent bonds between pairs of electrons. The bond between two carbon atoms has an energy of about 83 kcal/g-mol. The bond energy between a carbon and a hydrogen is about 99 kcal/g-mol and between a carbon and chlorine is about 81 kcal/g-mol. Some pairs of carbon atoms in Fig. 1.9 have two covalent bonds between them. This double bond has an approximate energy of 146 kcal/g-mol [11].

Hydrocarbons with only single bonds have no open structural positions where additional atoms can bond and are thus referred to as <u>saturated.</u> The paraffins are good examples. They have a general formula of C_nH_{2n+2}. Methane is n = 1 and ethane is n = 2. These, as well as compositions with n up to 15, are either liquid or gas at room temperature and are used as fuels. As the size of the molecules increase, the melting temperature increases, such that paraffins with about 30 carbon atoms per molecule are relatively rigid at room temperature. The increase in melting temperature with molecular size is partially due to decreased mobility, but mostly to increased van der Waals bonding between molecules. The larger molecules have more sites available for van der Waals bonds.

Hydrocarbons with double or triple bonds between a pair of carbon atoms are referred to as unsaturated. Under the appropriate conditions,

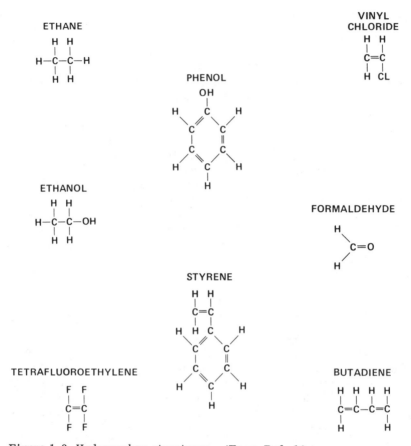

Figure 1.9 Hydrocarbon structures. (From Ref. 10.)

these bonds can be broken and replaced by single bonds that can link small molecules together to form large molecules. This is referred to as poly-merization [12,13].

Addition Polymerization

When a double bond is broken, it provides two sites at which new bonds may form, and the molecule is referred to as bifunctional. Ethylene, vinyl chloride, tetrafluoroethylene, styrene, and methyl methacrylate are all bi-functional. Addition polymerization can be achieved with bifunctional mole-cules by applying enough energy to break the double carbon bond. This energy can be in the form of heat, pressure, light, or a catalyst. Once the double bonds of a group of molecules are broken, an unstable electron structure will be present and the separate molecular units, called mers, will bond together to form a long chain (or polymer). The energy released during addition polymerization is greater than the energy that was required to start the reaction. The following illustrates addition polymerization of vinyl chloride to form polyvinyl chloride:

$$\left(\begin{matrix} H & H \\ \| & \| \\ C = C \\ \| & \| \\ H & Cl \end{matrix}\right)_n \xrightarrow[\text{light, or catalyst}]{\text{heat, pressure,}} \left(\begin{matrix} H & H & H & H & H & H \\ \| & \| & \| & \| & \| & \| \\ -C-C-C-C-C-C- \\ \| & \| & \| & \| & \| & \| \\ H & Cl & H & Cl & H & Cl \end{matrix}\right)_n \quad (1.1)$$

In more general terms, addition polymerization can be represented by

$$nA \longrightarrow (-A-)_n \quad (1.2)$$

Addition polymerization can also occur if more than one double bond is present (as in the polymerization of butadiene to make unvulcanized rubber), but only one of the bonds is broken and the resulting molecule is still linear.

Addition polymerization can be achieved with mixtures of two or more different monomers to achieve modified properties. This is called copolymerization and the resulting structure is referred to as a copolymer. This is analogous to solid solution in metals and ceramics and can be represented by

$$nA + mB \xrightarrow{\text{heat, pressure, etc.}} (-A_n B_m -) \quad (1.3)$$

The polymers produced by addition polymerization are typically thermoplastic; i.e., they soften when heated and can be plastically worked to produce a shape and then return to their initial properties upon cooling. Complex shapes of thermoplastic polymers can be produced in large quantity by such low-cost approaches as injection molding. Because of the reversible nature of plasticity, the thermoplastic polymers can be recycled.

Condensation Polymerization

Condensation polymerization involves reaction of two different organic molecules to form a new molecule, accompanied by release of a by-product:

$$pC + pD \xrightarrow{\text{heat, pressure, etc.}} (-E-) + pF \quad (1.4)$$

Either a linear or framework polymer can result from condensation polymerization, depending on whether one double bond or more than one double
bond is broken. The by-product is often water, but can also be other simple
molecules, such as an alcohol or an acid.

Dacron is a linear polymer produced by condensation polymerization.
It is synthesized from dimethyl terephthalate and ethylene glycol and forms
methyl alcohol (CH_3OH) as the by-product [10]. The reaction is shown in
Fig. 1.10. Note that no carbon double bonds were broken in this case, just
two C—O bonds on each dimethyl terephthalate and two C—OH bonds on each
ethylene glycol.

Phenol (C_6H_5OH) and formaldehyde (CH_2O) combine by condensation
polymerization to form a network structure as shown in Fig. 1.11. The
C=O bond in the formaldehyde is broken and a C—H bond in two adjacent
phenol molecules is broken. The remaining CH_2 of the formaldehyde then
has two unsatisfied carbon bonds and acts as a bridge between the two phenol
molecules. The O from the formaldehyde and the two H from the phenols
combine to form water as a by-product. This reaction occurs at several
C—H bonds in each phenol and results in the network structure [10]. The
phenol-formaldehyde polymer is known by several commercial names,
Bakelite and Texalite being two. Other condensation polymers include nylon
(hexamethylamine-adipic acid) and Melmac (melamine-formaldehyde).

Most of the condensation polymers are thermosetting resins. Once
polymerization has occurred, especially for the framework structures, the

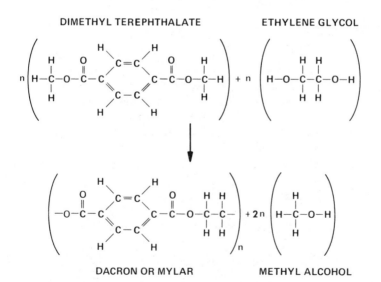

Figure 1.10 Formation of dacron by condensation polymerization.

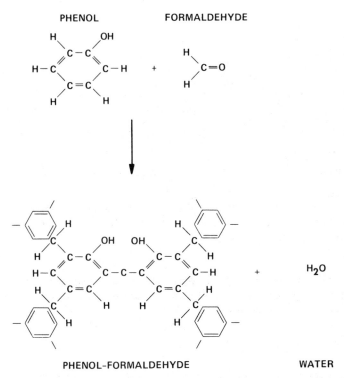

PHENOL-FORMALDEHYDE   WATER

Figure 1.11 Formation of a network structure by condensation polymerization of phenol and formaldehyde.

material is relatively rigid and does not increase in plasticity with increase in temperature. In general, the thermosetting resins have higher strength and higher-temperature capability than the thermoplastic resins, but are not as economical to fabricate.

Polymer Crystallization

The large molecules in a polymer can be oriented to produce a degree of crystallinity, resulting usually in modified properties. For instance, if the linear molecules in a polymer are random, van der Waals bonding between molecules will occur only in the limited number of positions where the appropriate atoms are adjacent. However, as alignment of the molecules increases, more atoms are in suitable positions for van der Waals bonds to form. Therefore, as the crystallinity increases, the strength tends to increase and the rate of creep decreases.

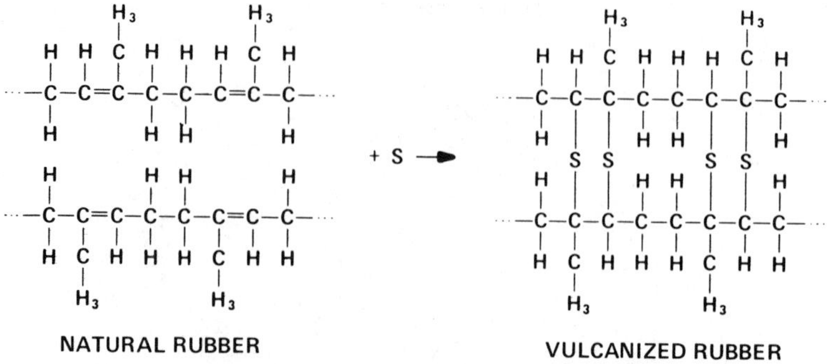

Figure 1.12 Cross-linking with sulfur in the vulcanization process for
natural rubber. (From L. H. Van Vlack, Elements of Materials Science,
2nd ed., © 1964, Addison-Wesley Publishing Co., Reading, Mass., Fig.
7.20. Reprinted with permission.)

The shape of the polymer molecules affects the ease and degree of crystallization and also the properties. Crystallization occurs most easily if the
individual monomers all have identical ordering.

Cross-Linking and Branching

Crystallization causes moderate changes in properties. Major changes can
occur in linear polymers by cross-linking or branching. In cross-linking,
adjacent chains are bonded together, usually by bridges between unsaturated
carbon atoms. The vulcanization of natural rubber with sulfur is a classic
example. The reaction is shown in Fig. 1.12. The degree of cross-linking
can be controlled by the amount of S added. Both the hardness and strength
increase as the amount of cross-linking increases.

PROBLEMS

1.1 Explain why hydrogen and oxygen are present in the atmosphere as H_2
and O_2 (diatomic molecules) and helium and argon are present as He and Ar
(monatomic molecules).

1.2 Show the complete electron notation for Co. Why does the 4s shell fill
before the 3d shell?

1.3 How many 3d electrons are in each of the following: (a) Fe, (b) Fe^{3+},
(c) Fe^{2+}, (d) Cu^+, (e) Cu^{2+}?

1.4 What chemical formula would result when yttrium and oxygen combine? What is the most likely coordination number? What is the relative percent ionic character? What relative hardness and melting temperature would be expected for yttrium oxide?

1.5 Copper, nickel, and gold have a face-centered cubic close-packed structure (i.e., unit cubic structure with atoms at the corners and center of the faces of a cube). Assuming that each atom can be modeled by a hard sphere, how many nearest neighbors does each atom have in the structure? Calculate the percent open space present in such a close-packed structure. The remaining space is filled by the spheres and is referred to as the atomic packing factor.

1.6 The atomic packing factor can also be estimated knowing the specific gravity of the material. Copper and diamond both have cubic structures. Calculate the packing factor for each. Explain the difference in terms of the nature of the interatomic bonding. Assume spherical atoms and the following data: specific gravity of copper and diamond, 8.96 g/cm^3 and 3.51 g/cm^3; atomic radii, 1.278 Å and 0.77 Å; and Avogadro's number, 6.02×10^{23} atoms per gram atom.

1.7 A ceramic composition can exist either as a crystalline structure or a metastable glass. Which form of this composition would you expect to have the higher strength at high temperature? Why?

1.8 Most materials expand when heated. What type of bonding would you expect to result in the lowest thermal expansion: metallic, ionic, or covalent? Why?

ANSWERS TO PROBLEMS

1.1 The outer shells of hydrogen and oxygen are not filled, so two atoms bond together covalently to produce H_2 and O_2, which are stable. He and Ar have complete outer shells and are stable as monatomic molecules.

1.2 Co $1s^2 2s^2 2p^6 3s^2 3p^6 3d^7 4s^2$. There is an overlap in energy between the 3d and 4s shells in the transition metals.

1.3 (a) Fe, 6; (b) Fe^{3+}, 5; (c) Fe^{2+}, 6; (d) Cu^+, 10; (e) Cu^{2+}, 9.

1.4 Y_2O_3 $\qquad \dfrac{Y^{3+}}{O^{2-}} = \dfrac{0.89}{1.40} = 0.636 \qquad CN = 6$

From Fig. 1.4,

$$\left| E_Y - E_O \right| \sim 2.2 \qquad \text{fraction ionic } 70\%$$

Because of the 3+ charge on the Y, we would expect properties similar to Al_2O_3, i.e., high hardness and high melting temperature.

1.5 12 nearest neighbors.

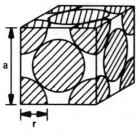

$a \cdot \sqrt{2} = 4r$

$a = \dfrac{4r}{\sqrt{2}}$

Filled space $= \dfrac{\text{volume spheres}}{\text{total volume}}$

There are 6 half spheres and 8 one-eighth spheres = total of $3 + 1 = 4$ spheres in one face-centered cubic unit.

Volume of 1 sphere $= \dfrac{4}{3}\pi r^3$

Volume of cube $= a^3 = \left(\dfrac{4r}{\sqrt{2}}\right)^3 = \dfrac{64r^3}{2\sqrt{2}}$

Filled space $= \dfrac{4(4\pi r^3/3)}{a^3} = \dfrac{16\pi r^3(2\sqrt{2})}{(3)(64r^3)}$

$= 0.74$

Open space $= 1.00 - 0.74 = 0.26 = 26\%$

1.6 Atomic packing factor $= \left(\dfrac{\text{atoms}}{\text{cm}^3}\right)\left(\dfrac{\text{volume}}{\text{atom}}\right)$

$\dfrac{\text{Atoms}}{\text{cm}^3} = \dfrac{\text{(S.G.)(Avogadro's number)}}{\text{atomic weight}} \quad \dfrac{\text{vol.}}{\text{atom}} = \dfrac{4\pi r^3}{3}$

$\dfrac{\text{Cu atoms}}{\text{cm}^3} = \dfrac{(8.96\ \text{g/cm}^3)(6.02 \times 10^{23}\ \text{atoms/g-atom})}{63.54\ \text{g/g-atom}} = 8.49 \times 10^{22}$

$\dfrac{\text{Vol.}}{\text{Cu atom}} = [(1.278\ \text{Å})(10^{-8}\ \text{cm})/\text{Å}]^3 \dfrac{4\pi}{3} = (1.278 \times 10^{-8})^3 \dfrac{4\pi}{3}$

$= 3.7 \times 10^{-24}$

$(\text{Packing factor})_{Cu} = (8.49 \times 10^{22})(3.7 \times 10^{-24}) \simeq 0.77$

This is in reasonable agreement with the packing factor estimated in Problem 1.5 for a face-centered cubic material.

$(\text{Atomic packing factor})_{diamond} \sim \dfrac{(3.51)(6.02 \times 10^{23})}{12.01} \dfrac{4\pi}{3} (0.77 \times 10^{-8})^3$

~ 0.34

This is much lower than for Cu and points out how much more open a co-valent structure is than a metallic structure.

1.7 The crystalline form would be expected to have higher strength at high temperature. The glass would soften at a lower temperature than the crystalline form and would be less able to sustain a load.

1.8 A covalent-bonded ceramic framework structure would probably have a lower level of expansion than a metallic or ionic bonded structure. Metallic and ionic structures are close packed. The expansion of each atom would be additive throughout the structure, yielding a high level of expansion. In the covalent ceramic, much of the expansion would be absorbed by the directional bonds and the open spaces in the structure so that the total expansion of the material would be minimized.

REFERENCES

1. R. D. Shannon and C. T. Prewitt, Effective ionic radii in oxides and fluorides, Acta Crystallogr. B25, 925 (1969).
2. W. D. Kingery, H. K. Bowen, and D. R. Uhlmann, Introduction to Ceramics, 2nd ed., John Wiley & Sons, Inc., New York, 1976, Chap. 2.
3. L. Pauling, The Nature of the Chemical Bond, 3rd ed., Cornell University Press, Ithaca, N.Y., 1960.
4. L. V. Azaroff, Introduction to Solids, McGraw-Hill Book Company, New York, 1960.
5. A. F. Wells, Structural Inorganic Chemistry, 3rd ed., Clarendon Press, Oxford, 1962.
6. R. C. Evans, An Introduction to Crystal Chemistry, 2nd ed., Cambridge University Press, London, 1964.
7. R. W. G. Wyckoff, Crystal Structures, Vols. 1-4, Interscience Publications, New York, 1948-1953.
8. R. H. Doremus, Glass Science, John Wiley & Sons, Inc., New York, 1973.
9. H. Rawson, Inorganic Glass-Forming Systems, Academic Press, Inc., New York, 1967.
10. L. H. Van Vlack, Elements of Materials Science, 2nd ed., Addison-Wesley Publishing Co., Reading, Mass., 1964, Chaps. 3, 7, and Appendix F.
11. F. W. Billmeyer, Jr., Textbook of Polymer Science, Interscience Publications, New York, 1962.
12. A. X. Schmidt and C. A. Marlies, Principles of High Polymer Theory and Practice, McGraw-Hill Book Company, New York, 1948.
13. G. F. D'Alelio, Fundamental Principles of Polymerization, John Wiley & Sons, Inc., New York, 1952.

2
Physical, Thermal, Electrical, Magnetic, and Optical Properties

Chapter 1 introduced the basic concepts of atomic bonding and crystal structure and pointed out property trends for classes of materials. This chapter and the next two chapters explore the interrelationships among bonding, crystal structure, and properties in greater detail and compare properties of individual materials, with the objective of giving the reader the understanding necessary to make a judicious ceramic material selection for the required application.

2.1 PHYSICAL PROPERTIES

Density

Density is a measure of the mass per unit volume and is reported in units such as grams per cubic centimeter or pounds per square inch. The term "density" used alone can be interpreted in various ways, so to be fully descriptive, modifiers assuring the intended interpretation should be used:

1. Crystallographic density—the ideal density that would be calculated from the continuous defect-free crystal lattice of the composition
2. Specific gravity—the same as crystallographic density
3. Theoretical density—the same as crystallographic density, but taking into account solid solutions and multiple phases
4. Bulk density—the measured density of a ceramic body, which includes all lattice defects, phases, and fabrication porosity

In this book the term "density" is used for the first three items. These all have the characteristic of zero fabrication porosity, so that the only open space is between the atoms in the structure. For all cases where fabrication porosity is present, the terms "bulk density" or "percent of theoretical density" are used.

The density is determined by the size and weight of the elements and by the tightness of packing of the structure. Elements of low atomic number and atomic weight (H, Be, C, Si, etc.) result in materials with low crystallographic density. Conversely, elements of high atomic number and atomic weight (W, Zr, Th, U, etc.) result in materials with high crystallographic density. Examples are listed in Table 2.1. Note that tungsten carbide (WC) has a density that is five times that of silicon carbide (SiC).

Organic materials have a low density because they are essentially structured from C and H and other low-atomic-weight elements, such as Cl and F.

Atomic packing has less effect on density than the factors noted above. Close packing in metals and ionic-bonded ceramics results in higher density than for the relatively open structures of covalent-bonded ceramics. For instance, considering only atomic weight in zirconium oxide (ZrO_2) and zircon ($ZrSiO_4$), one would expect zircon to have the higher density. However, the zircon structure is quite open due to the Si—O covalent bonding and has a lower density than ZrO_2 (4.65 g/cm^3 compared to 5.8 g/cm^3).

High-temperature polymorphs typically have lower density than the lower-temperature polymorph. Glasses have a lower density than the crystallized structure of the same composition.

Theoretical density can be calculated from crystal structure data and the atomic weights of the elements involved. Bulk density of a ceramic body can be determined by several approaches. If the body has a simple uniform geometrical shape (such as a solid cylinder or a rectangular bar) that can be measured accurately, the bulk density (B) can be determined easily by calculating the volume from the physical measurements and dividing this into the dry weight. The following are examples for simple shapes:

Solid cylinder
$$B_c = \frac{D}{V_c} = \frac{D}{\pi r^2 h} \qquad (2.1)$$

Rectangular bar or plate
$$B_r = \frac{D}{V_r} = \frac{D}{\ell wh} \qquad (2.2)$$

where B_c and B_r and V_c and V_r are, respectively, the bulk density and volume of the solid cylinder and the rectangular bar, D is the measured weight of the shape, r is the radius, h is the height, ℓ is the length, w is the width, and π is 3.14.

The bulk density of complex shapes is determined by Archimedes' principle, where the difference in the weight of the part in air compared to its weight suspended in water permits calculation of the volume. Parts with no surface-connected porosity can be immersed directly in water. Parts containing surface-connected porosity must either be coated with a wax or other impervious material of known density or boiled as defined in American

Table 2.1 Specific Gravity of Ceramic, Metallic, and Organic Materials

Material	Composition	Density (g/cm^3)
	Ceramic materials	
α-Aluminum oxide	α-Al_2O_3	3.95
γ-Aluminum oxide	γ-Al_2O_3	3.47
Aluminum nitride	AlN	3.26
Mullite	$Al_6Si_2O_{13}$	3.23
Boron carbide	B_4C	2.51
Boron nitride	BN	2.20
Berylium oxide	BeO	3.06
Barium titanate	$BaTiO_3$	5.80
Diamond	C	3.52
Graphite	C	2.1-2.3
Fluorite	CaF_2	3.18
Cerium oxide	CeO_2	7.30
Chromium oxide	Cr_2O_3	5.21
Spinel	$MgAl_2O_4$	3.55
Iron aluminum spinel	$FeAl_2O_4$	4.20
Magnetite	$FeFe_2O_4$	5.20
Hafnium oxide	HfO_2	9.68
Spodumene	$LiAlSi_2O_6$	3.20
Cordierite	$Mg_2Al_4Si_5O_{18}$	2.65
Magnesium oxide	MgO	3.75
Forsterite	Mg_2SiO_4	3.20
Quartz	SiO_2	2.65
Tridymite	SiO_2	2.27
Cristobalite	SiO_2	2.32
Silicon carbide	SiC	3.17
Silicon nitride	Si_3N_4	3.19

Table 2.1 (continued)

Material	Composition	Density (g/cm^3)
Titanium dioxide	TiO_2	4.26
Tungsten carbide	WC	15.70
Zirconium oxide	ZrO_2	5.80
Zircon	$ZrSiO_4$	4.65
	Metals	
Aluminum	Al	2.7
Iron	Fe	7.87
Magnesium	Mg	1.74
1040 Steel	Fe-base alloy	7.85
Hastelloy X	Ni-base alloy	8.23
HS-25 (L605)	Co-base alloy	9.13
Brass	70 Cu-30 Zn	8.5
Bronze	95 Cu-5 Sn	8.8
Silver	Ag	10.4
Tungsten	W	19.4
	Organic materials	
Polystyrene	Styrene polymer	1.05
Teflon	Polytetrafluoroethylene	2.2
Plexiglass	Polymethyl methacrylate	1.2
Polyethylene	Ethylene polymer	0.9

Society for Testing and Materials Specification ASTM C373 [1]. The latter technique permits direct measurement of bulk density, open porosity, water absorption, and apparent specific gravity and indirect assessment of closed porosity. The procedure involves first measuring the dry weight (D). The part is then boiled in water for 5 hr and then allowed to cool in the water for 24 hr. The wet weight in air W and the wet weight suspended in water S are then measured. The following can then be calculated:

Exterior volume $V = W - S$ $\hspace{6cm}$ (2.3)

Bulk density $B = \dfrac{D}{V}$ $\hspace{6cm}$ (2.4)

Apparent porosity $P = \dfrac{W - D}{V}$ $\hspace{5cm}$ (2.5)

Volume of impervious material $= D - S$ $\hspace{3.5cm}$ (2.6)

Apparent specific gravity $T = \dfrac{D}{D - S}$ $\hspace{4cm}$ (2.7)

Water absorption $A = \dfrac{W - D}{D}$ $\hspace{4.5cm}$ (2.8)

The apparent specific gravity can then be compared with the true specific gravity (usually available from handbooks or by crystallographic calculations), to estimate the amount of closed porosity.

Measurement of bulk density is necessary in the characterization of the properties of a ceramic material. The amount and distribution of porosity strongly influences strength, elastic modulus, oxidation resistance, wear resistance, and other important properties.

Melting Temperature

Many ceramic applications result directly from the high melting temperature of some ceramic materials. However, not all ceramic materials have high melting temperatures. For instance, B_2O_3 melts at only 460°C (860°F) and NaCl melts at 801°C (1474°F).

Table 2.2 lists approximate melting temperatures for some ceramic and metallic materials. High melting temperature is a result of high bond strength. Weakly bonded alkali metals and monovalent ionic ceramics have low melting temperatures. More strongly bonded transition metals (Fe, Ni, Co, etc.) and multivalent ionic ceramics have much higher melting temperatures. Very strongly bonded covalent ceramics have the highest melting temperatures.

Organic materials have low melting or decomposition temperatures because of the weak van der Waals bonding between molecules. Linear structures such as thermoplastics melt, whereas network structures such as thermosetting resins tend to decompose or degrade. Cross-linking or chain branching tends to increase the melting temperature of thermoplastic compositions.

Table 2.2 Melting Temperatures of Ceramic, Metallic, and Organic Materials

Material	Approximate melting temperature	
	°C	°F
Polystyrene (GP grade)	65–75	150–170[a]
Polymethyl methacrylate	60–90	140–200[a]
Na metal	98	208
Polyethylene	120	250[a]
Nylon 6[b]	135–150	275–300[a]
Polyimides	260	500[a]
Teflon	290	550[a]
B_2O_3	460	860
Al metal	660	1220
NaCl	801	1474
Ni-base superalloy (Hastelloy X)	1300	~2370
Co-base superalloy (Haynes 25)	1330–1410	2425–2570
Stainless steel (304)	1400–1450	2550–2650
CaF_2		
Fused SiO_2	~1650	3000
Si_3N_4[b]	~1750–1900	3180–3450
Mullite	1850	3360
Al_2O_3	2050	3720
Spinel	2135	3875
B_4C	2425	4220
SiC[b]	2300–2500	4170–4530
BeO	2570	4660
ZrO_2 (stabilized	2500–2600	4530–4710
MgO	2620	4750
WC	2775	5030
UO_2	2800	5070
TiC	3100	5520
ThO_2	3300	5880
W metal	3370	6010
C[a]	3500	6240
HfC	3890	6940

[a] Maximum temperature for continuous use.
[b] Sublimes.

2.2 THERMAL PROPERTIES

Heat Capacity

The heat capacity is the energy required to raise the temperature of a material and is reported in units of cal/mol·°C. It is also often reported in terms of "specific heat" with units of cal/g·°C. The heat capacity of ceramic materials increases with temperature up to around 1000°C, above which little further increase occurs.

Thermal Conductivity

Thermal conductivity (k) is the rate of heat flow through a material [2,3] and is usually reported in units of cal/sec·cm^2·°C·cm, where calories are the amount of heat, cm^2 is the cross section through which the heat is traveling, and cm is the distance the heat is traveling. Figure 2.1 shows the thermal conductivity of a variety of ceramic, metallic, and organic materials as a function of temperature [4,5]. In the following paragraphs we explore the parameters that affect thermal conductivity.

The amount of heat transfer is controlled by the amount of heat energy present, the nature of the heat carrier in the material, and the amount of dissipation. The heat energy present is a function of the volumetric heat capacity c. The carriers are electrons or phonons, where phonons can be thought of simply as quantized lattice vibrations. The amount of dissipation is a function of scattering effects and can be thought of in terms of attenuation distance for the lattice waves (sometimes referred to as the mean free path).

The thermal conductivity k is directly proportional to the heat capacity c, the quantity and velocity of the carrier v, and the mean free path λ [6].

$$k \propto cv\lambda \qquad\qquad\qquad\qquad (2.9)$$

Increasing the heat capacity, increasing the number of carriers and their velocity, and increasing the mean free path (i.e., decreasing attenuation or scattering) results in increased thermal conductivity.

In metals the carriers are electrons. Because of the nature of the metallic bond, these electrons are relatively free to move throughout the structure. This large number of carriers plus the large mean free path results in the high thermal conductivity of pure metals. Alloying reduces the mean free path, resulting in decreased thermal conductivity for metals such as the superalloys.

Most organic materials have low thermal conductivity due to the covalent bonding, the large molecule size, and the lack of crystallinity. Rubber, polystyrene, polyethylene, nylon, polymethyl methacrylate, Teflon, and most other common commercial organic materials have thermal conductivity

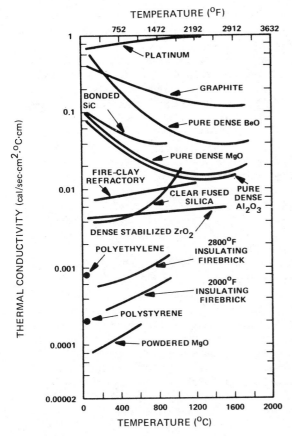

Figure 2.1 Thermal conductivity versus temperature for ceramic, metallic, and organic materials. (From Refs. 4 and 5.)

values at room temperature in the range 0.0002 to 0.0008 cal/sec·cm^2·°C·cm. Foamed polymers provide very good thermal insulation. The thermal conductivity of solid polymers can be increased by adding conductive fillers such as metals or graphite, but the other properties are also substantially modified.

In ceramics the primary carriers of thermal energy are phonons and radiation. The highest conductivities are achieved in the least cluttered structures, i.e., structures consisting of a single element, structures made up of elements of similar atomic weight, and structures with no extraneous atoms in solid solution.

Diamond and graphite are good examples of single-element ceramic structures having high thermal conductivity. Diamond has a thermal conductivity at room temperature of 9 W/cm·K, which is more than double that of copper. Because of its layer structure, graphite is anisotropic. Within the layers the bonding is strong and periodic and does not result in severe scattering of thermally induced lattice vibrations, resulting in high thermal conductivity in this direction (8.4 W/cm·K). Only weak van der Waals bonding occurs between layers, and lattice vibrations are quickly attenuated, resulting in much lower thermal conductivity in this direction (2.5 W/cm·K) [7].

BeO, SiC, and B_4C are good examples of ceramic materials composed of elements of similar atomic weight and size and that have high thermal conductivity. Lattice vibrations can move relatively easily through these structures because the lattice scattering is small. In materials such as UO_2 and ThO_2, where there is a large difference in the size and atomic weight of the anions and cations, much greater lattice scattering occurs and the thermal conductivity is low. UO_2 and ThO_2 have less than $1/10$ of the thermal conductivity of BeO and SiC. Materials such as MgO, Al_2O_3, and TiO_2 have intermediate values.

Solid solution decreases thermal conductivity. Solid solution refers to the substitution of an atom into the lattice position of another atom without resulting in a change to a new crystal structure. Solid solution occurs mostly with atoms of similar size. Si^{4+}, Al^{3+}, and Ge^{4+} having ionic radii (based on a coordination number of 4) of 0.26, 0.39, and 0.40 Å, respectively, typically replace each other in solid solution. Mg^{2+} and Ni^{2+} are so close in size (0.72 Å versus 0.69 Å based on 6 coordination) that they can replace each other in any proportion in MgO and NiO without a change in structure. However, even though the crystal structure does not change, the slight difference in ionic size and electron distributions results in enough lattice distortion to increase lattice wave scattering, as shown for MgO and NiO in Fig. 2.2 [4]. Other examples include $Al_6Si_2O_{13}$ (mullite), which has a much lower thermal conductivity than $MgAl_2O_4$ (magnesium aluminate spinel), which in turn has a lower thermal conductivity than Al_2O_3 or MgO.

Temperature has a strong effect on the thermal conductivity of ceramic materials. To understand the mechanisms we need to examine the relationship $k \propto cv\lambda$ as a function of temperature. The heat capacity c initially increases, but approaches a constant. The velocity v remains relatively constant. The mean free path λ is inversely proportional to the temperature T (i.e., $\lambda \propto 1/T$).

For crystalline ceramics, where lattice vibrations are the primary mode of heat conduction, the effect of λ dominates and thermal conductivity decreases as temperature increases. For glasses the structure is disordered even at room temperature and λ is low to start with and does not change significantly as the temperature increases. Thus for glasses the heat capacity c has a major effect and the thermal conductivity typically increases with temperature, as shown for fused silica (SiO_2 glass) in Fig. 2.1.

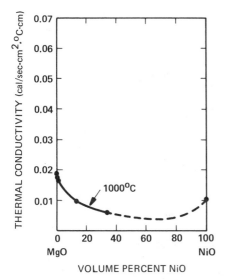

Figure 2.2 Effect of solid solution on thermal conductivity in the system MgO-NiO. (From Ref. 4.)

Radiation can significantly increase thermal conductivity as temperature is increased, especially for glass, transparent crystalline ceramics, and porous ceramics. Radiation is proportional to a power function of T and is usually in the range $T^{3.5}$ to T^5 for ceramic materials. The increase in conductivity with temperature for powdered MgO and insulating firebrick in Fig. 2.1 is due to radiation across the pores. In spite of this increase, the thermal conductivity of these porous materials is still very low and is an important factor in their application as high-temperature thermal insulation.

Thermal conductivity is an important property in determining the candidate material for many applications. Low thermal conductivity and high temperature stability are required for furnace linings used in numerous applications vital to industrial technology and our current standard of living, i.e., metals refining (including steelmaking), glass melting, electronics component processing, and dinnerware manufacturing, to name a few. High thermal conductivity is required for heat exchangers for energy conservation, for some braze fixtures, and for heat engine components where thermal stresses must be minimized. Most of these applications and the specific materials are discussed in later chapters, primarily in Chap. 13.

Thermal Expansion

The amplitude of atomic vibration within a structure increases as the temperature increases. For close-packed structures such as metals and ionic

ceramics, the amplitudes for each atom accumulate to produce a relatively
high level of expansion of the overall component. For covalent ceramics
some of the amplitude of vibration of individual atoms is absorbed by the
open space within the structure and by bond angle shifts. This results in a
much lower level of expansion of the overall component.

Thermal expansion data are reported in terms of the linear thermal
expansion coefficient α:

$$\alpha = \frac{\Delta \ell}{\ell_0 \, \Delta T} \qquad\qquad (2.10)$$

where ℓ_0 is the length at room temperature and $\Delta \ell$ is the change in length
for ΔT temperature increase. Units are typically in./in.·°C or cm/cm·°C.
Frequently, data are plotted in percent expansion versus temperature or in
parts per million expansion versus temperature. Thermal expansion coefficient versus temperature is plotted for important ceramic, metallic, and
organic materials in Fig. 2.3.

For cubic (isometric) single crystals and polycrystalline ceramics
(ceramics made up of many crystals with random orientation), the thermal
expansion coefficient is the same in all directions. For noncubic (anisometric) single crystals the thermal expansion coefficient is different for the
different crystallographic axes. Anisotropic expansion coefficients for
some anisometric crystals are listed in Table 2.3 [4].

Thermal expansion anisotropy is relatively small for $Al_6Si_2O_{13}$ (mullite),
Al_2O_3 (alumina), and TiO_2 (titania) compared to C (graphite), Al_2TiO_5
(aluminum titanate), and $CaCO_3$ (calcite). This can be explained in terms
of the crystal structures. Mullite, alumina, and titania all have structures
in which the bond strength is relatively uniform in each direction and the
atomic spacing varies only moderately. This results in only a small variation in thermal expansion along the different crystallographic axes. Graphite,
on the other hand, is a layer structure with very strong covalent bonding
within the layers but weak van der Waals bonding between layers. The thermal expansion within the layers is very low, as would be expected for a
strongly covalently bonded material, but is very high between the weakly
bonded layers. Aluminum titanate and calcite are not layer structures like
graphite, but do have structures with much different bonding and atomic
packing in the different crystallographic directions and thus also have highly
anisotropic thermal expansion.

When anisometric materials are fabricated into a polycrystalline ceramic
body the net thermal expansion can be very low, as shown for lithium aluminum silicate, $LiAlSi_2O_6$ (spodumene; also frequently called LAS) in Fig.
2.3. Such a material has very little dimensional change as a function of
temperature and can therefore withstand extreme thermal cycling or thermal shock without fracturing. Thermal shock resistance is discussed in
detail in Chap. 4.

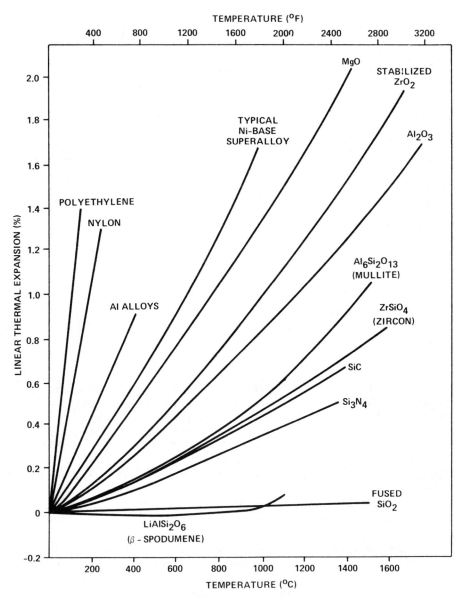

Figure 2.3 Thermal expansion characteristics of typical metals, polymers, and polycrystalline ceramics (from numerous sources.)

Table 2.3 Thermal Expansion Coefficients for Some Nonisotropic
Ceramics

	Linear thermal expansion coefficient	
Material	Normal to C axis $(\times 10^6$ per °C)	Parallel to C axis $(\times 10^6$ per °C)
Graphite	1	27
Al_2TiO_5 (aluminum titanate)	-2.6	11.5
$CaCO_3$ (calcite)	-6	25
Al_2O_3	8.3	9.0
$3Al_2O_3 \cdot 2SiO_2$ (mullite)	4.5	5.7
TiO_2	6.8	8.3
$ZrSiO_4$	3.7	6.2
SiO_2 (quartz)	14	9
$LiAlSi_2O_6$ (β-spodumene)	6.5	-2.0
$LiAlSiO_4$ (eucryptite)	8.2	-17.6

Source: Refs. 4, 26, and 27.

Low-thermal-expansion ceramic materials have broad potential for
both domestic and industrial applications. Perhaps the most frequently
encountered domestic application is the use of LAS-based polycrystalline
ceramics for heat-resistant cooking ware (Corningware) and stove tops.
Because of the low expansion of the LAS, the cooking ware can be removed
directly from an oven and immersed in cold water without breaking. Similarly, the stove top can withstand the very high temperature gradients between the position of the heating element and adjacent areas or when a pan
of frozen vegetables is placed directly on the preheated, red-hot burner.

Industrial applications for low-thermal-expansion ceramics are discussed in Chap. 13.

Displacive polymorphic transformations in which a rapid volume change
occurs show up on thermal expansion curves. Volume expansion curves for
the low cristobalite to high cristobalite, the low quartz to high quartz, and
the monoclinic to tetragonal ZrO_2 (zirconia) transformations are shown in
Fig. 2.4 [4]. It should be noted that the zirconia transformation results in
a decrease in volume rather than an expected increase. The significance
of this to material strength is discussed in Chap. 3.

Glasses and other noncrystalline solids range in thermal expansion behavior much like crystalline solids. However, in glasses the thermal expansion characteristics are controlled not only by the composition and the resulting structural aspects, but are also affected by the thermal history, i.e., temperature of the initial melt, rate of cooling, and subsequent heat treatments.

Low-thermal-expansion glasses are important in many applications which require a transparent material that is resistant to thermal shock. Some borosilicate glasses are in this category. Fused silica is also, and is one of the best thermal-shock-resistant materials available. Fused silica has also been fabricated in a porous foam which is used for lining critical surfaces of the space shuttle that are exposed to high temperatures during reentry. The low thermal expansion prevents thermal shock damage and the very low thermal conductivity protects underlying structures which are less thermal resistant.

Organic materials, especially the linear structures, have very high thermal expansion, as shown in Fig. 2.4. Polyethylene has a thermal expansion at room temperature of 180×10^{-6} cm/cm·°C. Vulcanized rubber, Teflon, and polystyrene have values of 81×10^{-6}, 99×10^{-6}, and 63×10^{-6} cm/cm·°C, respectively. The weak bonding between molecules results in the high thermal expansion. Polymers with network structures typically have lower thermal expansion because the bonding is stronger. The values for melamine-formaldehyde, urea-formaldehyde, and phenol-formaldehyde are 27×10^{-6}, 27×10^{-6}, and 72×10^{-6} cm/cm·°C, respectively [5]. These values are still substantially higher than the values for most ceramics and metals.

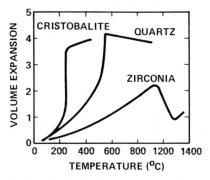

Figure 2.4 Volume expansion curves for cristobalite, quartz, and ZrO_2 showing the effects of displacive polymorphic transformations. (From Ref. 4.)

2.3 ELECTRICAL PROPERTIES

Ceramic materials have a wide range of electrical properties. Some do not allow passage of an electric current even in a very strong electric field and thus are excellent insulators. Others allow an electric current to pass only under certain conditions or when an energy threshold has been reached and thus are useful semiconductors. Still others do allow passage of an electric current and have application as electrical conductors. Some ceramics do not conduct electricity but undergo internal charge polarization which allows the material to be used for storage of an electrical charge in capacitors. The following sections describe why ceramics have these electrical properties and discuss some of the important applications.

Electrical Conductivity

The electrical conductivity of a material is somewhat analogous to thermal conductivity. The amount of conduction is a function of the amount of energy present (in this case, the size of the electric field), the number of carriers, and the amount of dissipation.

The carriers in metals are electrons. Because of the nature of the metallic bond, these electrons are relatively free to move throughout the structure and result in high electrical conductivity. This is especially true for pure metals, where atom size and packing are uniform and nothing is present to dissipate the free motion of the electrons. Alloying disrupts the uniformity of the structure and reduces the electrical conductivity. An increase in temperature also disrupts the structure (due to lattice vibration) and results in a decrease in electrical conductivity.

Most organic materials have poor electrical conductivity because they lack carriers and are thus used for electrical insulation or dielectric applications. Resistivities of most polymers are greater than 10^{10} Ω-cm. The resistivity can be reduced by addition of a filler such as graphite or powdered metal. In some cases, polymer structures have been modified to produce donor-acceptor sites and resistivities as low as 10^2 Ω-cm have been achieved.

Ceramic materials show a broad range of electrical conductivity behavior. Table 2.4 compares the electrical resistivity (reciprocal of electrical conductivity) of some metals, polymers, and ceramics at room temperature.

Like metals, some transition metal oxides, such as ReO_3, CrO_2, VO, TiO, and ReO_2, conduct by electrons [4]. This results from an overlap of electron orbitals such that wide unfilled d or f bands are present. Under the influence of an electric field, these electrons are relatively free to move and carry charge through the material. The presence of impurities can reduce the conductivity slightly by scattering. An increase in temperature has the same effect.

Table 2.4 Electrical Resistivity of Some Metals,
Polymers, and Ceramics at Room Temperature

Material	Resistivity (Ω-cm)
Metallic conduction	
Copper	1.7×10^{-6}
Iron	10×10^{-6}
Tungsten	5.5×10^{-6}
ReO_3	2×10^{-6}
CrO_2	3×10^{-5}
Semiconductors	
SiC	10
B_4C	0.5
Ge	40
Fe_3O_4	10^{-2}
Insulators	
SiO_2	$>10^{14}$
Steatite porcelain	$>10^{14}$
Fire-clay brick	10^8
Low-voltage porcelain	$10^{12}-10^{14}$
Al_2O_3	$>10^{14}$
Si_3N_4	$>10^{14}$
MgO	$>10^{14}$
Phenol-formaldehyde	10^{12}
Vulcanized rubber	10^{14}
Teflon	10^{16}
Polystyrene	10^{18}
Nylon	10^{14}

Source: Compiled from Refs. 4-7.

In some ceramic materials, especially oxides and halides, ions can be carriers to provide electrical conduction. The degree of conductivity is largely dependent on the energy barrier that must be overcome for the ion to move from one lattice position to the next. At low temperature, conductivity is low. However, when the temperature is high enough to overcome the barrier to lattice diffusion, the conductivity increases. The presence of lattice defects such as vacancies and interstitials in the structure aids conduction. Controlled impurities can be added to increase the concentration of defects and thus increase the electrical conductivity.

Ionically conductive ZrO_2 is used in sensitive oxygen detection devices. The conduction is due to diffusion of oxygen ions through the crystal lattice. The rate of diffusion is dependent on the temperature and the oxygen partial pressure in the environment. The oxygen diffusion is measured in the device either in terms of voltage or current, depending on the construction of the device. Response time is less than 1 sec. Such devices are used in automobile and truck engines to monitor air-fuel mixtures and the resulting combustion efficiency. They are used in a similar fashion in high-temperature furnaces, but as a means of controlling the fuel-air mixture to assure that excess oxygen is present and that fuel is not wasted by burning too rich a fuel mixture. Obviously, this also helps control air pollution.

Semiconductors

Other metal oxides, such as CoO, NiO, Cu_2O, and Fe_2O_3, have an energy gap between the filled and empty electron bands such that conduction will occur only when external energy is supplied to bridge the energy gap. These materials are semiconductors. An increase in temperature can provide the energy. For instance, NiO, Fe_2O_3, and CoO are insulators at low temperature with conductivities of less than 10^{-16} $(\Omega\text{-cm})^{-1}$. In the range 250 to 1000 K the conductivity increases nearly linearly to values in the range 10^{-4} to 10^{-2} $(\Omega\text{-cm})^{-1}$.

Semiconducting properties can be achieved in many ceramics by doping or by forming lattice vacancies through nonstoichiometry. Examples of ceramics responsive to this approach include TiO_2, ZnO, CdS, $BaTiO_3$, Cr_2O_3, Al_2O_3, and SiC.

Ceramics with semiconductor properties have many important applications. Cu_2O is used as a rectifier, a device that allows voltage to flow only in one direction and can thus convert alternating current to direct current. Semiconductor spinel materials such as Fe_3O_4, diluted in controlled solid solution with nonconducting spinel materials such as $MgAl_2O_4$, $MgCr_2O_4$, and Zn_2TiO_4, can be used as a thermistor, a device that has carefully controlled electrical resistance as a function of temperatures [8].

SiC can be doped to yield a semiconductor material with high-temperature stability that is used for a variety of resistance-heating-element applications. By control of the resistivity of the SiC and the cross section, almost any desired operating conditions up to about 1500°C (2732°F) can be achieved. The electricity has to work to get through the material, resulting in the production of heat. Increasing the resistivity, decreasing the cross section of the heating element, or increasing the applied voltage increases the amount of heat produced.

Figure 2.5 shows some configurations of SiC heating elements. The larger heating elements are used primarily for industrial applications where a controlled, clean heat source is required. The smaller ones are igniters. They have been developed for use in home appliances such as gas clothes dryers and gas stoves to replace the pilot light. The igniters are safer and save on fuel consumption. Prior to development of these igniters for home appliances, approximately 35 to 40% of the gas used was wasted by the pilot light. It has been estimated that this wastage for appliances in the United States was over 6 million cubic feet of natural gas per day prior to widespread installation of igniters. Initial igniters were developed for gas clothes dryers and became available on a limited basis in 1968. Currently, about 750,000[*] per year are being manufactured. Marketing of igniters for gas ranges started in 1974 and has now reached a volume of 1,700,000[*] per year.

Semiconduction properties can also be achieved or controlled by mixing a conducting or semiconducting material with a nonconducting material, often without significantly affecting other properties. Table 2.5 shows the variation in resistivity for near-theoretical density Si_3N_4 containing additions of SiC [9]. Table 2.6 shows the heating response of some of these variations. The electrical resistance R was calculated using the formula

$$R = \frac{E}{I} \qquad\qquad\qquad (2.11)$$

where E is the applied voltage and I the measured current. The resistivity ρ was calculated using the formula

$$\rho = \frac{AR}{\ell} \qquad\qquad\qquad (2.12)$$

where A is the cross-sectional area of the specimen and ℓ the gauge length of the specimen.

Electrical Insulators

As shown in Table 2.4, most pure oxide and silicate ceramics are very resistant to passage of electricity and are thus good insulators. This high

---

[*]Approximations supplied by Duncan McKeown, Carborundum Co., Niagara Falls, N.Y.

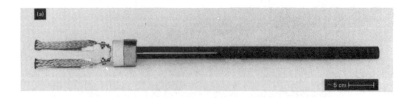

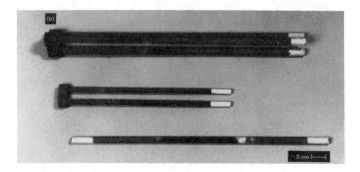

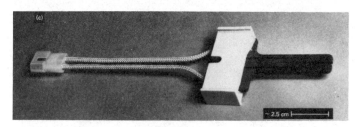

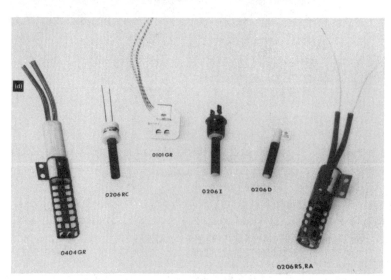

Figure 2.5 Variety of SiC heating elements. (Parts a and d courtesy of Carborundum Co., Niagara Falls, N.Y. Parts b and c courtesy of Norton Co., Worcester, Mass.)

Table 2.5 Mechanical and Electrical Properties of Si_3N_4-SiC
            Compositions

Composition (Si_3N_4/SiC)	Bulk density (g/cm^3)	MOR[a] at 20°C (psi)	MOR[b] at 1375°C (psi)	Resistivity (Ω-cm)
100/0	3.25	104,000	53,000	10^{10}
90/10	3.27	80,200	--	10^6
85/15	3.33	105,300	--	50,000
82.5/17.5	3.32	115,200	--	720
80/20	3.05	121,800[b]	58,700	136
77.5/22.5	3.31	97,400	--	35
75/25	3.30	125,100	46,600	13
72.5/27.5	3.34	103,400	52,000	8.2
70/30	3.18	105,400[b]	46,700	4.5
67.5/32.5	3.39	108,800	--	3.3
65/35	3.39	120,700	46,000	2.0
60/40	3.03	106,500[b]	46,800	1.9

[a] 4-point flexure test.
[b] 3-point flexure test.
Source: Ref. 9.

electrical resistivity, combined with properties such as chemical inertness
and high temperature stability, has led to many important applications.
    Different applications require different characteristics. For instance,
substrates (thin plates of insulator material that other electrical components
and circuits are attached to) must have acceptably high strength, chemical
inertness, and electrical resistance, but also must have high thermal conductivity and fine surface finish. The high thermal conductivity is required
to remove heat built up by resistors. Under some conditions 250°C (482°F)
can be generated, which is higher than most organic substrates can withstand and high enough to result in chemical changes in some glass compositions. For this type of application, Al_2O_3 works best. The Al_2O_3 is more
resistant to the temperature, but also has a higher thermal conductivity
and reduces the temperature by dissipating the heat. The thermal conductivity of Al_2O_3 is about 25 times that of typical glass compositions. Al_2O_3
also has the advantage that it can be trimmed to size and have holes drilled
using a laser [10].

Table 2.6 Heating Response of Si_3N_4–SiC Composites

(Si_3N_4–SiC) Composition (%)	Temperature (°C)	Time (min)	Current (A)	AC voltage (V)	Calculated resistance (Ω)	Calculated resistivity (Ω–cm)
80–20	Under 200	No change in 4 min	<0.1	45	--	--
70–30	R.T.	0	0.1	45	450	17.8
	375	3	0.2	45	225	8.9
	600	7	0.3	45	150	5.9
60–40	R.T.	0	1.0	45	45.0	1.8
	600	0.75	1.6	45	19.6	0.8
	850	3	3.1	45	14.5	0.6
80–20	R.T.	0	--	120	300	11.9
	600	1.5	0.4	120	200	7.9
	750	5	0.6	120		
70–30	200	0.15	1	120	120	4.7
	600	0.35	2	120	60	2.4
	1200	1	3	120	40	1.6
60–40	R.T.	0	2.4	80	33	1.3
	600	0.15	2.7	80	30	1.2
	1200	0.5	5	80	16	0.7
	1300	1	5.8	80	14	0.5
70–30	R.T.	0	0.3	80	266	10.5
	600	1.75	1.1	80	73	2.9
	700	3	1.5	80	53	2.1
60–40	R.T.	0	1.5	80	53	2.1
	200	0.1	2	80	40	1.6
	600	0.3	4	80	20	0.8
	1400	0.7	7	80	11	0.5

[a]R.T., room temperature.
Source: Adapted from Ref. 9.

Close control over surface finish is also necessary for substrate applications. Electrical leads and components applied to the substrates are often quite thin for our modern miniaturized devices. For example, in one case Ta is anodized on an Al_2O_3 substrate to form an insulating oxide layer 2000 Å thick. A discontinuity roughly 250 Å (1 μin.) thick can cause a short in the coating and result in malfunction of the whole device [10].

Another interesting application for ceramics is insulators for high-voltage power lines. Most people are aware that ceramics are used for these insulators, but do not realize how demanding the application really is. First, the insulators must be very strong because they support the weight of the power lines. Next time you drive through the countryside, notice how far apart the power line towers are and imagine the load the insulators must endure during a heavy wind or snow storm. Then consider how long these insulators are designed to last. They must be highly reliable. Second, the insulator must be resistant to weather damage and to absorption of water. Internal absorption can result in arcing at the high voltages involved.

Another demanding application that we are all aware of but take for granted is the spark plug insulator for the internal combustion engine [11]. These are made of high-alumina compositions that have very high electrical resistivity. In most spark plugs the alumina insulator must withstand several thousand volts at each spark discharge, a pressure pulse of about 10.4 MPa (1500 psi), and the thermal radiation from 2400°C (4350°F) combustion temperature. The spark plug must survive many hours of these conditions repeated at a frequency of about 25 to 50 cycles per second.

Dielectric Properties

Ceramic insulator materials are often useful due to their dielectric properties. Dielectric refers to the polarization that occurs when the material is placed in an electric field. The negative charge shifts toward the positive electrode and the positive charge toward the negative electrode. The total polarizability resulting from a combination of electronic, ionic, and dipole orientation effects is referred to as the dielectric constant (K'). Dielectric constants for some ceramic and organic materials are listed in Table 2.7.

Another important dielectric property is the dielectric strength, defined as the capability of the material to withstand an electric field without breaking down and has units of volts per unit of thickness of the dielectric material. Although accurate values are hard to obtain experimentally, the following general ranges may be cited: titanates, 100 to 300 V/mil; micas, 125 to 5500 V/mil; and phenolics, ~2000 V/mil. In general, single-crystal ceramics have higher dielectric strength than polycrystalline ceramics. Organic materials have relatively high dielectric strength, accounting for their widespread use as electrical insulation.

Table 2.7 Dielectric Constants for Ceramic and
Organic Materials

Material	K'
NaCl	5.9
Mica	2.5-7.3
MgO	9.6
BeO	6.5
Al_2O_3	8.6-10.6
TiO_2	15-170
Porcelain	5.0
Fused SiO_2	3.8
High-lead glass	19.0
$BaTiO_3$	1600
$BaTiO_3 + 10\%\ CaZrO_3 + 1\%\ MgZrO_3$	5000
$BaTiO_3 + 10\%\ CaZrO_3 + 10\%\ SrTiO_3$	9500
Rubber	2.0-3.5
Phenolic	7.5

Source: Compiled from Refs. 7, 8, 10, and 12.

One important application of dielectric materials is in the fabrication of
capacitors. The charge that can be stored in a capacitor is dependent on the
dielectric constant of the material between the capacitor plates. For example, if the material has $K' = 10$, the capacitor can store 10 times as much
charge as it could if only vacuum were present between the plates. Materials with a high dielectric constant, such as $BaTiO_3$, make possible substantial miniaturization of capacitors. The dielectric properties are dependent
on the strength of the electric field and the temperature; pure $BaTiO_3$ is not
optimum for all applications. However, Pb^{2+}, Sr^{2+}, Ca^{2+}, and Cd^{2+} can be
substituted for some of the Ba^{2+} ions and Zr^{4+}, Ce^{4+}, Th^{4+}, Sn^{4+}, and Hf^{4+}
can be substituted for a portion of the Ti^{4+} to produce a wide range of solid
solution compositions from which a composition having the required properties can be selected.

The availability of ceramic dielectric materials that allow miniaturization of capacitors is extremely important in our modern transistorized society.

E. C. Henry provides a dramatic yet amusing example [12]. A typical transistorized pocket radio contains several 3/8-in.-diameter $BaTiO_3$-type capacitors. The best dielectric material available in 1910 for capacitor fabrication was naturally-occurring mica, with a dielectric constant of less than 10. To supply the needs of our pocket radio, each capacitor made with mica would have to be about 40 cm (15.5 in.) across. In 1943 TiO_2 was the best capacitor dielectric, with a dielectric constant of about 80. With TiO_2, a capacitor for our pocket radio could now be reduced to about 9 cm (3.5 in.) across.

Some dielectric materials, including $BaTiO_3$, have ferroelectric behavior [12]. Ferroelectricity is defined as the spontaneous alignment of electric dipoles within the material and is analogous to the alignment of magnetic dipoles in ferromagnetic materials. In $BaTiO_3$ the spontaneous alignment involves the Ti^{4+} ions, which are in octahedral sites in the perovskite crystal structure. Because of the large size of the Ba^{2+} ions, the Ti^{4+} ion is almost too small for the octahedral site and tends to position itself in an off-center position, resulting in an electric dipole. This is only true for the tetragonal polymorph of $BaTiO_3$, which is the stable polymorph below 120°C (248°F). $BaTiO_3$ is cubic between 120°C (248°F) and 1460°C (2660°F) and hexagonal between 1460°C (2660°F) and 1618°C (2940°F), but neither of the polymorphs is ferroelectric. The temperature at which ferroelectric properties cease is referred to as the Curie temperature, which is 120°C (248°F) for $BaTiO_3$.

Other ceramic materials that are ferroelectric include $PbTiO_3$, $NaTaO_3$, $LiTaO_3$, WO_3, KH_2PO_4, and Rochelle salt (potassium-sodium tartrate tetrahydrate). Each has a different Curie temperature and different response to an electric field. $PbTiO_3$ has a Curie temperature of 480°C (896°F) and can thus be used at a higher temperature than $BaTiO_3$.

The ferroelectric materials are anisotropic and therefore have different electrical properties in the different crystallographic directions. When an electric voltage is applied, a mechanical distortion results. This piezoelectric property has led to widespread use of $BaTiO_3$ and other ferroelectric ceramics as transducers in ultrasonic devices, microphones, phonograph pickups, accelerometers, strain gauges, and sonar devices.

Rochelle salt initially made the commercial phonograph possible by converting vibrations from the needle into electrical signals which could be amplified. Modern stereo phonographs use the more environmentally stable ceramics such as lead zirconate titanate.

Ferroelectric ceramics are used in communication bandpass filters which can be tuned according to the resonance of the ferroelectric material to pass or receive a specific frequency and reject others. Similarly, they can filter out coded signals in the presence of noise or jamming.

Ceramic transducers are used widely in ultrasonic cleaners, ultrasonic machining, and ultrasonic nondestructive evaluation. They are also used in delay lines such as those required for radar systems. A transducer that is connected to a bar or rod (waveguide) receives an electrical input which it converts to an acoustic wave which passes along the waveguide. A transducer at the opposite end of the waveguide converts the acoustic wave back into the original electric input, but delayed according to the material and length of the waveguide. This is used in radar systems to compare information from one echo with the next echo and for range calibration. It is also used in color television sets and many other applications.

2.4 MAGNETIC PROPERTIES

Magnetic ceramics play an essential role in modern technology. Important applications include permanent magnets, memory units with rapid switching times in digital computers, and circuit elements in radio, television, microwave, and other electronic devices.

The intrinsic magnetic properties of a material are determined by the electronic structure and the crystal structure [8]. According to the Pauli exclusion principle, only two electrons can occupy a given energy level and these two electrons must have opposite spins. These spins produce a magnetic moment. If the energy level is occupied by both electrons, the magnetic moments cancel because of the opposite spins. A net magnetic moment occurs only when electron energy levels are half filled, such as in the transition elements, rare earth elements, and actinide elements and in some of the structures derived from these elements. Magnetic moments can exist in individual ions within a structure but be cancelled out by opposite moments in surrounding ions in the structure. Such is the case for FeO, for instance. Fe^{2+} ions in one plane all contain electrons with parallel spins but in the adjacent plane have electrons with opposite spin.

Ceramic materials having magnetic properties are commonly referred to as ferrites [13-15]. The ferrites are divided into classes according to crystal structure. Cubic ferrites have either the spinel structure or the garnet structure, hexagonal ferrites have the magnetoplumbite or related structures, and orthorhombic ferrites have the perovskite structure. Table 2.8 lists the composition ranges for each of these classes. All contain either transition elements with partially filled d electron shells or rare earth elements with partially filled f electron shells such that a net magnetic moment is present.

The hexagonal ferrites (especially barium, strontium, and lead hexaferrites) are frequently used for permanent magnets because of their high magnetization, compact size, and low cost.

Most household magnets used to latch cupboards, hang memos, and so on, are hexagonal ferrites. Ferrite powder is added to rubber or plastic as a filler and is used for the strip on refrigerator doors and acts as a latch and also provides sealing. Hexagonal ferrite magnets are also used in the motors of electric toothbrushes, electric knives, and automobile windshield

Table 2.8 Ferrite Composition and Structures

Cubic ferrites

   Spinel   General structure MFe_2O_4, where Fe is tri-
                                      valent and M is divalent Ni, Mn, Mg, Zn, Cu,
                                      Co, or a mixture

   Garnet   General structure $R_3Fe_5O_{12}$, where Fe is tri-
                                        valent and R is a trivalent rare earth, typical-
                                      ly Y or Gd

Hexagonal ferrites

   Various $BaFe_{12}O_{19}$, $Ba_2MFe_{12}O_{22}$, $BaM_2Fe_{16}O_{27}$,
    structures $Ba_3M_2Fe_{24}O_{41}$, $Ba_2M_2Fe_{28}O_{46}$,
                                   $Ba_4M_2Fe_{36}O_{60}$, where M is divalent Ni, Co,
                                    Zn, or Mg, and Ba can be replaced by Sr and
                                    Pb

Orthorhombic ferrites

   Perovskite   General structure $RFeO_3$, where R is a trivalent
                                      rare earth and Fe is trivalent and can be par-
                                      tially replaced by trivalent Ni, Mn, Cr, Co,
                                      Al, Ca, or V^{5+}

Source: Adapted from Ref. 13.

wipers, heater blowers, air conditioners, power windows, seat adjustors, and convertible-top raisers. In 1968, each automobile manufactured had approximately four ferrite motors [12].

Ferrite permanent magnets are also used in such diverse applications as in loudspeakers, the cyclotron particle accelerator, and in electricity meters (to support the weight of the rotor to reduce wear in the bearings).

Spinel and garnet ferrites have different properties than the hexagonal ferrites and are therefore used in different types of applications. One of the most important applications in the past was in digital computers for data storage. A square loop magnetic response is required for data storage, i.e., the material has to have two stable states of magnetization that can be switched from one to the other. One state represents the digit "0" and the other the digit "1," allowing storage of data. Data are retrieved by switching. The cycle time for switching is about 5 μsec, which allows for very rapid data handling. Each storage unit or core consists of a ferrite magnet ring or toroid less than 0.5 mm (0.020 in.) in diameter. Each computer of this type had many of these cores. Over 6 billion cores were manufactured in 1969. The use of ferrite memory cores has been decreasing since 1969 because of the development of alternative technologies more amenable to miniaturization and cost reduction. These technologies also utilize ceramics. Currently the predominant memory and data storage systems are MOS (metal oxide/silicon), disks, and tapes. The latter two use dispersed ferrite coatings and magnetic recording heads.

Rare earth garnet ferrites have also been used for data storage. One particularly interesting approach involved depositing a thin film (~ 5 μm thick) of the rare earth garnet material on a nonmagnetic substrate having a slightly different thermal expansion coefficient. During cooling, the mismatch in thermal contraction induced preferred magnetization perpendicular to the plane of the rare earth garnet layer and separated into alternating microscopic domains of opposite spin. These domains, which look like bubbles under a polarizing microscope, can provide the binary memory input for a digital computer [16]. Similar bubble domain structures have been achieved in thin single-crystal plates of orthorhombic ferrites [17].

The magnesium-zinc and nickel-zinc spinel ferrites are important in transformer and inductor applications, where a material with high permeability and low loss is required. The ceramic compositions have been very successful at competing with silicon-iron and nickel-iron metallic alloys for these applications. Some of the specific applications include television, radio, electronic ignition systems, high-frequency fluorescent lighting systems, Touch-Tone telephones, communication interference filters, high-frequency welding, submarine communications, and high-speed tape and disk recording heads.

The ferrites are also important for microwave and a variety of other applications. The reader is encouraged to refer to Ref. 13, which further describes the ferrites and their applications and contains an extensive list of other references where more detailed information is available.

The technology of magnetic ceramics is well developed. The magnetic properties can be controlled over a broad range by control of composition and processing. Conversely, lack of careful control of compositions and processing can prevent achieving the required properties.

2.5 OPTICAL PROPERTIES

The optical properties of a material include absorption, transparency, index of refraction, color, and phosphorescence. These properties are determined primarily by the level of interaction between the incident electromagnetic radiation and the electrons within the material.

Absorption and Transparency

Absorption and transparency are closely related optical properties. If the incident electromagnetic radiation stimulates electrons to move from their

initial energy level to a different energy level, the radiation is absorbed
and the material is opaque to this particular wavelength of radiation. Metals
have many open energy levels for electron movement and thus are opaque
to most wavelengths of electromagnetic radiation.

Ionic ceramics have filled outer electron shells comparable to inert gas
electron configurations and do not have energy levels available for electron
movement. Most single crystals of ionic ceramics are transparent to most
electromagnetic wavelengths. Covalent ceramics vary in their level of optical transmission. Ones that are good insulators and have a large band gap
transmit. Those that are semiconductors and have a small band gap can
transmit under some conditions but become opaque as soon as enough energy is present for electrons to enter the conduction band.

Optical absorption can also occur due to resonance. This results when
the frequency of the electromagnetic radiation is comparable to the natural
frequency of the material. The resulting oscillations in the material absorb the radiation and the material is optically opaque.

Absorption due to electron transition and resonance is intrinsic. Absorption can also result from extrinsic effects due to scattering by inclusions,
pores, grain boundaries, or other internal flaws. Such absorption is undesirable in most optical applications. For instance, much effort was expended
to develop a transparent polycrystalline Al_2O_3 material. The first problem
was to control the starting materials and processing to eliminate all porosity,
both in the grains and between grains. Until this was achieved, the transparency was severely reduced by scattering from pores. Even when the
pores were removed, though, additional scattering resulted from the grain
boundaries. This resulted because Al_2O_3 is anisotropic and the light passing
through the material was affected differently by grains oriented in different
directions.

This would not be a problem with polycrystalline Y_2O_3, which is isotropic .

Optical transparency is important in many applications. Glass and a
variety of ionic ceramics are transparent in the visible range of the spectrum.
This is the range between wavelengths of 0.4 and 0.7 μm. There are many
applications in this range for windows, lenses, prisms, and filters. Transparency in other wavelength ranges is important for electrooptical and
electromagnetic window materials for tactical and strategic missiles, aircraft, remotely piloted vehicles, spacecraft, battlefield optics, and high-
energy lasers. MgO, Al_2O_3, and fused SiO_2 are transparent in the ultraviolet (0.2 to 0.4 μm), a portion of the infrared (0.7 to 3.0 μm), and the
radar (>1000 μm) ranges. MgF_2, ZnS, ZnSe, and CdTe are transparent to
infrared and radar wavelengths [18].

Color

Color is another optical property which leads to many ceramic applications.
Color results from the absorption of a relatively narrow wavelength of radiation

within the visible region of the spectrum (0.4 to 0.7 μm). For this type of absorption, transition of electrons must occur. This occurs primarily where transition elements having an incomplete d shell (V, Cr, Mn, Fe, Co, Ni, Cu) or f shell (rare earth elements) are present. It can also result from nonstoichiometry.

The oxidation state and bond field are also important in color formation. For instance, neither S^{2-} nor Cd^{2+} cause visible absorption, but CdS produces strong yellow. Similarly, Fe^{3+} and S^{2-} are responsible for the color of amber glass.

Ceramic colorants are widely used as pigments in paints and other materials produced and used at low temperatures. They are especially important where processing is done at elevated temperature where other types of pigments are destroyed. For instance, porcelain enamels which are fired in the range 750 to 850°C require ceramic colorants. Ceramic having the spinel structure AB_2O_4 (such as blue $CoAl_2O_4$) are often used in this temperature range. Doped ZrO_2 and $ZrSiO_4$ are used at higher temperatures (1000 to 1250°C) because of their increased resistance to attack by the glass in which they are dispersed. Dopants include vanadium (blue), praseodymium (yellow), and iron (pink).

Phosphorescence

Phosphorescence is another important optical property displayed by some ceramic compositions. Phosphorescence is the emission of light resulting from the excitation of the material by the appropriate energy source. Ceramic phosphors are used in fluorescent lights, oscilloscope screens, TV screens, photocopy lamps, and other applications [19].

The fluorescent light consists of a sealed glass tube coated on the inside with a halogen phosphate [such as $Ca_5(PO_4)_3(Cl, F)$ or $Sr_5(PO_4)_3(Cl, F)$ doped with Sb and Mn] and filled with mercury vapor and argon. A capacitor provides an electric discharge which stimulates radiation of the mercury vapor at a wavelength of 2537 Å. This ultraviolet radiation excites a broad band of radiation in the visible range from the phosphor, producing the light source.

For oscilloscopes and TV sets the phosphor is excited by an electron beam which sweeps across the phosphor-coated screen. The decay time of the light emission of the phosphor is important. For color TV, decay occurs in approximately 1/10 to 1/100 of a second. A phosphor having slower decay is necessary for a radar screen. More than one phosphor is required for a color TV set, each being selected to emit a narrow wavelength of radiation corresponding to one of the primary colors. The most difficult color to achieve was red.

Initially $Zn_3(PO_4)_2$ doped with Mn was used for red, but it had low luminous efficiency and required high electron-beam currents. YVO_4 and

Y_2O_2S, both doped with Eu, have been developed more recently and have improved efficiency. The phosphorescence results from transitions of 4f electrons in the Eu [20].

Photocopy lamps also make use of phosphorescent ceramic coatings. $MgGa_2O_4$ doped with Mn is typically used, resulting in a narrow band of light emission in the green wavelength range. Strontium magnesium pyrophosphate doped with Eu^{2+} has also been used. It gives off light in the ultraviolet and blue range.

Lasers

LASER is the acronym for "light amplification by the stimulated emission of radiation" [21-23]. Ceramic materials that are currently most important for laser applications include Al_2O_3 doped with Cr^{3+} (the ruby laser), $Y_3Al_5O_{12}$ doped with Nd^{3+} (the yttrium aluminum garnet or YAG laser), and glass doped with Nd^{3+}. All emit radiation of a specific wavelength, which is determined by the dopant: 0.694 μm for the Cr^{3+}-doped Al_2O_3 and 1.06 μm for the Nd^{3+}-doped materials.

The ceramic component of the laser consists of a cylindrical rod typically 0.3 to 1.5 cm in diameter by 5 to 15 cm long, with the ends polished to a flatness of $\lambda/10$ (where $\lambda = 0.59$ μm) and parallelism of ±5 sec of arc. The rod must be as flaw free as possible to avoid losses due to scattering, and the dopant must be uniformly dispersed. Usually, a tungsten-iodine filament lamp or a rare-gas arc lamp is used to stimulate the rod. A small portion of the lamp output is absorbed by the dopant ions (the rest is dissipated as heat), resulting in electron transitions to a higher-energy state. As these electrons drop back into their initial energy state, light of a single wavelength specific to the dopant is emitted. Mirrors placed at the ends of the ceramic rod reflect the stimulated light back into the rod, where further coherent light is emitted and amplification results. The intensity can be built up and then released as a "pulse" by removing the mirror as in the case of a pulsed laser. Alternatively, the mirror can be only partially reflective and allow emission of a portion of the coherent light as a continuous beam, as in the case of a continuous laser.

Index of Refraction

As discussed earlier in the section on absorption and transparency, light interacts with the electrons in the material. For transparent materials, the degree of interaction varies, but results in a decrease in the velocity (v) of the light compared to its velocity in vacuum. The index of refraction η is defined as the ratio of the velocity of light in vacuum to the velocity in the material:

$$\eta = \frac{v_{\text{vacuum}}}{v_{\text{material}}} \tag{2.13}$$

The index of refraction for a material varies according to the wavelength of incident radiation. This is called <u>dispersion</u> and normally results in a decrease in η as wavelength increases. Therefore, when comparing the index of refraction of various materials, make sure that the wavelength was constant or that dispersion curves are available to make the necessary corrections.

The change in velocity when light passes from one material to another causes the light to bend or change direction. This is called <u>refraction</u> and is the effect that makes a fish in the water look like it is in a different position than it really is. The angle of refraction r is related to the angle of incidence i and the index of refraction by the equation

$$\sin r = \frac{\sin i}{\eta} \tag{2.14}$$

for the case where one medium is air or vacuum. This is shown schematically in Fig. 2.6. Note that a portion of the light is also reflected.

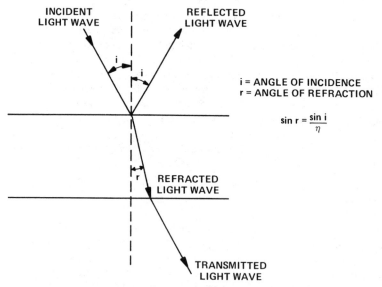

Figure 2.6 Primary refraction, reflection, and transmission for a plate of transparent material in air.

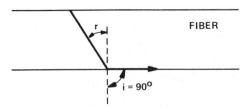

Figure 2.7 Minimum angular requirement for r to achieve total internal reflection within a fiber.

Equation (2.14) is used extensively in the design of optical devices, sometimes as a correction (as in lenses, to make sure that the image is of the appropriate size and at the desired focal plane) and sometimes to achieve special effects. For instance, it is desirable with gemstones to optimize the color and brilliance of the material. The stones are cut at angles that allow maximum light to enter and be retained to accentuate the color and brilliancy.

Another important example is fiber optics. For communications, the fiber must carry coherent light from a laser for large distances with minimal loss. To minimize losses, a refractive index and angle of refraction are selected such that total internal reflection is achieved. Figure 2.7 shows how total internal reflection is achieved. First, let us imagine that the fiber is being lighted internally from the left. We want to calculate the value of the internal angle that will result in refraction parallel to the length of the fiber. The minimum angle of r occurs when i = 90°, and the equation for this critical value of r becomes

$$\sin r_{crit} = \frac{\sin i}{\eta} = \frac{1}{\eta} \tag{2.15}$$

The value of r_{crit} for a typical glass (with $\eta = 1.5$) is about 42°. Any angle larger than this will result in total internal reflection. Because the critical angle for total reflection is relatively low, the light can be transmitted around corners without loss.

As discussed earlier, losses in transmission of light through a transparent material can result from scattering and absorption due to internal defects, inclusions, or inhomogeneity. Similar losses can occur in fibers due to surface defects. Scratches, grease, or dust on the surface can cause scattering or change the critical angle for total internal reflection. Interesting engineering approaches have been devised to avoid these losses. One approach is to clad (or encase) the fiber in a thin layer of glass having a lower index of refraction. Another approach is to provide a gradient in the index of refraction from the surface to the interior using ion exchange techniques. This results in the actual focusing of the light in a sinusoidal path along the fiber axis [24, 25].

PROBLEMS

2.1 An engineer is designing a furnace for hot pressing. He wishes to apply the heat circumferentially and the pressure axially. He would like to
obtain as uniform as possible a temperature in the interior of the furnace,
especially across the diameter, yet he needs to avoid excessive heat loss
axially, which might damage the hydraulic press. He could line the top and
bottom of the furnace with porous refractory, but available porous refractories are not strong enough to withstand the hot pressing pressures. What
material could he select that would help him achieve a uniform temperature
in the furnace, yet minimize axial heat loss? Why?

2.2 Porosity has a strong influence on the properties of ceramic materials.
Therefore, bulk density is routinely measured and used in the material
specification to determine if the material is acceptable or should be rejected.
What effect would you expect porosity to have on strength? on elastic
modulus? on oxidation or corrosion resistance? on thermal conductivity?
on optical transparency?

2.3 Fiber refractories have become important for lining furnaces. They
are used in the form of loose blankets (like fiberglass insulation), cotton-
like bundles, and fiber boards. What functions do they fulfill? What advantages do they have over conventional brick refractories?

2.4 Compare the mechanisms of thermal conductivity at room temperature
and high temperature for a glass, a polycrystalline ceramic, and a porous
refractory brick.

2.5 Sodium chloride (NaCl) single crystals are transparent and colorless,
but become colored when annealed at a suitable temperature in sodium vapor.
What is a possible mechanism?

2.6 Figure 2.3 shows the percent linear thermal expansion as a function
of temperature. The average coefficient of thermal expansion can be calculated from the curves for each material over any selected temperature
range. Calculate the average thermal expansion coefficient of Al_2O_3 from
room temperature to 1000°C. How does it compare with the values listed
in Table 2.3? Explain.

2.7 According to Table 2.6, a 70 Si_3N_4-30 SiC composition reached 1200°C
in 1 min at 120 V and 3 A. This material would be acceptable as the resistance heating element of an igniter for a household appliance. Assuming
that material could be fabricated into 4-mm-diameter rods, what length
would be required? If it could be fabricated only into flat rectangular plates,
suggest an alternative configuration and dimensions.

2.8 $BaTiO_3$ has piezoelectric properties and is used as a transducer.
What limits the use temperature of a $BaTiO_3$ transducer?

2.9 Explain why organic materials are such effective electrical insulators. Why are they also not good capacitor dielectrics?

2.10 Explain the source of magnetism in ceramic materials.

ANSWERS TO PROBLEMS

2.1 Graphite which has been processed in such a way that the crystals are all oriented in the same direction, so that the graphite has high thermal conductivity in one plane and low thermal conductivity in the direction perpendicular to this plane. Such an orientation results when graphite is produced by extrusion. The top and bottom blocks of the furnace can be machined from the extruded graphite so that the plane of high conductivity is perpendicular to the direction of hot pressing and will allow rapid heat transfer radially from the heat source to the furnace interior. The direction of low conductivity would then be parallel to the pressing direction and would minimize heat loss to the platens of the press.

2.2 Porosity increases oxidation or corrosion and decreases strength, elastic modulus, thermal conductivity, and optical transparency.

2.3 Much air space is present between fibers. This results in a very low density aggregate that has the low thermal conductivity necessary for furnace insulation. The advantages are the following:

Fiber blankets are flexible and permit easy furnace fabrication and repair. Resulting furnaces are lightweight and of minimum size.
Fiber-lined furnaces have low thermal inertia and can be heated or cooled
    very rapidly. Brick-lined furnaces are bulky and the brick acts as a
    heat sink that takes much longer to heat and cool.

2.4 The glass does not have a periodic lattice structure and thus cannot transmit heat by phonon conduction. It has no free electrons, so cannot transmit heat by electron conduction. The glass thus has low thermal conductivity. As the temperature is increased, the conductivity increases slightly because as the temperature continues to increase, further heat transfer occurs by radiation.

Polycrystalline ceramics conduct heat primarily by phonon conduction. As the temperature increases, scattering effects in the lattice increase and the thermal conductivity decreases.

Porous refractories are made of low-conductivity ceramics and contain a high volume of porosity to decrease the thermal conductivity further. The thermal conductivity increases with temperature due to radiation across the pores.

2.5 Excess sodium is incorporated into the structure, producing nonstoichiometry. This modified structure interacts differently with the incident light. A specific wavelength is absorbed, resulting in the coloration. Specifically, the excess sodium is compensated by the formation of chlorine ion vacancies which have each trapped an electron. These are referred to as **F centers**.

2.6 $\Delta T = 980°C$; from Fig. 2.3, $\Delta \ell \simeq 0.84\%$, or on a unit basis $\Delta \ell / \ell \simeq 0.0084$. Coefficient of thermal expansion $= \Delta \ell / \ell\ \Delta T = 0.0084/980°C = 8.6 \times 10^{-6}$ per °C. This is intermediate between the values 8.3 and 9.10 $\times 10^{-6}$ per °C listed in Table 2.3 for the coefficient in different crystallographic directions for Al_2O_3.

2.7 ρ = resistivity = $\dfrac{AR}{\ell}$ $R = \dfrac{E}{I}$

Combining these equations gives

$$\ell = \frac{AE}{\rho I} = \frac{\pi r^2 E}{\rho I} = \frac{\pi (0.2)^2 (120)}{(1.6)(3)} = 3.14 \text{ cm}$$

Possible rectangular configurations:

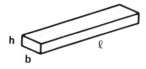

A = bh $\dfrac{A}{\ell} = \dfrac{\rho I}{E} = 0.04$

Theoretically, any combination of $A/\ell = 0.04$ would work. The best combination would be the one most easily manufactured and compatible with the size constraints of the device.

Alternative configuration:

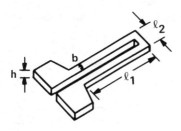

A = bh $\ell = 2\ell_1 + \ell_2$

$\dfrac{A}{\ell} = \dfrac{\rho I}{E} = 0.04$

Many possible combinations of b, h, ℓ_1, and ℓ_2.

2.8 Only the tetragonal polymorph of $BaTiO_3$ has ferroelectric properties. This polymorph is stable only up to 120°C (248°F).

2.9 Organic materials have very low electrical conductivity plus a very
high dielectric strength and are thus excellent electrical insulators. However, because of the large molecular size, organic materials do not undergo much polarization in an electric field and therefore have a low dielectric
constant.

REFERENCES

1. ASTM Specification C373, ASTM Standards, Part 13, American Society
   for Testing and Materials, Philadelphia, 1969.
2. C. Kittel, Introduction to Solid State Physics, 3rd ed., John Wiley &
   Sons, Inc., New York, 1968.
3. M.I.T., Thermal conductivity, J. Am. Ceram. Soc. 37[2], 67-110(1954).
4. W. D. Kingery, H. K. Bowen, and D. R. Uhlmann, Introduction to
   Ceramics, 2nd ed., John Wiley & Sons, Inc., New York, 1976,
   Chap. 12.
5. L. H. Van Vlack, Elements of Materials Science, 2nd ed., Addison-
   Wesley Publishing Co., Reading, Mass., 1964, p. 420.
6. L. H. Van Vlack, Physical Ceramics for Engineers, Addison-Wesley
   Publishing Co., Reading, Mass., 1964, Chap. 9.
7. Kirk-Othmer Encyclopedia of Chemical Technology, Vol. 4, 3rd ed.
   (H. F. Mark, D. F. Othmer, C. G. Overberger, and G. T. Seaborg,
   eds.), John Wiley & Sons, Inc., New York, 1978.
8. W. D. Kingery, H. K. Bowen, and D. R. Uhlmann, Introduction to
   Ceramics, 2nd ed., John Wiley & Sons, Inc., New York, 1976,
   Chaps. 17-19.
9. D. W. Richerson, U.S. Patent 3,890,250, Hot Pressed Silicon Nitride
   Containing Finely Dispersed Silicon Carbide or Silicon Aluminum Oxynitride (1975).
10. D. G. Thomas, in Physics of Electronic Ceramics, Part B (L. L.
    Hench and D. B. Dove, eds.), Marcel Dekker, Inc., New York, 1972,
    pp. 1057-1090.
11. E. Ryshkewitch, Oxide Ceramics, Academic Press, Inc., New York,
    1960, p. 251.
12. E. C. Henry, Electronic Ceramics, Doubleday & Company, New York,
    1969.
13. E. E. Riches, Ferrites, A Review of Materials and Applications,
    Mills & Boon Ltd., London, 1972.
14. E. C. Snelling, Soft Ferrites, ILIFFE Books Ltd., London, 1969.
15. Yasushi Hoshino, Shuichi Iida, and Mitsuo Sugimoto, eds., Ferrites,
    University Park Press, Tokyo, 1971.
16. A. H. Bobeck, E. G. Spencer, L. G. Van Uitert, S. C. Abrahams,
    R. L. Barns, W. H. Grodkiewicz, R. C. Sherwood, P. H. Schmidt,

D. H. Smith, and E. M. Walters, Uniaxial magnetic garnets for domain wall "bubble" devices, Appl. Phys. Lett. 17, 131 (1970).

17. A. H. Bobeck, R. F. Fischer, A. J. Perneski, J. P. Remeika, and
    L. G. van Uitert, Trans. IEEE MAG-5, 544 (1969).

18. S. Musikant, R. A. Tanzilli, R. J. Charles, G. A. Slack, W. White,
    and R. M. Cannon, Advanced Optical Ceramics, General Electric Co.,
    Document No. 78SDR2195, prepared as a final report for ONR contract
    N00014-77-C-0649, 1978.

19. H. L. Burrus, Lamp Phosphors, Mills & Boon Ltd., London, 1972.

20. M. J. Taylor, in Modern Oxide Materials (B. Cockayne and D. W.
    Jones, eds.), Academic Press, Inc., London, 1972, pp. 120-146.

21. A. F. Harvey, ed., Coherent Light, Wiley-Interscience, London,
    1970.

22. A. Sona, ed., Lasers and Their Applications, Gordon and Breach
    Science Publishers, New York, 1976.

23. B. Cockayne, in Modern Oxide Materials (B. Cockayne and D. W.
    Jones, eds.), Academic Press, Inc., London, 1972, pp. 1-28.

24. W. D. Kingery, H. K. Bowen, and D. R. Uhlmann, Introduction to
    Ceramics, 2nd ed., John Wiley & Sons, Inc., New York, 1976,
    Chap. 13.

25. J. A. Cole, Communications through a glass wire, Ceramic Industry,
    April 1977, pp. 28-30.

26. W. Ostertag, G. R. Fischer, and J. P. Williams, Thermal expansion
    of synthetic β-spodumene and β-spodumene-silica solid solutions,
    J. Amer. Ceram. Soc. 51(11), 651-654 (1968).

27. F. H. Gillery and E. A. Bush, Thermal contraction of β-eucryptite
    ($Li_2O \cdot Al_2O_3 \cdot 2SiO_2$) by x-ray and dilatometer methods, J. Am. Ceram.
    Soc. 42(4), 175-177 (1959).

Mechanical Properties
and Their Measurement

The mechanical properties of a material determine its limitations for structural applications where the material is required to sustain a load. To make a judicious material selection for such applications, it is useful to understand mechanical properties terminology, theory, and test approaches as well as to obtain the specific property data for the candidate materials.

The objective of this chapter is to review the basic principles of elasticity and strength [1-5] and to develop an understanding of why the actual strength of a ceramic component is far below its theoretical strength. This information is applied concurrently to an evaluation of the property data measurement techniques and the limitations that must be considered when using these data for material selection and component design.

3.1 ELASTICITY

When a load is applied to a material, deformation occurs due to a slight change in the atomic spacing. The load is defined in terms of stress σ, which is typically in units of pounds per square inch (psi) or megapascals (MPa). The deformation is defined in terms of strain ϵ, which is typically in units of inches (or centimeters) of deformation per inches (or centimeters) of the initial length or in percent.

The amount and type of strain is dependent on the atomic bond strength of the material, the stress, and the temperature. Up to a certain stress limit for each material the strain is reversible; i.e., when the stress is removed, the atomic spacing returns to its original state and the strain disappears. This is referred to as elastic deformation, and the stress and strain are related by a simple proportionality constant. For tensile stress

$$\sigma = E \epsilon \tag{3.1}$$

and the proportionality constant E is called the modulus of elasticity or Young's modulus.

For shear loading

$$\tau = G\gamma \tag{3.2}$$

where τ is the shear stress, γ the shear strain, and G the proportionality constant, referred to as the shear modulus or the modulus of rigidity.

At ambient and intermediate temperatures for short-term loading, most ceramics behave elastically with no plastic deformation up to fracture, as illustrated in Fig. 3.1a. This is known as brittle fracture and is one of the most critical characteristics of a ceramic that must be considered in design for structural applications.

Metals also behave elastically up to a certain stress, but rather than fracture in a brittle manner like ceramics, most metals deform in a ductile manner as the stress is further increased. This is referred to as plastic deformation or plastic strain and is not reversible. Some metals, aluminum for instance, have a smooth transition from elastic strain to plastic strain, as shown in Fig. 3.1b. Others, low-carbon steels for instance, have a discontinuity at the outset of plastic strain. This is called the yield point and is illustrated in Fig. 3.1c.

Not all ceramics behave in a brittle fashion and not all metals behave in a ductile fashion. Most ceramic materials undergo plastic deformation at high temperature. Even at room temperature ceramics such as LiF, NaCl, and MgO undergo plastic deformation, especially under sustained loading. These ceramics all have the rock salt structure, which has cubic symmetry and thus has many slip systems available for plastic deformation by dislocation movement.

Pure metals have the greatest degree of ductile behavior. Addition of alloying elements reduces ductility to the point where some metals are brittle at room temperature. Cast iron is a good example. Intermetallic compounds also have little or no ductility.

Modulus of Elasticity

As shown in equation (3.1), the modulus of elasticity E is the proportionality constant between elastic stress and elastic strain and can be thought of simply as the amount of stress σ required to produce unit elastic strain ϵ.

$$E = \frac{\sigma}{\epsilon} \tag{3.3}$$

Considering E on the atomic level under conditions of tensile stress which tend to pull the material apart, it will be recognized that as the load increases, the interatomic spacing increases. The stronger the bonding, the greater the stress required to increase the interatomic spacing, and thus the higher the value of the modulus of elasticity. Ceramics with strong

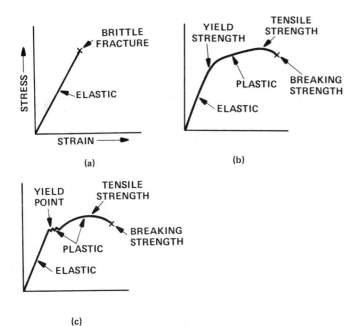

Figure 3.1 Types of stress-strain behavior. (a) Brittle fracture typical of ceramics. (b) Plastic deformation with no distinct yield point. (c) Plastic deformation with yield point.

covalent bonding have the highest E values. Diamond has an E of 150×10^6 psi. Ceramics with weak ionic bonding have much lower E values. NaCl has an E of 6.4×10^6 psi. Metals are intermediate. Iron-base alloys and superalloys typically have an E of about 30×10^6 psi. Typical values for other ceramics, metals, and organics are listed in Table 3.1 [6,7,8].

Bond strength, and thus E, varies in different crystallographic directions. This anisotropy must be considered when dealing with single crystals and the crystallographic orientation defined when reporting elastic modulus data. However, most ceramic products are made up of many crystals in random orientation. These polycrystalline ceramics have a uniform elastic modulus in all directions which is equal to the average of the moduli for the various crystallographic directions. The elastic modulus values most commonly reported for ceramic materials are average values for polycrystalline bodies. Even though these polycrystalline ceramics have an apparent single elastic modulus, the engineer must be aware that the individual crystals within the microstructure are anisotropic and that internal stresses may be present which can affect the application of the material.

Table 3.1 Typical Room-temperature Elastic Modulus Values for
Important Engineering Materials

| | Average elastic modulus, E | |
Material	GPa	psi
Rubber	0.0035-3.5	$5 \times 10^2 - 5 \times 10^5$
Nylon	2.8	0.4×10^6
Polymethyl methacrylate	3.5	0.5×10^6
Urea-formaldehyde	10.4	1.5×10^6
Bulk graphite	6.9	1×10^6
Concrete	13.8	2×10^6
NaCl	44.2	6.4×10^6
Aluminum alloys	69	10×10^6
Fused SiO_2	69	10×10^6
Typical glass	69	10×10^6
ZrO_2	138	20×10^6
Mullite ($Al_6Si_2O_{13}$)	145	21×10^6
UO_2	173	25×10^6
Iron	197	28.5×10^6
MgO	207	30×10^6
Ni-base superalloy (IN-100)	210	30.4×10^6
Spinel ($MgAl_2O_4$)	284	36×10^6
Si_3N_4	304	44×10^6
BeO	311	45×10^6
Al_2O_3	380	55×10^6
SiC	414	60×10^6
TiC	462	67×10^6
Diamond	1035	150×10^6

Source: Data from Refs. 6, 7, and 8.

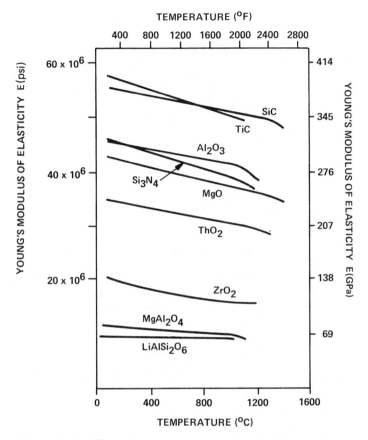

Figure 3.2 Effect of temperature on the elastic modulus.

Figure 3.2 shows the effect of temperature on the elastic moduli of various ceramics. In each case E decreases slightly as the temperature increases. This results from the increase in interatomic spacing due to thermal expansion. As the interatomic spacing increases, less force is necessary for further separation.

Many materials that will be encountered by an engineer are made up of more than one composition or phase and will have an elastic modulus intermediate between the moduli of the two constituent phases. Examples are cermets (Co-bonded WC, for instance), glass- or carbon-reinforced organics, oxide-dispersion-strengthened metals, and glass-bonded ceramics (machinable glass-bonded mica, for instance). In cases where the elastic modulus value is not available, it can be estimated using the law of mixtures:

$$E = E_a V_a + E_b V_b \tag{3.4}$$

where E_a and E_b are the elastic moduli of the constituents, V_a and V_b the volume fractions, and E the estimated elastic modulus of the mixture. This is a simplified relationship and is suitable only for rough estimates.

Porosity also affects the elastic modulus, always resulting in a decrease. MacKenzie [9] has derived a relationship for estimating the elastic modulus of porous materials:

$$E = E_0 (1 - 1.9P + 0.9P^2) \tag{3.5}$$

where E_0 is the elastic modulus of nonporous material and P the volume fraction of pores. This relationship is valid for materials containing up to 50% porosity and having a Poisson's ratio of 0.3.

Elastic Modulus Measurement

Two techniques are commonly used to measure the elastic modulus. The first involves direct measurement of strain as a function of stress, plotting the data graphically, and measuring the slope of the elastic portion of the curve. This technique can be conducted accurately at room temperature using strain gauges, but is limited at temperatures above which strain gauges can be reliably attached. Some individuals have calculated the elastic modulus from load-deflection curves obtained from tensile strength testing. Such data may provide a rough approximation, but it is not likely to be accurate, because the deflection curve typically contains other components in addition to the elastic strain of the material, i.e., deflection of fixtures and looseness in both the load train and data recording systems.

A second method for determining elastic modulus is based on measurement of the resonant frequency of the material and calculation of E from the equation

$$E = CMf^2 \tag{3.6}$$

where C is a constant depending on the specimen size and shape and on Poisson's ratio, M is the mass of the specimen, and f is the frequency of the fundamental transverse (flexural) mode of vibration. E can also be determined using the longitudinal or torsional vibration modes, but the equations will be different. These equations plus tables that give values for C can be found in ASTM Specification C747 [10].

This technique can be used accurately over the complete temperature range and for the various crystallographic directions of single crystals as well as for the average elastic modulus of a polycrystalline material.

Poisson's Ratio

When a tensile load is applied to a material, the length of the sample increases slightly and the thickness decreases slightly. The ratio of the thickness decrease to the length increase is referred to as Poisson's ratio v:

$$v = - \frac{\Delta d/d}{\Delta \ell/\ell} \qquad (3.7)$$

This is shown schematically in Fig. 3.3.

Poisson's ratio typically varies from 0.1 to 0.5. Values for various materials at room temperature are listed in Table 3.2.

For isotropic and polycrystalline ceramics, Poisson's ratio, Young's modulus, and the shear modulus are related by the equation

$$E = 2G(1 + v) \qquad (3.8)$$

3.2 STRENGTH

The term "strength" is ambiguous for both metals and ceramics. One must use modifiers to be specific, i.e., yield strength, tensile strength, compressive strength, flexural strength, ultimate strength, fracture (or breaking) strength, and theoretical strength. The following sections discuss the

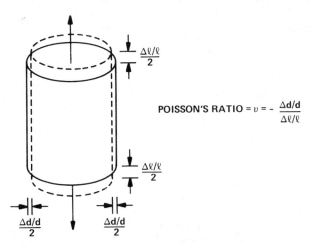

$$\text{POISSON'S RATIO} = v = - \frac{\Delta d/d}{\Delta \ell/\ell}$$

Figure 3.3 Physical definition of Poisson's ratio.

Table 3.2 Poisson's Ratio at Room
Temperature for Various Materials

Material	Approximate Poisson's ratio
SiC	0.14
$MoSi_2$	0.17
HfC	0.17
Concrete	0.20
B_4C	0.21
Si_3N_4	0.24
SiO_2	0.25
Al_2O_3	0.26
Steels	0.25-0.30
Most metals	0.33
BeO	0.34
MgO	0.36

types of strength and attempt to provide the reader with an understanding of
the strength characteristics of ceramics and the criteria that must be considered when selecting a ceramic material for a structural application.

Theoretical Strength

Theoretical strength can be defined as the tensile stress required to break
atomic bonds and pull a structure apart. The equation

$$\sigma_{th} = \left(\frac{E\gamma}{a_0}\right)^{1/2}$$

(3.9)

has been derived for estimating the theoretical strength under tensile loading,
where σ_{th} is the theoretical strength, E the elastic modulus, a_0 the interatomic spacing, and γ the fracture surface energy. The theoretical strength
for ceramic materials typically ranges from 1/10 to 1/5 of the elastic modulus.
    Aluminum oxide (Al_2O_3), for instance, has an average elastic modulus
of 380 GPa (55×10^6 psi) and would thus have a theoretical strength in the

range 38 GPa (5.5×10^6 psi) to 76 GPa (11×10^6 psi). However, the theoretical strength of a ceramic material has not been achieved. This is due to the presence of fabrication flaws and structural flaws in the material, which result in stress concentration and fracture at a load well below the theoretical strength.

Table 3.3 compares the theoretical strengths of Al_2O_3 and silicon carbide (SiC) with typical tensile strengths reported for specimens fabricated by different approaches. Most ceramic products are fabricated from the polycrystalline approach. The fracture strengths of the polycrystalline versions of SiC and Al_2O_3 are only about $1/100$ of the theoretical strength.

Effects of Flaw Size

The presence of a flaw such as a crack, pore, or inclusion in a ceramic material results in stress concentration. Inglis [11] showed that the stress concentration at the tip of an elliptical crack in a nonductile material is

$$\frac{\sigma_m}{\sigma_a} = 2\left(\frac{c}{\rho}\right)^{1/2} \tag{3.10}$$

where σ_m is the maximum stress at the crack tip, σ_a the applied stress, $2c$ the length of the major axis of the crack, and ρ the radius of the crack tip.

To obtain an idea of the effect of flaws on stress concentration, one can assume that the crack tip radius is approximately equal to the atomic spacing

Table 3.3 Comparison of Theoretical Strength and Actual Strength

Material	E [GPa (psi)]	Estimated theoretical strength [GPa (psi)]	Measured strength of fibers [GPa (psi)]	Measured strength of polycrystalline specimen [GPa (psi)]
Al_2O_3 [a]	380 (55×10^6)	38 (5.5×10^6)	16 (2.3×10^6)	0.4 (60×10^3)
SiC	440 (64×10^6)	44 (6.4×10^6)	21 (3.0×10^6)	0.7 (100×10^3)

[a] From R. J. Stokes, in The Science of Ceramic Machining and Surface Finishing, NBS Special Publication 348, 1972, U.S. Government Printing Office, Washington, D.C., p. 347.

a_0 (~2 Å) and use some recent flaw size and strength data for reaction-bonded silicon nitride (Si_3N_4). For a flaw size c of 170 μm (~0.007 in.) fracture occurred at 21.7×10^3 psi (150 MPa). Substituting in equation (3.10) the stress concentration factor is 1840. It is apparent that even a small flaw in ceramics is extremely critical and leads to substantial stress concentration.

Griffith [12] proposed an equation of the form

$$\sigma_f = A\left(\frac{E\gamma}{c}\right)^{1/2} \tag{3.11}$$

for relating the fracture stress to the material properties and the flaw size, where σ_f is the fracture stress, E the elastic modulus, γ the fracture energy, c the flaw size, and A is a constant that depends on the specimen and flaw geometries.

Evans and Tappin [13] have presented a more general relationship:

$$\sigma_f = \frac{Z}{Y}\left(\frac{2E\gamma}{c}\right)^{1/2} \tag{3.12}$$

where Y is a dimensionless term that depends on the flaw depth and the test geometry, Z is another dimensionless term that depends on the flaw configuration, and c is the depth of a surface flaw (or half the flaw size for an internal flaw) and E and γ are defined as above. For an internal flaw that is less than 1/10 of the size of the cross section under tensile loading, $Y = 1.77$. For a surface flaw that is much less than 1/10 of the thickness of a cross section under bend loading, Y approaches 2.0. Z varies according to the flaw shape, but is usually between 1.0 and 2.0.

The effect of a planar elliptical crack at the surface of a ceramic specimen is the easiest to analyze. This type of crack results commonly from machining, but can also occur due to impact, thermal shock, glaze crazing, or a number of other causes. The Z value depends on the ratio of the depth of the flaw (c) to the length of the flaw (ℓ), as illustrated in Fig. 3.4.

The effects of three-dimensional flaws such as pores and inclusions have not been analyzed as rigorously. However, it is evident that the severity of strength reduction is affected by a combination of factors:

1. The shape of a pore
2. The presence of cracks or grain boundary cusps adjacent to a pore
3. The distance between pores and between a pore and the surface
4. The size and shape of an inclusion
5. The differences in elastic moduli and coefficients of thermal expansion between the inclusion and the matrix

These factors are discussed individually in the following paragraphs.

(a)

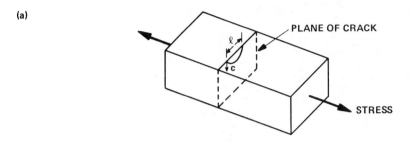

(b)

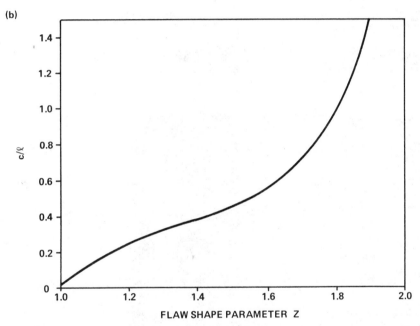

Figure 3.4 Flaw shape parameter Z values for elliptical surface crack
morphologies. (a) Geometry. (b) Z versus c/ℓ. (Reprinted with permission from Progress in Materials Science, Vol. 21, Pts. 3/4: Structural
Ceramics, A. G. Evans and T. G. Langdon, © 1976, Pergamon Press, Ltd.

Pore Shape A simple spherical pore theoretically would have less stress
concentration effect than a sharp crack. However, pores in ceramics are
not perfectly spherical. Some are roughly spherical but most are highly
irregular. Roughly spherical pores can result due to air entrapment during
processing as shown in Fig. 3.5a for a specimen of reaction-bonded silicon
nitride (Si_3N_4) fabricated by slip casting. This specimen fractured at 150
MPa (21,800 psi) during 4-point bend testing. Several models were evaluated to relate the flaw size to the fracture stress [14]. The best correlation

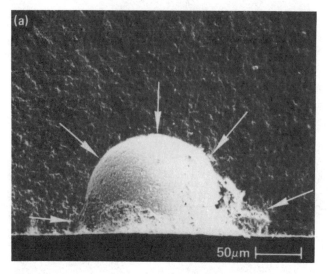

(a)

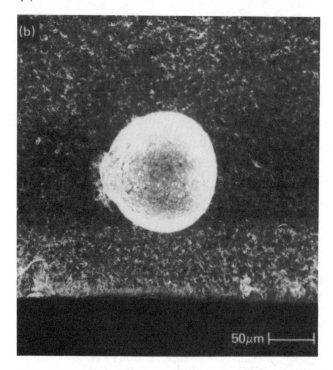

(b)

Figure 3.5 Scanning electron photomicrographs of fracture surfaces of reaction-bonded silicon nitride containing nearly spherical pores resulting from air entrapment during processing. Arrows outline flaw dimensions used to calculate fracture stress.

between measured and calculated fracture stress was obtained by assuming that the pore was equivalent to an elliptical crack of the cross section outlined by the arrows in Fig. 3.5a.

For this calculation, $Y = 2$, $Z = 1.58$ (from Fig. 3.4), $E = 219 \times 10^9$ N/m^2 (32×10^6 psi), and $\gamma = 11.9$ J/m^2, resulting in a calculated σ_f of 157 MPa (22,800 psi). This is very close to the value measured by 4–point bend testing.

Pore-Crack Combinations   The simplest and most common combination between a pore and a crack involves the intersection of the pore with grain boundaries of the ceramic material. If the pore is much larger than the grain size of the material (as in the previous example), the extremities of the pore provide a good approximation of the critical flaw size. If the size of the pore approaches the size of the grains, the effect of cracks along the grain boundaries probably will predominate, and the effective flaw size will be larger than that of the pore. Evans and Langdon [14] provide a more detailed treatment of pores, cracks, and pore crack combinations and provide an extensive list of references.

Internal Pores   The effect on strength of an internal pore depends on the shape of the pore and the position of the pore with respect to the surface. If the pore is close to the surface, the bridge of material separating it from the surface may break first and result in a critical flaw whose dimensions would be the size of the pore plus the bridge. In this case, the measured strength is likely to be less than would be predicted simply by the dimensions of the pore and less than if the pore intersected the surface [13]. Experimental data by the author for internal spherical pores which were within about half a radius of the surface correlated well with this ligament theory, resulting in calculated stresses that were within 10% of the measured values [15]. However, for pores progressively farther from the surface, the ligament theory did not apply. In fact, other theories proposed in the literature also did not give acceptable correlation. It appears that further study is necessary if we are to understand more fully the stress concentration effects of internal pores and to derive mathematical relationships that provide accurate predictions of fracture stress.

Figure 3.5b shows an internal nearly-spherical pore in reaction-bonded Si_3N_4 that was approximately one radius from the surface. The measured bend strength was 215 MPa (31,200 psi). This represents the peak tensile stress at the specimen surface at the time of fracture. Since the tensile stress in a bend specimen decreases linearly inward from the surface (reaching zero at the midplane of the bar), the stress at the plane of the flaw can easily be calculated. At the plane intersecting the deepest portion of the pore, the corrected stress was 182 MPa (26,400 psi). At the plane intersecting the centerline of the pore, the corrected stress was 193 MPa (28,000 psi). Using the ligament theory, the calculated fracture stress was

126 MPa (18,300 psi), well below the measured fracture stress. Assuming a flaw size equal to the pore diameter and Y = 1.77 (for an internal flaw) and the same E and γ values as used earlier, the calculated stress was 182 MPa (26,400 psi). This appears to be excellent correlation. However, of the eight specimens evaluated in this way, the average difference between measurement and calculation was 16%, with the highest being 37%. This further illustrates that the effect of internal flaws is not well enough understood for accurate quantitative analysis and that additional experimental and analytical effort is required.

Pore Clusters   If a group of pores are close together, the material bridges between them can crack first, linking the pores together and producing a much larger flaw that results in much lower strength. The probability is high that pores separated by less than one pore radius will link.

Inclusions   Inclusions typically occur in ceramic materials due to contamination of the ceramic powders during processing. Sources of contamination are discussed in detail in Part II. The degree of strength reduction due to an inclusion depends on the thermal and elastic properties of the inclusion compared to the matrix material. Thermal expansion differences can result in cracks forming adjacent to the inclusion during cooling from the fabrication temperature. Elastic modulus difference can result in the formation of cracks when a stress is applied. The worst decrease in strength occurs when the inclusion has a low coefficient of thermal expansion and a low elastic modulus compared to the matrix material. In this case, the effective flaw size is larger than the visible inclusion size. It is equivalent to the inclusion size plus the length of the adjacent cracks.

Inclusions with a high thermal expansion coefficient or a high elastic modulus have less effect on strength. These conditions produce circumferential cracks rather than radial cracks, and the effective flaw size approaches that of a flat elliptical crack equivalent in elliptical dimensions to the inclusion.

Strength Measurement

Tensile Strength   Strength can be measured in a number of different ways, as illustrated in Fig. 3.6. Tensile strength testing is typically used for characterizing ductile metals. A metal tensile test specimen is attached to threaded fixtures of any universal test machine that can provide a calibrated pull load at controlled rate. Yield strength, breaking strength, and elongation are measured in a single test. The tensile strength is defined as the maximum load P (the stress at fracture for a ceramic) divided by the original cross-sectional area A:

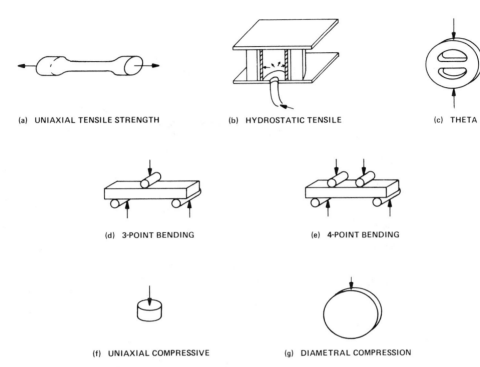

(a) UNIAXIAL TENSILE STRENGTH   (b) HYDROSTATIC TENSILE   (c) THETA

(d) 3-POINT BENDING   (e) 4-POINT BENDING

(f) UNIAXIAL COMPRESSIVE   (g) DIAMETRAL COMPRESSION

Figure 3.6 Schematics of strength tests.

$$\sigma_t = \frac{P}{A} \tag{3.13}$$

Ceramics are not normally characterized by tensile testing due to the high cost of test specimen fabrication and the requirement for extremely good alignment of the load train during testing. Any misalignment introduces bending and thus stress concentration at surface flaws, which results in uncertainty in the tensile strength measurement. For accurate tensile strength measurement with ceramics, strain gauges must be used to determine the amount of bending and stress analysis must be conducted to determine the stress distribution within the test specimen. A ceramic tensile test specimen is shown in Fig. 3.7. This specimen was designed for testing in a very sophisticated test facility at Southern Research Institute, Birmingham, Alabama, in which close alignment was achieved through the use of gas bearings [16].

Tensile strength can also be measured by applying a hydrostatic load to the inside of a thin-walled hollow cylinder specimen configuration (Fig. 3.6b) [17]. This has been used for room-temperature strength measurement

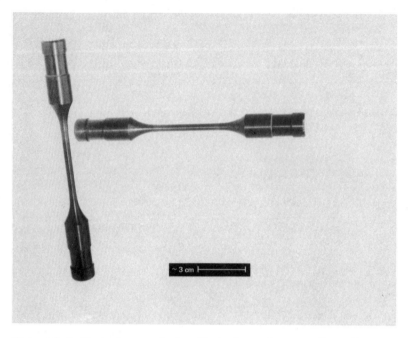

Figure 3.7 Typical ceramic tensile test specimen configuration.

but has not been adapted for elevated temperatures. Achieving adequate
seal of the pressurizing fluid at high temperature would be difficult. Another
limitation of this test is the likelihood for fracture to occur at flaws on the
corners at the ends of the hollow cylinder. It is difficult to machine these
corners without producing chips or cracks. The chips or cracks can be
removed by radiusing the edges, but this results in a reduced specimen
thickness which adversely modifies the stress distribution during testing.

Another method of obtaining tensile strength of a ceramic material is
known as the theta test [18]. The configuration is shown in Fig. 3.6c.
Application of a compressive load to the two arches produces a uniaxial
tensile stress in the crossbeam. Very little testing has been conducted with
this configuration owing largely to difficulty in specimen fabrication.

Compressive Strength Compressive strength is the crushing strength of
a material, as shown in Fig. 3.6d. It is rarely measured for metals, but
is commonly measured for ceramics, especially those that must support
structural loads, such as refractory brick or building brick. The compressive strength of a ceramic material is usually much higher than the
tensile strength so that it is often beneficial to design a ceramic component
so that it supports heavy loads in compression rather than tension. In fact,

Table 3.4 Comparison of Hardness and Compressive Strength for Polycrystalline Ceramic Materials

Material	Vickers hardness		Calculated stress $Hv/3$ yield		Measured compressive strength
	kg/mm^2	kpsi	kg/mm^2	kpsi	kg/mm^2 kpsi
Al_2O_3	2370	3360	790	1120	650
BeO	1140	1620	380	540	360
MgO	660	930	220	310	200
$MgAl_2O_4$	1650	2340	550	780	400
Fused SiO_2	540	780	180	260	190
ZrO_2 (+ CaO)	1410	1980	470	660	290
$ZrSiO_4$	710	1140	270	380	210
SiC	3300	4680	1100	1560	--
Diamond	9000	13,780	3000	4260	910
NaCl	21	30	7	10	6
B_4C	4980	7080	1660	2360	414

Source: Ref. 19.

in some applications the ceramic material is prestressed in a state of compression to give it increased resistance to tensile loads that will be imposed during service. The residual compressive stresses must first be overcome by tensile stresses before additional tensile stress can build up to break the ceramic. Concrete prestressed with steel bars is one example. Safety glass is another example.

Rice [19] has conducted an extensive review of the literature and has proposed that the upper limit of compressive strength is the stress at which microplastic yielding (deformation involving slip along crystallographic planes) occurs. He suggests that the microplastic yield stress can be estimated by dividing the measured microhardness, either Vickers or Knoop, by 3, and that the compressive strength of current well-fabricated ceramic materials ranges from about one-half to three-fourths of the yield stress. Table 3.4 lists some of the hardness and compressive strength data reported by Rice. There does appear to be some correlation between the microhardness and compressive strength.

The following factors probably contribute to reduction of the measured compressive strength below the microplastic yield stress for the material: flaws in the ceramic, such as cracks, voids, and impurities; twinning; elastic and thermal expansion anisotropy; and misalignment during testing. Grain size also appears to have a large effect. In general, the compressive strength increases as the grain size decreases.

Bend Strength   The strength of ceramic materials is generally characterized by bend testing (also referred to as flexure testing), as illustrated in Fig. 3.6e and f. The test specimen can have a circular, square, or rectangular cross section and is uniform along the complete length. Such a specimen is much less expensive to fabricate than a tensile specimen.

Bend testing is conducted with the same kind of universal test machine used for tensile and compressive strength measurements. As shown in Fig. 3.6e and f, the test specimen is supported at the ends and the load is applied either at the center (3-point loading) or at two positions (4-point loading). The bend strength is defined as the maximum tensile stress at failure and is often referred to as the modulus of rupture, or MOR. The bend strength for a rectangular test specimen can be calculated using the general flexure stress formula:

$$S = \frac{Mc}{I} \tag{3.14}$$

where M is the moment, c the distance from the neutral axis to the tensile surface, and I the moment of inertia. For a rectangular test specimen I = $bd^3/12$ and c = $d/2$, where d is the thickness of the specimen and b is the width. Figure 3.8 illustrates the derivation of the 3-point and 4-point flexure formulas for rectangular bars.

The strength characterization data for ceramics are reported in terms of MOR or bend strength. Specimens are relatively inexpensive and testing is straightforward and quick. However, there is a severe limitation on the usability of MOR data for ceramics: the measured strength will vary significantly depending on the size of the specimen tested and whether it is loaded in 3-point or 4-point. To understand the magnitude and reason for this variation, data for hot-pressed Si_3N_4[*] may be used as an example.

For specimens having a rectangular cross section of 0.32 × 0.64 cm (0.125 × 0.250 in.), 3-point bend testing over a 3.8-cm (1.5-in.) span resulted in an average MOR of about 930 MPa (135 kpsi). Four-point bend testing of bars from the same batch resulted in an average MOR of only 724

---

[*]Norton Company NC-132 hot-pressed Si_3N_4.

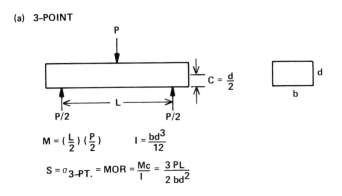

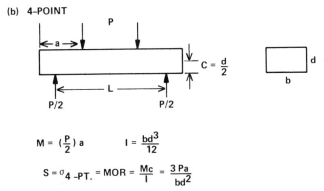

Figure 3.8 Derivation of 3- and 4-point flexure formulas for a rectangular bar.

MPa (105 kpsi). Uniaxial tensile testing of a comparable cross section of the same hot-pressed Si_3N_4 yielded a strength of only 552 MPa (80 kpsi). Which of these strengths should an engineer use? Why are they different? The engineer can answer the first question only if he or she understands the answer to the second question. The answer to the second question can best be visualized by referring to Fig. 3.9.

The stress distribution for 3-point bending is shown in Fig. 3.9a. The peak stress occurs only along a single line on the surface of the test bar opposite the point of loading. The stress decreases linearly along the length of the bar and into the thickness of the bar, reaching zero at the bottom supports and at the neutral axis, respectively. The probability of the largest flaw in the specimen being at the surface along the line of peak stress is very

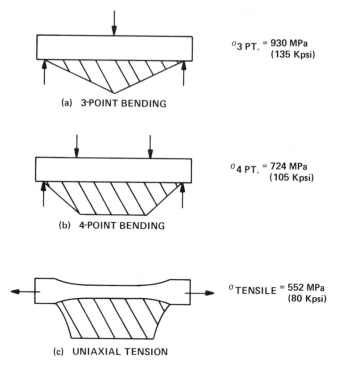

$\sigma_{3 \ PT.} = 930 \ MPa$
(135 Kpsi)

(a) 3-POINT BENDING

$\sigma_{4 \ PT.} = 724 \ MPa$
(105 Kpsi)

(b) 4-POINT BENDING

$\sigma_{TENSILE} = 552 \ MPa$
(80 Kpsi)

(c) UNIAXIAL TENSION

Figure 3.9 Comparison of the tensile stress distributions for 3-point, 4-point, and uniaxial tensile test specimens along with typical average strengths as measured by each technique for Norton NC-132 hot-pressed Si_3N_4. Shaded area represents the tensile stress, ranging from zero at the supports of the bend specimens to maximum at midspan, and being uniformly maximum along the whole gauge length of the tensile specimen.

low. Therefore, the specimen will either fracture at a flaw smaller than the largest flaw or in a region of lower stress, whichever one satisfies equation (3.11) first. For the case of hot-pressed Si_3N_4, where $\sigma_{3\text{-pt.}} = 930$ MPa (135,000 psi) with the assumptions that $E = 303 \times 10^9$ N/m2, $\gamma = 30$ J/m2, $Z = 1.5$, and $Y = 2$, c from equation (3.11) equals 10 μm (0.0004 in.). In other words, a flaw of 10 μm depth on the surface at midspan would result in fracture at 930 MPa. Halfway between midspan and the bottom support, the load at fracture would be $(1/2)(930) = 465$ MPa (67,400 psi). The critical flaw size to cause fracture at this point would be 41 μm. If the 41-μm flaw had been at midspan, the bar would have fractured at 326 MPa (47,200 psi) rather than 930 MPa. The point of this example is that the MOR or bend strength measured in a 3-point test does not reveal the strength limit of the material or even the local stress and flaw size that caused fracture. All it tells is the peak stress on the tensile surface at the time of fracture, as provided by equation (3.14).

Three-point bend testing results in synthetically high strength values. These values can be used for design only if treated statistically or probabilistically. This point is discussed in Chap. 11.

Four-point bend testing results in lower strength values for a given ceramic material than does 3-point bending. Figure 3.9b illustrates the approximate stress distribution in a 4-point bend specimen. The peak stress is present over the area of the tensile face between the load points. The tensile stress decreases linearly from the surface to zero at the neutral axis and from the load points to zero at the bottom supports. The area and volume under peak tensile stress or near peak tensile stress is much greater for 4-point bending than for 3-point bending, and thus the probability of a larger flaw being exposed to high stress is increased. As a result, the MOR or bend strength measured in 4-point is lower than that measured in 3-point. For the Si_3N_4, the average 4-point bend strength was 724 MPa (105,000 psi), which would correspond to a critical flaw size of 17 μm (0.0007 in.) at the tensile surface in the region of peak stress. As was shown for 3-point bending, fracture could also occur at lower stress regions from larger flaws.

Uniaxial tensile strength testing results in lower strength values for a given ceramic than does bend testing. Figure 3.9c illustrates that the complete volume of the gauge section of a tensile test specimen is exposed to the peak stress. Therefore, the largest flaw in this volume will be the critical flaw and result in fracture when the critical stress as defined by equation (3.11) is reached. To continue the Si_3N_4 example, the average fracture stress in uniaxial tensile loading was 552 MPa (80,000 psi) and the critical flaw size was 29 μm (0.0011 in.)

In summary, the observed strength value is dependent on the type of test conducted. More specifically, it is dependent simultaneously on the flaw size distribution of the material and the stress distribution in the test specimen. As the uniformity of the flaws within a material increases, the strength values measured by bend and tensile testing approach each other. Such is the case for most metals. For most ceramic materials, the apparent strength will decrease when going from 3-point to 4-point to tensile testing and as specimen size increases.

Biaxial Strength   The tensile, compressive, and bend tests discussed in the previous paragraphs all involve loading in a single direction and thus produce uniaxial stress fields. Many applications for materials impose multiaxial stress fields. Very few data are available for the response of ceramics to multiaxial stress fields.

The specimen test shown in Fig. 3.6g provides strength data under a biaxial stress condition. A flaw in the material is exposed simultaneously to both tensile and shear stresses [20,21].

Biaxial loading frequently occurs at the contact zone between two ceramic parts or between a ceramic and a metal part, especially during relative motion due to mechanical sliding or thermal cycling. Under certain conditions, very localized surface tensile stresses much higher than the applied load can result [22,23]. This is illustrated in Fig. 3.10.

Application of only a normal force N results in compressive stresses. Simultaneous application of a tangential force T results in localized tensile stress at the edge of the contact zone opposite the direction of the tangential force. This tensile stress is a maximum at the surface of the ceramic and rapidly decreases inward from the surface. The magnitude of the tensile stress increases as the coefficient of friction increases. It reaches a peak when the static friction is highest, but is immediately reduced once sliding begins because the dynamic coefficient of friction is lower.

Most engineers are not aware of this mechanism of tensile stress generation, yet it is probably a common cause of chipping, spalling, cracking, and fracture of ceramic components.

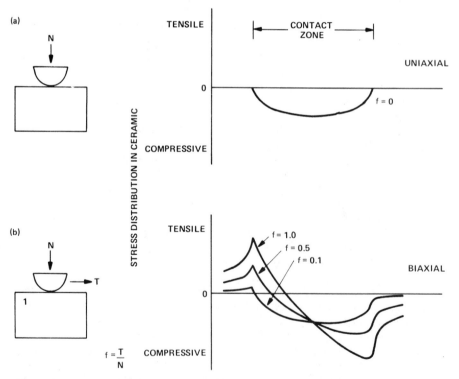

Figure 3.10 Contact loading showing uniaxial and biaxial effects on stress distribution. (Adapted from Ref. 22.)

Strength Data for Ceramic Materials

Table 3.5 summarizes strength data for a variety of ceramic materials. The purpose of this table is to provide the reader with a general indication of the strengths typically measured for the different types of ceramic materials. The data are adequate for making an initial material selection for an application, but are not suitable for analytical design or life prediction calculations. The best procedure is to select candidate materials from Table 3.5 and then contact current suppliers and obtain their most up-to-date data. As discussed earlier, to understand the data adequately, one must know the porosity content of the material, the size of the test specimen, and the geometry of the strength test (i.e., 3-point versus 4-point, and span of fixture). The data in Table 3.5 were obtained from many sources and were undoubtedly measured on a wide variety of specimen sizes and fixture configurations.

For many applications the high-temperature strength is more important than the room-temperature strength. The strength of nearly all ceramic materials decreases as the temperature is increased (graphite is an exception). Figures 3.11 and 3.12 compare strength versus temperature for some ceramic and metallic materials.

The strength would be expected to decrease with increasing temperature in proportion to the decrease in elastic modulus. This appears to occur for most ceramic materials at intermediate temperatures. At still higher temperatures, the rate of strength decrease is more rapid, generally due to nonelastic effects. For instance, most ceramics have secondary chemical compositions concentrated at grain boundaries that soften at high temperature and decrease the load-bearing capability of the bulk ceramic. This is discussed in more detail in Sec. 4.1 and in Chap. 7.

3.3 FRACTURE TOUGHNESS

Discussions so far have considered strength and fracture in terms of critical flaw size. An alternative approach considers fracture in terms of crack surface displacement and the stresses at the tip of the crack. This is the fracture mechanics approach.

The stress concentration at a crack tip is denoted in terms of the stress intensity factors K_I, K_{II}, and K_{III}. The subscripts refer to the direction of load application with respect to the position of the crack. If the load is perpendicular to the crack, as is typically the case in a tensile or bend test, the displacement is referred to as mode I and is represented by K_I. This is also called the opening mode as is most frequently operational for ceramics. Similarly, shear loading is referred to as mode II and mode III and represented by K_{II} and K_{III} [24]. The load directions for the three modes are illustrated in Fig. 3.13.

Table 3.5 Typical Room Temperature Strengths of Ceramic Materials

Material	MOR		Tensile	
	MPa	kpsi	MPa	kpsi
Sapphire (single-crystal Al_2O_3)	620	90	--	--
Al_2O_3 (0-2% porosity)	350-580	50-80	200-310	30-45
Sintered Al_2O_3 (<5% porosity)	200-350	30-50	--	--
Alumina porcelain (90-95% Al_2O_3)	275-350	40-50	172-240	25-35
Sintered BeO (3.5% porosity)	172-275	25-40	90-133	13-20
Sintered MgO (<5% porosity)	100	15	--	--
Sintered stabilized ZrO_2 (<5% porosity)	138-240	20-35	138	20
Sintered mullite (<5% porosity)	175	25	100	15
Sintered spinel (<5% porosity)	83-220	12-32	--	19
Hot-pressed Si_3N_4 (<1% porosity)	620-965	90-140	350-580	50-80
Sintered Si_3N_4 (~5% porosity)	414-580	60-80	--	--
Reaction-bonded Si_3N_4 (15-25% porosity)	200-350	30-50	100-200	15-30
Hot-pressed SiC (<1% porosity)	621-825	90-120	--	--
Sintered SiC (~2% porosity)	450-520	65-75	--	--
Reaction-sintered SiC (10-15% free Si)	240-450	35-65	--	--
Bonded SiC (~20% porosity)	14	2	--	--

Table 3.5 (continued)

Material	MOR		Tensile	
	MPa	kpsi	MPa	kpsi
Fused SiO_2	110	16	69	10
Vycor or Pyrex glass	69	10	--	--
Glass-ceramic	245	10-35	--	--
Machinable glass-ceramic	100	15	--	--
Hot-pressed BN (<5% porosity)	48-100	7-15	--	--
Hot-pressed B_4C (<5% porosity)	310-350	45-50	--	--
Hot-pressed TiC (<2% porosity)	275-450	40-65	240-275	35-40
Sintered WC (2% porosity)	790-825	115-120	--	--
Mullite porcelain	69	10	--	--
Steatite porcelain	138	20	--	--
Fire-clay brick	5.2	0.75	--	--
Magnesite brick	28	4	--	--
Insulating firebrick (80-85% porosity)	0.28	0.04	--	--
2600°F insulating firebrick (75% porosity)	1.4	0.2	--	--
3000°F insulating firebrick (60% porosity)	2	0.3	--	--
Graphite (grade ATJ)	28	4	12	1.8

Mode I is most frequently encountered for ceramic materials. Stress analysis solutions for K_I for a variety of crack locations within simple geometries have been reported by Paris and Sih [25]. Experimental data for a variety of materials have been obtained for the critical stress intensity K_{I_C}. This is the stress intensity factor at which the crack will propagate and lead to fracture. It is also referred to as <u>fracture toughness</u> and is

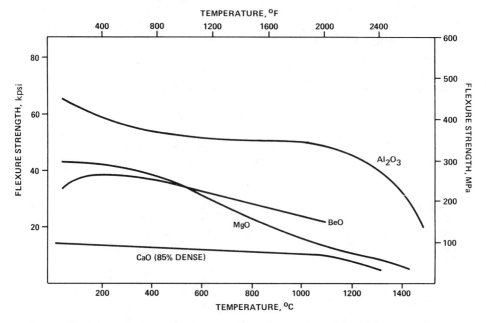

Figure 3.11 Strength versus temperature for oxide and silicate ceramics.

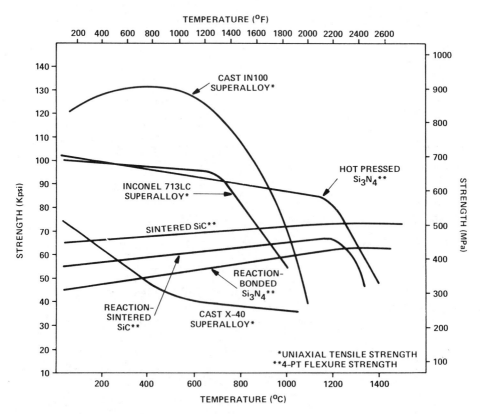

Figure 3.12 Strength versus temperature for carbide and nitride ceramics and superalloy metals.

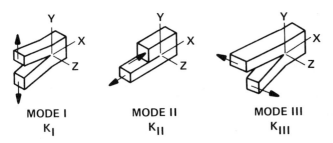

MODE I   MODE II   MODE III
K_I K_{II} K_{III}

Figure 3.13 Stress intensity factor notations for various displacement modes.

considered a basic property of the material. The higher the fracture toughness, the more difficult it is to initiate and propagate a crack.

The mode I stress intensity factor is related to other parameters by the following equations [25]:

For plane strain,

$$K_I = \left(\frac{2\gamma E}{1 - \nu^2} \right)^{1/2} \tag{3.15}$$

For plane stress,

$$K_I = (2\gamma E)^{1/2} \tag{3.16}$$

and for applied stress σ_a and crack length 2c,

$$K_I = \sigma_a Y c^{1/2} \tag{3.17}$$

where γ is the fracture energy, E the elastic modulus, ν is Poisson's ratio, and Y is a dimensionless term determined by the crack configuration and loading geometry.

A variety of experimental techniques for measuring fracture toughness are available and are discussed by Evans [27]. Schematics of the specimen geometries and equations for the calculation of K_I are included in Fig. 3.14.

(a) SINGLE-EDGE NOTCHED BEAM (SENB) SPECIMEN

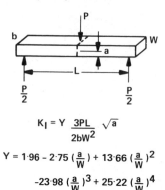

$$K_I = Y \; \frac{3PL}{2bW^2} \; \sqrt{a}$$

$$Y = 1 \cdot 96 - 2 \cdot 75 \left(\frac{a}{W}\right) + 13 \cdot 66 \left(\frac{a}{W}\right)^2$$
$$-23 \cdot 98 \left(\frac{a}{W}\right)^3 + 25 \cdot 22 \left(\frac{a}{W}\right)^4$$

(b) DOUBLE TORSION (DT) SPECIMEN

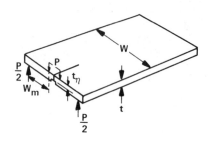

$$K_I = PW_m \left[\frac{3(1 + \nu)}{Wt^3 t_\eta}\right]^{\frac{1}{2}}$$

(c) CONSTANT-MOMENT SPECIMEN

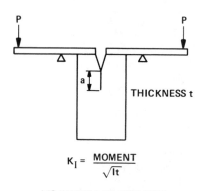

THICKNESS t

$$K_I = \frac{MOMENT}{\sqrt{It}}$$

I IS INERTIA OF ONE ARM.

(d) TAPERED DOUBLE CANTILEVER BEAM (TCB) SPECIMEN

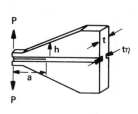

$$K_I = 2P\left(\frac{m}{t t_\eta}\right) \qquad M = \frac{1}{t} + \frac{3a^2}{h^3}$$

Figure 3.14 Techniques for experimental determination of mode I stress intensity factor. (Compiled from A. G. Evans, <u>Fracture Mechanics of Ceramics</u>, Vol. 1, Plenum Publishing Corp., New York, 1974, pp. 25–26.)

PROBLEMS

3.1 The average thermal expansion coefficient of MgO is 5×10^{-6} in./in.$\cdot$°F. What temperature change is required to produce the same linear change as a stress of 15,000 psi?

3.2 Fused silica has a thermal expansion coefficient of about 2.8×10^{-7} in./in.$\cdot$°F. What temperature change is required to produce the same linear change as a stress of 15,000 psi? How does this compare with the answer to Problem 3.1? What degree of thermal shock resistance would you expect for each material?

3.3 What would have the greater effect on the elastic modulus of an Al_2O_3 material, 5 vol % intergranular glass or 5 vol % porosity?

3.4 Ten rectangular bars of BeO 0.5 in. wide by 0.25 in. thick were tested in 3-point bending over a 2.0-in. span. The failure loads for each (in ascending order) were 280, 292, 296, 299, 308, 317, 319, 330, 338, and 360 lb. Calculate the modulus of rupture (MOR) for each and the mean MOR for the group.

3.5 A ceramic material is tested in 3-point and 4-point bending and in uniaxial tension. The resulting MOR and strength values are 80,000, 60,000, and 25,000 psi. What can we deduce from these data about the flaw distribution and uniformity of this material?

ANSWERS TO PROBLEMS

3.1 $\sigma = E\epsilon$ ϵ = linear change = $\dfrac{\sigma}{E}$

$\sigma = 15,000$ psi $E = 30 \times 10^6$ psi $\epsilon = \dfrac{15 \times 10^3}{30 \times 10^6} = 5 \times 10^{-4}$ in./in.

$\alpha = \dfrac{\Delta\ell}{\ell \Delta T}$ $\Delta T = \dfrac{\Delta\ell}{\ell\alpha}$ $\dfrac{\Delta\ell}{\ell} = \epsilon$

$\Delta T = \dfrac{5 \times 10^{-4} \text{ in./in.}}{5 \times 10^{-6} \text{ in./in.}\cdot°F} = 100°F$

3.2 $\epsilon = 5 \times 10^{-4}$ in./in.

$\Delta T = \dfrac{5 \times 10^{-4}}{2.8 \times 10^{-7}} = 1800°F$

This is much higher than for MgO. MgO would have much less resistance to thermal shock than would fused silica.

3.3 $E = E_a V_a + E_b V_b = (55 \times 10^6 \text{ psi})(0.95) + (10 \times 10^6)(0.05)$

$= 52.7 \times 10^6 \text{ psi}$

$E = E_0(1 - 1.9\rho + 0.9\rho^2) = (55 \times 10^6)[1 - (1.9)(0.05) + 0.9(0.05)^2]$

$= 49.8 \times 10^6 \text{ psi}$

The porosity would have the greater effect.

3.4

Load	$3 PL/2bd^2$ (psi)
280	26,900
292	28,000
296	28,400
299	28,700
308	29,600
317	30,400
319	30,600
330	31,700
338	32,400
360	34,600

$\bar{\sigma} = \text{mean} = \dfrac{\Sigma}{N} = \dfrac{301,000}{10}$

$\bar{\sigma} = 30,100 \text{ psi}$

3.5 The material contains a broad size distribution of flaws and is probably not very homogeneous.

REFERENCES

1. F. A. McClintock and A. S. Argon, Mechanical Behavior of Materials, Addison-Wesley Publishing Co., Inc., Reading, Mass., 1966.
2. J. B. Wachtman, ed., Mechanical and Thermal Properties of Ceramics, NBS Special Publication 303, U.S. Government Printing Office, Washington, D.C., 1969.
3. A. S. Tetelman and A. J. McEvily, Fracture of Structural Materials, John Wiley & Sons, Inc., New York, 1967.
4. R. C. Bradt, D. P. H. Hasselman, and F. F. Lange, eds., Fracture Mechanics of Ceramics, Vols. 1 and 2, Plenum Publishing Corp., New York, 1974.
5. R. C. Bradt, D. P. H. Hasselman, and F. F. Lange, eds., Fracture Mechanics of Ceramics, Vols. 3 and 4, Plenum Publishing Corp., New York, 1978.
6. W. D. Kingery, H. K. Bowen, and D. R. Uhlmann, Introduction to Ceramics, 2nd ed., John Wiley & Sons, Inc., New York, 1976, Chap. 15.

7. L. H. Van Vlack, Elements of Materials Science, Addison-Wesley Publishing Co., Inc., Reading, Mass., 1964, pp. 418-420.

8. L. H. Van Vlack, Physical Ceramics for Engineers, Addison-Wesley Publishing Co., Inc., Reading, Mass., 1964, p. 118.

9. J. K. MacKenzie, Proc. Phys. Soc. (Lond.) B63, 2 (1950).

10. ASTM Specification C747, Annual Book of ASTM Standards, American Society for Testing and Materials, Philadelphia, pp. 1064-1074.

11. C. E. Inglis, Stresses in a plate due to the presence of cracks and sharp corners, Trans. Inst. Nav. Arch. 55, 219 (1913).

12. A. A. Griffith, The phenomenon of rupture and flow in solids, Philos. Trans. R. Soc. Lond. Ser. A 221(4), 163 (1920).

13. A. G. Evans and G. Tappin, Proc. Br. Ceram. Soc. 20, 275-297 (1972).

14. A. G. Evans and T. G. Langdon, Structural ceramics, in Progress in Materials Science, Vol. 21, Pts. 3/4: Structural Ceramics, Pergamon Press, Inc., Elmsford, N.Y., 1976.

15. K. M. Johansen, D. W. Richerson, and J. J. Schuldies, Ceramic Components for Turbine Engines, Phase II Technical Report under Air Force Contract F33615-77-C-5171, Feb. 29, 1980.

16. C. D. Pears and H. W. Starrett, An Experimental Study of the Weibull Volume Theory, AFML-TR-66-228, Mar. 1967.

17. R. Sedlacek and F. A. Halden, Method of tensile testing of brittle materials, Rev. Sci. Instrum. 33(3), 298-300 (1962).

18. W. B. Shook, Critical Survey of Mechanical Property Test Methods for Brittle Materials, ASD-TDR-63-491 (July 1963).

19. R. W. Rice, The compressive strength of ceramics, in Materials Science Research, Vol. 5: Ceramics in Severe Environments (W. W. Kriegel and H. Palmour III, eds.), Plenum Publishing Corp., New York, 1971, pp. 195-229.

20. A. Rudnick, A. R. Hunter, and F. C. Holden, An analysis of the diametral compression test, Mater. Res. Std. 3(4), 283-289 (1963).

21. A. Rudnick, C. W. Marschall, W. H. Duckworth, and B. R. Emrich, The Evaluation and Interpretation of Mechanical Properties of Brittle Materials, AFML-TR-67-316, DCIC 68-3 (Mar. 1968).

22. D. G. Finger, Contact Stress Analysis of Ceramic-to-Metal Interfaces, Final Report, Contract N00014-78-C-0547, Sept. 1979.

23. D. W. Richerson, W. D. Carruthers, and L. J. Lindberg, Contact stress and coefficient of friction effects on ceramic interfaces, in Surfaces and Interfaces in Ceramic and Ceramic-Metal Systems, Materials Science Research, Vol. 14 (J. Pask and A. Evans, eds.), Plenum Publishing Corp., New York, 1981, pp. 661-676.

24. P. M. Braiden, in Mechanical Properties of Ceramics for High Temperature Applications, AGARD Report No. 651, NTIS No. AD-A034 262, Dec. 1976.

25. P. C. Paris and G. C. Sih, Stress analysis of cracks, <u>ASTM STP No. 381</u>, 1965, p. 30.
26. J. B. Wachtman, Jr., Highlights of progress in the science of fracture of ceramics and glass, J. Am. Ceram. Soc. <u>57</u>(12), 509-519 (1974).
27. A. G. Evans, in <u>Fracture Mechanics of Ceramics</u>, Vol. 1 (R. C. Bradt, D. P. H. Hasselman, and F. F. Lange, eds.), Plenum Publishing Corp., New York, 1974, pp. 25-26.

Time, Temperature,
and Environmental
Effects on Properties

The elastic properties of ceramics and relationships between flaw size and strength as determined by flexure and tensile testing were discussed in Chap. 3. In flexure and tensile strength tests, the load is applied rapidly and fracture occurs from single flaws or groups of flaws initially in the material. If the load is applied slowly or is held at a level below that required for instantaneous fracture, flaws initially in the material can grow to larger size and result in fracture at a much lower stress. This is known as slow crack growth, static fatigue, or stress rupture.

Ceramics are often used in severe environments that metals cannot tolerate. The life of a ceramic component in such an environment is often controlled by the formation of new flaws in the ceramic that are larger than those initially present. Such new flaws or extensions of existing flaws can result from chemical attack (oxidation, corrosion, etc.), erosion, impact, or thermal shock.

The objective of this chapter is to examine the susceptibility of important ceramic materials to slow crack growth and environmental effects as a function of time and temperature.

4.1 CREEP

The term creep is normally used to refer to deformation at a constant stress as a function of time and temperature. Creep is plastic deformation rather than elastic deformation and thus is not recovered after the stress is released. A typical creep curve has four distinct regions, as shown in Fig. 4.1. The secondary creep region is the most useful for predicting the life of a ceramic component. It is typified by a constant rate of deformation and is often referred to as steady-state creep.

Steady-state creep can be represented by the equation

$$\dot{\epsilon} = A\sigma^n e^{-Q_c/RT} \tag{4.1}$$

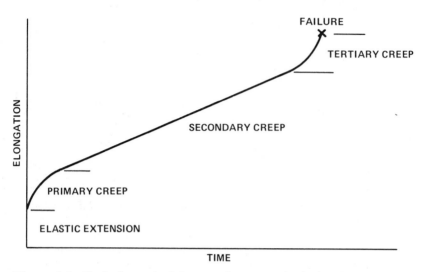

Figure 4.1 Typical constant-temperature, constant-stress creep curve.

where σ is the stress, T the absolute temperature, Q_c the activation energy for creep, and A and n are constants for the specific material. The constant n is usually referred to as the stress exponent.

The activation energy for creep Q_c can be obtained by measuring the slope of a plot of $\dot{\epsilon}$ versus $1/T$. The stress exponent and the activation energy provide information about the mechanism of creep, i.e., whether it is controlled by viscoelastic effects, diffusion, porosity, or some other mechanism. Understanding the mechanism provides information about the kinetics of flaw growth. This flaw growth information plus a knowledge of the initial flaw size distribution (as determined by statistical evaluation of room-temperature strength tests) can be used in conjunction with fracture mechanics theory and stress rupture testing to estimate the life of the ceramic component under similar stress and temperature conditions.

The creep rate of ceramic materials is affected by temperature, stress, crystal structure of single crystals, microstructure (grain size, porosity, grain boundary chemistry) of polycrystalline ceramics, viscosity of noncrystalline ceramics, composition, stoichiometry, and environment [1].

Effects of Temperature and Stress

Temperature and stress both have strong effects on creep, as would be expected from examination of equation (4.1). In general, as the temperature or stress increases, the creep rate increases and the duration of steady-state creep decreases. This is shown schematically in Fig. 4.2. When

comparing creep data for different materials under different conditions, it
is often useful to plot the data as steady-state creep   versus either $1/T$ or
log σ to produce the type of curves shown in Fig. 4.3.

Effects of Crystal Structure

The mechanism of creep in single crystals is the movement of dislocations
through the crystal structure. Such movement is accommodated by slip
along preferential crystal planes or by homogeneous shear (twinning).
Highly symmetrical cubic structures such as NaCl and MgO have many
planes available for slip [1]. At low temperature slip occurs along {110}
planes in the <$1\bar{1}0$> direction. At high temperature slip also occurs
along {001} and {111} planes in the <110> direction, resulting in five independent slip systems. Diamond, silicon, CaF_2, UO_2, TiC, UC, and
$MgAl_2O_4$ (spinel) all have five independent slip systems at high temperature.
However, the temperatures at which these systems become operative and
the stresses required for slip vary depending on the bond strength of the
material. Slip occurs at low temperature and low stress in weakly ionic
bonded NaCl, but requires much higher temperature and stress for strongly
covalent bonded diamond or TiC.

Single crystals with less symmetry than cubic crystals have fewer slip
systems available. Graphite, Al_2O_3, and BeO are hexagonal and have only
two independent slip systems. Slip of {0001} planes occurs in the <$11\bar{2}0$>
direction.

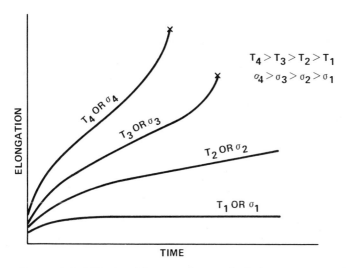

Figure 4.2 Effects of temperature and stress on creep rate.

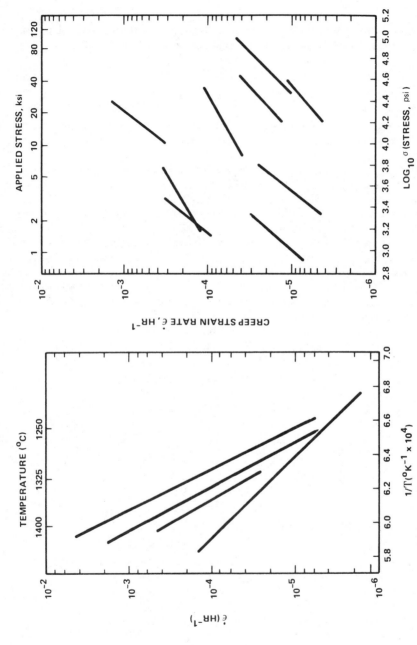

Figure 4.3 Additional methods for summarizing creep data and comparing differences between materials. Each line represents a different material.

104

For measurable creep of a single crystal to occur, dislocations must be present or created and then start moving. Energy is required to form dislocations, to initiate their movement, and to keep them moving. Increasing the stress or the temperature increases the energy available for forming and moving dislocations.

Defects in the lattice structure decrease creep by pinning dislocations. Dislocations oriented across the direction of slip and large precipitates have the largest effect on blocking slip motion. Solid solution and point defects have less effect. However, even their effect can be substantial. Chin et al. [2] reported that additions of 840 ppm of Sr^{2+} ions in solid solution in KCl increased the compressive stress required to produce yield from about 2 MPa to over 20 MPa.

Effects of Microstructure of Polycrystalline Ceramics

Creep of polycrystalline ceramics is usually controlled by different mechanisms than those which control single crystals. Dislocation movement is generally not a significant factor because of the random orientation of the individual crystals (more typically referred to as grains when in a polycrystalline structure) and the difficulty for a dislocation to travel from one grain through the grain boundary into the adjacent grain.

Creep in polycrystalline ceramics is usually controlled by the rate of diffusion or by the rate of grain boundary sliding. Diffusion involves the motion of ions, atoms, or vacancies through the crystal structure (bulk diffusion) or along the grain boundary (grain boundary diffusion). Grain boundary sliding often involves porosity or a different chemical composition at the grain boundary. Grain boundary sliding is an important (and undesirable) contribution to fracture in many ceramic materials densified by hot pressing or sintering. Additives are required to achieve densification. These additives concentrate at the grain boundaries together with impurities initially present in the material. If a glass is formed, it may soften at a temperature well below the temperature at which the matrix material would normally creep, allowing slip along grain boundaries. Grain boundary sliding is normally accompanied by the formation of pores at the grain boundaries, especially by cavitation at triple points (points at which three grains meet). The combination of reduced load-bearing capability due to softening of the grain boundary glass phase and formation of new flaws usually results in fracture before appreciable plastic deformation can occur.

The creep mechanism varies for materials at different temperatures and stresses. Ashby [3] has constructed a "deformation mechanism map" in which he plots shear stress versus temperature. His map for MgO is shown in Fig. 4.4. It indicates that at low temperature and high stress MgO creeps by dislocation motion and that at lower stress and increased temperature MgO creeps by diffusion mechanisms.

The creep mechanism for Al_2O_3 also changes according to stress and temperature, as described by Kingery et al. [1]. In single crystals Al_2O_3 and polycrystalline Al_2O_3 with grains larger than 60 μm slip can occur along {0001} planes at 2000°C at about 7 MPa (1000 psi) and at 1200°C at about 70 MPa (10,000 psi). Other planes require stresses greater than 140 MPa (20,000 psi) to cause slip even at 2000°C and thus are usually not a factor. In polycrystalline Al_2O_3 with grain size in the range 5 to 60 μm, the creep rate is controlled by Al^{3+} ion diffusion through the lattice. Below 1400°C and for finer grain sizes, Al^{3+} ion diffusion along the grain boundaries appear to be rate controlling.

The examples for MgO and Al_2O_3 show that creep mechanism and rate can vary according to temperature, stress, and grain size. Porosity also has

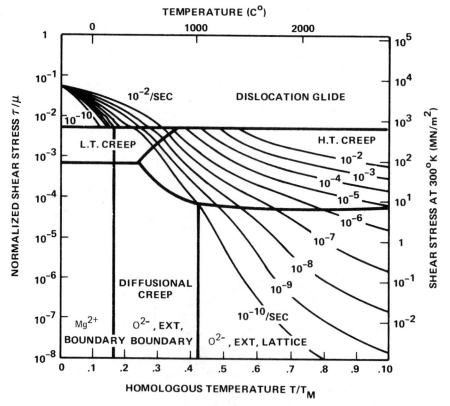

Figure 4.4 Deformation mechanism map for MgO. (Reprinted with permision from Acta Metall. 20, M. Ashby, "Deformation mechanism maps," © 1972, Pergamon Press Ltd.)

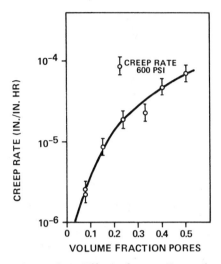

Figure 4.5 Effect of porosity on the creep rate of a polycrystalline Al_2O_3 ceramic. (From Ref. 1.)

has a substantial effect. Figure 4.5 shows the effect of porosity on the creep rate of a polycrystalline Al_2O_3 ceramic [1]. In this case, the creep rate was substantially increased by increased porosity, perhaps because of the decreased cross-sectional area available to resist creep. Similar results were reported by Kingery et al. [1] for MgO, where material with 12% porosity had six times the creep rate of comparable material with only 2% porosity.

Creep in Noncrystalline Ceramics

Creep in glasses is controlled by viscous flow and is a function of the viscosity of the glass at the temperature of interest. The viscosity of glass varies over a wide range with temperature. Soda-lime-silica glass (common container or window glass) has a viscosity of 10^{15} P at 400°C and 10^2 P at 1300°C. The softening point of glass is defined as $10^{7.6}$ P, which occurs for soda-lime-silica glass at about 700°C. Soda-lime-silica glass melts in the range 1300 to 1500°C. This glass creeps at low temperatures.

The viscous flow of glasses is an important mechanism of creep in many commercial polycrystalline ceramics. These polycrystalline ceramics contain secondary glass phases at the grain boundaries. The glass phases result from additions made during processing to achieve densification and from impurities initially present in the ceramic powders or picked up during processing. The viscosity characteristics of the glass are highly dependent on composition. For silicate glasses, the viscosity decreases with increased concentration of modifying cations. For example, the viscosity of fused silica is reduced by 10^4 P at 1700°C by the addition of 2.5 mole % K_2O. F^-, Ba^{2+}, and Pb^{2+} are particularly effective at reducing the viscosity of glasses. Additions of SiO_2 or Al_2O_3 typically increase viscosity.

The reasons for the viscosity changes in glass as a function of composition change are not well understood. It is currently thought that addition of cations produces breaks in the Si-O network structure, which reduces the overall bonding strength and decreases the viscosity.

Effects of Composition, Stoichiometry, and Environment

Composition governs the bonding and structure of single-crystal, polycrystalline, and noncrystalline ceramics and thus determines the baseline creep tendencies. Grain size, porosity, grain boundary glass phases, and second-phase dispersions all modify the baseline creep properties; so does stoichiometry. A stoichiometric ceramic is one that has all crystallographic lattice positions filled according to the normal chemical formula. A nonstoichiometric ceramic is one that has a deficiency of one type of atoms. For instance, $TiC_{0.75}$ is a nonstoichiometric form of TiC in which insufficient carbon atoms were available during formation. The structure is basically a TiC structure with vacancies to make up for the missing carbon atoms. Nonstoichiometric materials have different diffusion characteristics than their stoichiometric equivalent and thus have different creep characteristics.

Environment apparently can also affect creep, although very little work has been done to study or quantify the effects. Joffe et al. [4] observed that NaCl had brittle behavior in air but was ductile when immersed in water. Evidently, the NaCl tested in air fractured from surface defects before creep had a chance to occur. The surface of NaCl in water was dissolved, removing the surface defects and allowing plastic deformation by creep. Although it has not been studied in detail, it is likely that high-temperature oxidation and corrosion mechanisms can also alter the surface flaw size or provide crack blunting to change the creep characteristics of ceramic materials.

Measurement of Creep

Creep has been measured in tension, compression, torsion, and bending. The data from these different approaches may not be comparable. Therefore, it is important to know as many details as possible about how the data were generated, i.e., test configuration, specimen size, sensitivity of deflection measurement, and so on.

Creep testing consists of measurement of deflection at a constant load and constant temperature. The deflection is usually measured with a transducer. Corrections should be made for thermal expansion within the test

Table 4.1 Torsional Creep of Some Ceramic Materials

Material	Creep rate at 1300°C (in./in. · hr)
	At 1800 psi (12.4 MPa)
Polycrystalline Al_2O_3	0.13×10^{-5}
Polycrystalline BeO	(30×10^{-5})[a]
Polycrystalline MgO (slip cast)	33×10^{-5}
Polycrystalline MgO (hydrostatic pressed)	3.3×10^{-5}
Polycrystalline $MgAl_2O_4$ (2-5 μm)	26.3×10^{-5}
Polycrystalline $MgAl_2O_4$ (1-3 mm)	0.1×10^{-5}
Polycrystalline ThO_2	(100×10^{-5})[a]
Polycrystalline ZrO_2 (stabilized)	3×10^{-5}
Quartz glass	$20,000 \times 10^{-5}$
Soft glass	$1.9 \times 10^9 \times 10^{-5}$
Insulating firebrick	$100,000 \times 10^{-5}$
	At 10 psi
Quartz glass	0.001
Soft glass	8
Insulating firebrick	0.005
Chromium magnesite brick	0.0005
Magnesite brick	0.00002

[a] Extrapolated.
Source: Ref. 1.

specimen and deflection measurement system and for elastic and plastic deflection in the test fixture and deflection measurement system. Corrections can often be checked by comparing the cumulative creep measured by the deflection system with an actual physical measurement of the specimen dimensions before and after testing.

Table 4.1 summarizes torsional creep data for some ceramic materials. The polycrystalline oxides have much lower creep rates than the glasses or the insulating firebrick. The high creep rates of the firebrick can be attributed largely to the presence of glass phases and porosity.

Figure 4.6 shows the creep characteristics of several Si_3N_4 and SiC materials. Some of the data were measured in compression by Seltzer [5] and others were measured in tension by Larsen and Walther [6]. In general, the creep resistance of pure nitrides and carbides is very high because of their strong covalent bonding. However, other factors may decrease the creep resistance. Let us look at each curve in Fig. 4.6, starting at the upper left.

NC-132[*] is Si_3N_4 hot pressed with MgO additives. During the high-temperature hot-pressing operation, a complex glassy phase is formed at the grain boundaries. This glass phase is primarily a magnesium silicate modified by Ca, Fe, Al, and other impurities initially present in the Si_3N_4 powder. At temperatures roughly above 1100°C, grain boundary sliding occurs under loading, resulting in a mechanism for creep. Figure 4.7 shows an NC-132 hot-pressed Si_3N_4 specimen exposed in bending to 276 MPa (40,000 psi) for 50 hr at 1100°C (~2000°F).

NCX-34[*] is Si_3N_4 hot pressed with Y_2O_3 additives. A grain boundary phase is also present in this material, but it is more refractory than the complex magnesium silicate phase in NC-132 and requires a higher temperature to initiate grain boundary sliding. Thus NCX-34 has somewhat improved creep resistance.

NC-350 is Si_3N_4 prepared by the reaction-bonding process (described in Chap. 7). No additives are required to densify this material, so no grain boundary phase forms. The creep resistance is greatly improved.

NC-435[*] is a SiC-Si material prepared by a reaction sintering process (described in Chap. 7). Direct bonding occurs between SiC particles and SiC produced by the process. No glassy grain boundary phases are present and the creep resistance is relatively good.

The sintered α-SiC[†] is prepared by pressureless sintering at very high temperatures (>1900°C) with low levels of additives. A grain boundary glass phase does not form. The creep rate is very low.

The Si_3N_4 of Greskovich and Palm [7] was densified with additions of $BeSiN_2$ and SiO_2 and has a creep rate only slightly higher than Carborundum's α-silicon carbide. The low creep results from the thinness and refractoriness of the grain boundary phase. Transmission electron microscopy using high-resolution lattice imaging techniques has shown that the grain boundary phase is only about 10 Å thick.

Polycrystalline ceramics typically have stress and temperature thresholds below which slow crack growth does not occur or is negligible with respect to the life requirement of the component. This is not surprising. Energy is required to cause a flaw or crack to increase in size. The amount

---

[*]Manufactured by the Norton Company, Worcester, Mass.
[†]Manufactured by the Carborundum Co., Niagara Falls, N.Y.

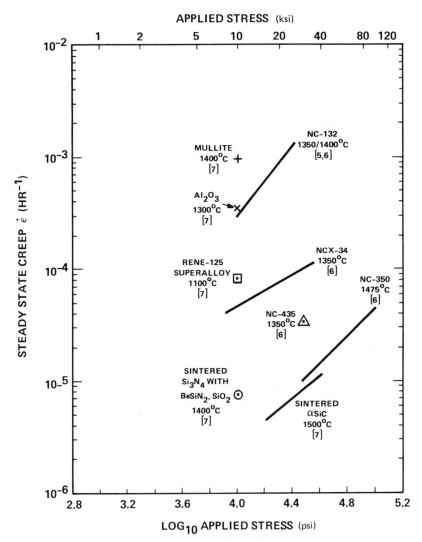

Figure 4.6 Creep of Si_3N_4 and SiC materials and comparison with mullite, Al_2O_3, and Rene-125 superalloy. (From Refs. 5 and 6.)

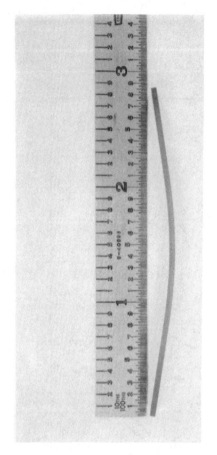

Figure 4.7 Hot-pressed Si_3N_4 specimen deformed by creep under a load of
276 MPa (40,000 psi) at 1100°C (~2200°F) for 50 hr.

of energy is dependent on the bond strength of the material and on the mechanisms available for crack growth. Crack growth is relatively easy if the
grain boundaries of the material are coated with a glass phase. At high
temperature, localized creep of this glass can occur, resulting in grain
boundary sliding. Figure 4.8a shows the fracture surface of an NC-132
hot-pressed Si_3N_4 specimen which fractured after 2.2 min under a static
bending load of 276 MPa (40,000 psi) at ~1100°C (~2000°F). The initial flaw
was probably a shallow (20 to 40 μm) machining crack. It linked up with
cracks formed by grain boundary sliding and separation and pores formed
by triple-point cavitation to produce the new flaw or structurally weakened
region seen in Fig. 4.8 as the large semicircular area extending inward

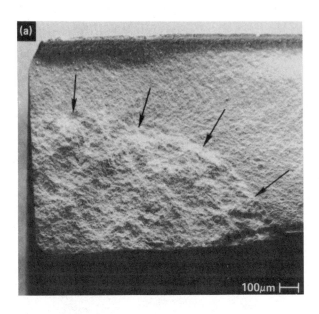

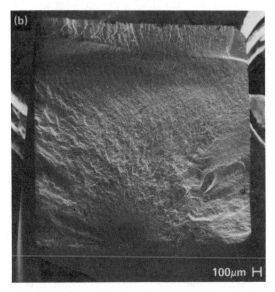

Figure 4.8 Comparison of a slow crack growth fracture versus a normal bend fracture for hot-pressed Si_3N_4. (From Ref. 9.)

from the tensile surface. This was the effective flaw size at fracture. Figure 4.8b shows the fracture surface of an NC-132 hot-pressed Si_3N_4 specimen from the same batch, but fractured under rapid loading (normal 4-point bend test at a cross-head speed of 0.02 in./min) at room temperature, where slow crack growth did not occur. This specimen fractured at 876 MPa (127,000 psi).

The example above illustrates the importance of knowing the static fatigue properties of the ceramic material. To use the material successfully in an application requires an understanding of the time-dependent properties in addition to the short-term properties.

Creep Consideration for Component Design

Creep is not normally a critical consideration for low-temperature applications. However, for intermediate- and high-temperature applications where structural loading is present, creep may be life limiting and must be carefully considered.

4.2 STATIC FATIGUE

Static fatigue, also known as stress rupture, involves subcritical crack growth at a stress which is lower than required for instantaneous fracture. Flaws initially in the material slowly grow under the effects of the stress to a size such that instantaneous fracture can occur according to the Griffith relationship [equation (3.11)] discussed in Chap. 3. Static fatigue is measured by applying a static tensile load at a constant temperature and recording the time to fracture. The experimental setup is similar to that used to measure creep, except that deformation measurements are not required.

Stress rupture is commonly used for predicting the life of metals at temperatures and stresses that simulate the application conditions. Stress rupture testing can also be useful in predicting the life of ceramic materials and in determining life-limiting characteristics that might be improved with further material development. However, stress rupture data for ceramics are usually not easy to interpret because of the large scatter, as illustrated in Table 4.2 for hot-pressed and reaction-bonded Si_3N_4 materials.

The factors discussed previously which influence creep also influence static fatigue. Slow crack growth tends to increase with increasing temperature or stress and with the presence of glassy phases. It is affected by grain size and by the presence of secondary phases. For instance, particulate dispersion tends to interrupt slow crack growth and increase the stress rupture life.

To characterize completely the stress rupture life of a ceramic requires many test repetitions at a variety of temperatures and stresses. When

Table 4.2 Stress Rupture Data for Hot-pressed Si_3N_4[a], Illustrating
Typical Scatter

| Temperature | | Stress | | Time of test (hr) | | Source of data |
| | | | | No failure | Failure | |
°C	°F	MPa	kpsi			
1066	1950	310	45	50		
1066	1950	310	45		0.46	
1066	1950	310	45		0.18	b
1066	1950	310	45	50		
1066	1950	310	45	50		
1200	2192	262	38		380	
1200	2192	276	40	480		
1200	2192	262	38		180	
1200	2192	262	38		105	c
1200	2192	262	38		52	
1200	2192	276	40		32	

[a] NC-132 from Norton Co., Worcester, Mass.
[b] AiResearch Report No. 76-212188(10), Ceramic Gas Turbine Engine Demonstration Program, Interim Report No. 10, Aug. 1978, prepared under contract N00024-76-C-5352.
[c] G. D. Quinn, Characterization of Turbine Ceramics After Long-Term Environmental Exposure, AMMRC TR 80-15, Apr. 1980, p. 18.

selecting candidate materials for an application, it is often desirable to obtain a quick determination of the susceptibility of the material to static fatigue. This can be done by variable-stressing rate experiments. Samples are loaded at three or four rates (as controlled by the cross-head speed of the test equipment) to fracture at the temperature of interest and the data are plotted as strength versus load rate, as shown in Fig. 4.9. If the strength changes as a function of load rate, the ceramic is susceptible to slow crack growth and can probably be removed from consideration if a long stress rupture life is required. If little or no strength change is detected as a function of load rate, the material is suitable for further evaluation by longer-term static fatigue tests.

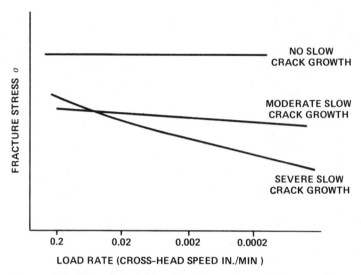

Figure 4.9 Use of variable stressing rate experiments for estimating the susceptibility of a ceramic to slow crack growth.

Static fatigue is not restricted to high temperatures. It also occurs at room temperature for silicate glasses and for many other ceramics. The mechanism involves localized corrosion at the crack tip or where stress is concentrated at a flaw. The corrosion is accelerated by the presence of water or, in some cases, other chemicals [8]. This mechanism is referred to as stress corrosion and should be considered during the design of glass components and polycrystalline ceramics known to contain a glass phase.

4.3 CHEMICAL EFFECTS

The resistance of a ceramic to chemical attack is largely a function of the strength of atomic bonding and the kinetics of thermochemical equilibrium for the ceramic and the surrounding environment at the temperature of exposure. Weakly ionic bonded ceramics have relatively low resistance to chemical attack. For instance, NaCl is soluble in water and $CaCO_3$ is dissolved by weak acids. Strongly ionic and covalent-bonded ceramics are much more stable. Al_2O_3 and $ZrSiO_4$ are stable enough that they can be used for crucible and mold linings for melting and casting metal alloys at temperatures above 1200°C (2200°F). Si_3N_4 and Si_2ON_2 (silicon oxynitride) have also been used in contact with molten metals and molten ceramics without appreciable reaction, especially in the processing of aluminum.

Alumina-, magnesia-, chromia-, and zirconia[*]-based ceramic refractories are indispensable for lining the high-temperature furnaces used for metals refining and glass manufacturing.

Ceramics are often selected instead of metals for applications because of their superior chemical stability over a broad temperature range. Al_2O_3, porcelain, and many other oxides, silicates, borides, carbides, and nitrides are resistant to acids and bases and are used in a variety of corrosive applications where metals do not survive. Al_2O_3 and porcelain are particularly important to consider for these applications because they are currently in high-volume production for liners, seals, laboratory ware, and a variety of specialty items and can be obtained for prototype evaluation at moderate cost with reasonable delivery.

Selection of ceramics for high-temperature corrosive applications requires a knowledge of chemical resistance plus other properties, such as thermal conductivity, thermal expansion coefficient, strength, and creep resistance. The following paragraphs review the response of some ceramics to high-temperature environments and define the combinations of properties required for some current and projected applications.

Gas-Solid Reactions

Oxidation Many engineering operations are conducted at high temperature in an air or oxygen environment. Oxides and stoichiometric silicates are typically stable at high temperature in an oxygen or mixed oxygen–nitrogen atmosphere. Most of the carbides, nitrides, and borides are not. Si_3N_4 and SiC, which are becoming important commercial ceramics, are used as examples.

At elevated temperature, oxygen interacts with any exposed surface of SiC or Si_3N_4. If the oxygen partial pressure is roughly 1 mm Hg or higher, SiO_2 will form a protective layer at the surface. This is referred to as passive oxidation. Formation of SiO_2 will initially be rapid, but will decrease as the thickness of the layer increases. Under these conditions, oxidation will be controlled by oxygen diffusion through the SiO_2 layer and will be parabolic. The oxidation rate increases as the temperature increases, as shown by data from Singhal summarized in Fig. 4.10.

At low partial pressure (1 mm Hg or below) of oxygen, inadequate oxygen is present to form a protective SiO_2 layer. Instead, SiO (silicon monoxide) gas forms. This is called active oxidation. It is roughly linear

---

[*]It is common practice to refer to the oxides by a contracted form ending in "ia" (e.g., Al_2O_3, alumina; MgO, magnesia; Cr_2O_3, chromia; ZrO_2, zirconia).

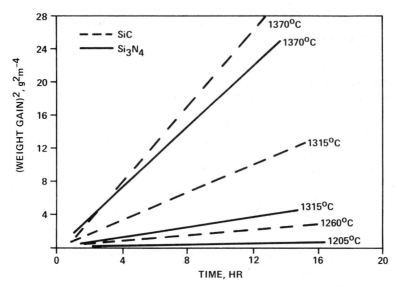

Figure 4.10 Oxidation rate versus temperature for Si_3N_4 and SiC materials. [After S. C. Singhal, in Ceramics for High Performance Applications (J. J. Burke, A. E. Gorum, and R. N. Katz, eds.), Brook Hill Publishing Co., Chestnut Hill, Mass., 1974, pp. 533-548.]

and continuous and the component can be completely consumed. Active oxidation conditions are not common but must be kept in mind.

The kinetics of passive oxidation were initially determined by weight gain and dimensional change measurements. More important, though, for some applications such as heat engines is the effect of oxidation on the strength of the material. Depending on the ceramic, the temperature, and the initial surface condition, oxidation can increase, decrease, or not affect the strength. This is illustrated for Norton NC-132 hot-pressed Si_3N_4 in Fig. 4.11. Specimens with grinding grooves parallel to the length (longitudinal) and thus parallel to the stressing direction were much stronger than those with grinding grooves perpendicular to both the length (transverse) and stressing direction. Low-temperature oxidation (1000°C) significantly increased the strength for the transverse machined specimens by blunting or reducing the severity of the surface flaws associated with the grinding damage. High-temperature oxidation completely removed the grinding grooves, but resulted in formation of surface pits, which had an equivalent or worse effect on the strength.

The oxidation characteristics of the NC-132 were found to be composition and impurity controlled. Oxygen from the atmosphere was reacting with the Si_3N_4 to form SiO_2, as expected, but impurities within the Si_3N_4 (magnesium, iron, calcium, etc.) were interacting with the SiO_2 to form a

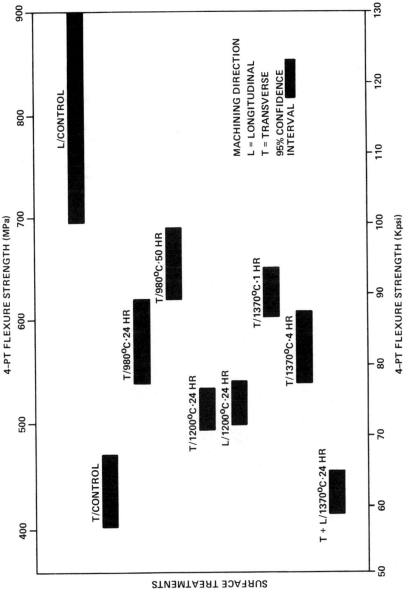

Figure 4.11 Oxidation of NC-132 hot-pressed silicon nitride, showing strength increase for low-temperature exposure and strength decrease for high-temperature exposure. (From Ref. 9.)

complex silicate having a much lower melting temperature and viscosity than pure SiO_2 and a different diffusion rate for oxygen. This resulted in increased oxidation of the Si_3N_4 and even possibly chemical corrosion due to limited solubility of Si_3N_4 in the complex silicate surface layer.

Increased purity and use of densification aids other than MgO (such as Y_2O_3, CeO_2, and $BeSiN_2$) in the Si_3N_4 result in improved oxidation resistance. This is illustrated by data reported by Greskovich and Palm [7] and shown in Fig. 4.12.

The strength of reaction-bonded Si_3N_4 is also strongly affected by high-temperature oxidation. Reaction-bonded Si_3N_4 does not contain the densification

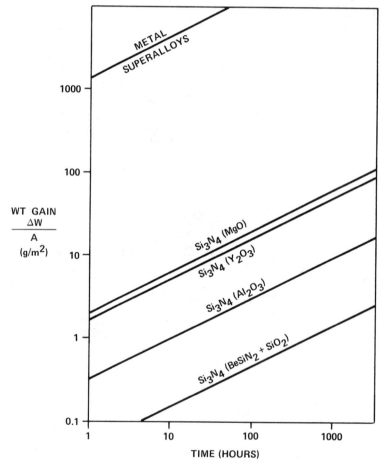

Figure 4.12 Oxidation resistance of hot-pressed Si_3N_4 containing various densification aids (shown in parentheses). (From Ref. 7.)

aids present in hot-pressed Si_3N_4 but does contain 15 to 25% porosity, depending on the grade and the manufacturer (the fabrication and characteristics of reaction-bonded Si_3N_4 are discussed in Chap. 7). High-porosity reaction-bonded Si_3N_4 undergoes oxidation over a broad temperature range along the interconnected internal pore channels, forming cristobalite or amorphous silica. Initially, a slight strength increase occurs as a result of blunting of strength-controlling defects. However, longer exposure results in substantial formation of internal SiO_2 and a decrease in strength. Godfrey and Pitman [10] reported that reaction-bonded Si_3N_4 with an initial density of 2.55 g/cm^3 ($\sim$20% porosity) had an average strength decrease greater than 30% after oxidation exposure for 250 hr at 1250°C. It is thought that the strength reduction is due partially to consumption of Si_3N_4 and replacement by weaker SiO_2 and partially to the differences in thermal expansion and elastic properties of Si_3N_4, cristobalite, and amorphous silica. The latter can result in crack formation, due especially to the approximately 5% shrinkage that occurs at about 250°C when high cristobalite undergoes a displacive phase transformation to low cristobalite.

Reaction-bonded Si_3N_4 with low porosity or very small pores can be improved by controlled oxidation. Rapid oxidation, especially at temperatures above 1300°C, can quickly seal the surface with coherent oxide layers and protect the interior from oxidation. Such treatment does not always result in a strength increase, but does significantly increase the long-term stability under further exposure to a high-temperature oxidizing atmosphere. This is illustrated in Fig. 4.13. The flash oxidation at 1350°C sealed the surface and prevented further oxidation. Specimens exposed at 900°C and 1100°C without a prior flash oxidation showed a continuous weight gain, indicating that internal oxidation was taking place [11].

Other Gas-Solid Reactions Many industrial processes are conducted under atmospheres other than oxygen or air. It is important for an engineer to understand the type of reactions that could occur and to know where to obtain specific reaction data for the particular case. Vaporization and thermal decomposition are the most common reactions. The rate of vaporization is dependent on the vapor pressure of the solid at the temperature of exposure. The vapor pressure increases with temperature. However, once the equilibrium vapor pressure is reached, no further vaporization will occur unless the temperature is increased or the atmosphere is changed. For this reason, greater material loss would be expected in vacuum or in a flowing gas than in a static atmosphere. For instance, oxides will volatilize much more rapidly in a vacuum than in air.

The application of silica-containing glass and refractories in reducing atmospheres is limited by loss of silica. Volatilization in vacuum or an inert gas can occur by the reaction

$$2SiO_2 (s) \longrightarrow 2SiO (g) + O_2 (g) \qquad\qquad (4.2)$$

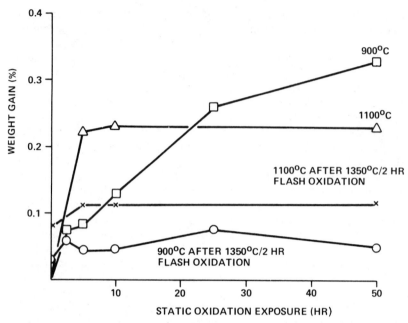

Figure 4.13 Benefit of flash oxidation treatment for minimizing further oxidation of reaction-bonded Si_3N_4.

or in hydrogen by the reaction

$$H_2 \text{ (g)} + SiO_2 \text{ (s)} \longrightarrow SiO \text{ (g)} + H_2O \qquad\qquad (4.3)$$

The amount of silica loss can be substantial, as shown in Fig. 4.14 for refractory brick.

Si_3N_4 and SiC refractories are also susceptible to volatilization at high temperature. These two materials do not melt, as do metals and most other ceramics. They dissociate or decompose, Si_3N_4 at about 1900°C and SiC at about 2500°C. However, in vacuum or in inert atmosphere, both materials can dissociate at much lower temperatures. This becomes an important factor in the processing of Si_3N_4 components by sintering or hot pressing. Temperatures between 1700 and 1800°C are required to achieve densification, but positive steps must be taken to minimize the dissociation reactions that are competing with the densification reactions. Use of vacuum or an inert atmosphere would favor the dissociation reactions. Use of a positive nitrogen pressure or packing the parts in a powder bed containing Si_3N_4 depresses the dissociation reactions and allows densification to occur.

Decomposition reactions are very important in processing of oxide ceramics and refractories. The oxides are usually derived from metal salts such as carbonates, hydroxides, nitrates, sulfates, acetates, oxalates, or alkoxides. These salts are either naturally occurring raw materials or the results of chemical refining operations. They are used as a portion of the raw materials in the fabrication of glass, refractories, and a wide variety of ceramic products. They decompose at low to intermediate temperatures to produce a solid oxide plus a gas. Heating rates and decomposition temperatures must be carefully controlled to avoid breaking the ceramic part during processing, as a result of too rapid evolution of the reactant gas. It is also important that complete decomposition occur before melting or densification of the part begins.

Liquid-Solid Reactions

Liquid-solid reactions are generally referred to as corrosion. Corrosion can result from direct contact of a reactive liquid with the ceramic as a simple dissolution reaction or can be more complex, such as interactions with impurities in the ceramic or surrounding gas or liquid environment.

The kinetics of a reaction are often more important for ceramics than whether or not a reaction occurs. A reaction can only occur if fresh reactants can get to the ceramic surface and if reaction products can get away

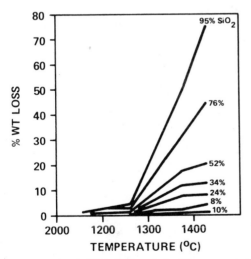

Figure 4.14 Weight loss of silicon-containing refractory brick after 50 hr in a 100% hydrogen atmosphere. (From Ref. 12.)

from the surface. Often the reaction products remain on the surface and act as a boundary layer, which limits further reaction. In this case diffusion through the boundary layer controls the rate of corrosion. For instance, in glass melting furnaces and in other industrial processes where silicate slags are present, the boundary layer may be on the order of 1 cm thick, due to the high viscosities and low fluid velocities involved. This protects the refractory linings and significantly extends their life.

Ambient Temperature Corrosion   Ceramics show a broad range of resistance to corrosion at room temperature. Strongly bonded ceramics such as Al_2O_3 and Si_3N_4 are virtually inert to attack by aqueous solutions, including most strong acids and bases. On the other hand, many of the weakly bonded ionic metal salts, including most of the nitrates, oxalates, chlorides, and sulfates, are soluble in water or weak acids.

Ceramic silicates are also very stable. A notable exception is attack by hydrofluoric acid (HF). HF readily dissolves most silicate glass compositions.

Lay [13] has prepared a review of the resistance of many engineering ceramics, including Al_2O_3, BeO, MgO, ZrO_2, spinel, Si_3N_4, mullite, and SiC, to corrosion by acids, alkalis, gases, fused salts, metals, and metal oxides.

High-Temperature Corrosion of Oxides   High-temperature corrosion of oxide ceramics is encountered in many cases where the ceramic is in contact either with a molten ceramic or a molten metal. The former case is especially important in the glassmaking industry, where it has been estimated that around 318 million gross of glass containers and 3.5 billion square feet of flat glass were manufactured in 1979 [14]. This represents a large-volume usage of oxide refractories for glass-melting furnace linings.

Corrosion of Al_2O_3 in a $CaO-Al_2O_3-SiO_2$ melt provides a good example for comparing some of the rate-controlling factors for the corrosion of an oxide by a glass or slag. Data of Cooper et al. [15] are summarized in Fig. 4.15. The corrosion rate increases as the temperature increases (curve B versus curve E). The corrosion rate is typically higher for a polycrystalline ceramic than a single crystal (curve D versus curve C), due to grain boundary effects. The corrosion rate is lower for natural convection than for forced convection or for cases either where the melt is flowing or the ceramic is moving (curves A, C, and E).

Environments for oxide ceramics in metal melting and refining processes are even more severe than the conditions in glass-melting furnaces. An important example is the Basic Oxygen Steelmaking Process (BOSP) used in a major portion of iron refining in the United States [16]. BOSP is a batch process conducted in a ceramic refractory brick-lined furnace referred to as the basic oxygen furnace (BOF). During a typical cycle the BOF is first tilted and charged with up to 100 tons of scrap steel and 250 tons of 1300°C

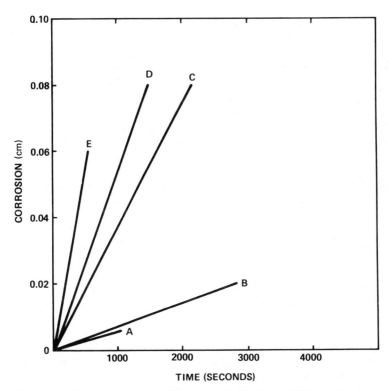

Figure 4.15 Corrosion of Al_2O_3 in CaO-Al_2O_3-SiO_2 melt. A, single crystal, 1550°C, natural convection; B, single-crystal disk rotating at 1200 rpm, 1410°C; C, single crystal, 1500°C, forced convection; D, polycrystalline, 1500°C, forced convection; E, single-crystal disk rotating at 1200 rpm, 1550°C. (From Ref. 15.)

molten crude iron. The BOF is then turned upright and oxygen is introduced through a water-cooled lance. CaO and CaF_2 are added as flux to remove Si, P, S, and other impurities by formation of a slag. The BOF cycle is approximately 1 hr long, with the temperature reaching 1600 to 1700°C (~2900 to 3100°F) and the furnace lining bathed in churning molten slag and steel.

The BOF service environment is extremely severe. The refractory lining is exposed to impact, abrasion, and thermal shock during charging of the scrap steel and the 1300°C crude iron and is exposed to both molten slag and molten metal corrosion during operation. Two primary types of refractories are used. One is referred to as pitch-bonded or tar-bonded and consists mostly of MgO particles coated with pitch and warm-bonded to form bricks. The other is referred to as tar-impregnated and consists of

porous MgO ceramic brick formed and then impregnated under vacuum by
molten pitch. These two types of bricks are installed in the furnace and
then "burned in" under controlled temperature and reducing conditions to
pyrolyze the pitch to elemental carbon. The carbon is resistant to wetting
by the molten slag and metal and inhibits penetration of the brick. A typical BOF ceramic lining ranges from 18 in. thick at the opening to 36 in.
thick in regions of maximum erosion and corrosion. Its life is usually less
than 1000 cycles, with a rate of refractory consumption between 0.08 and
0.15 cm (0.03 to 0.06 in.) average recession per cycle.

Corrosion in Heat Engines   During the past 10 years interest has grown in
the potential use of ceramics in heat engines as a means of achieving higher
operating temperatures and increased efficiency (decreased fuel consumption). Si_3N_4 and SiC currently appear to be the leading candidate materials
due to their unique combination of high strength and thermal conductivity,
low thermal expansion, and good high temperature stability.

Si_3N_4 and SiC are being evaluated in gas turbine engines for combustor
liners, stator vanes (nonmoving airfoil shapes that guide the hot-gas flow
in the optimum direction into the rotor), rotors, and a variety of liners for
hot-gas flow through the turbine. Under operating conditions, these components are exposed to oxidizing atmospheres flowing at moderate to high
velocity. Oxidation does not appear to be a problem, due to the formation
of a passive SiO_2 surface layer. However, corrosion can occur as a result
of the combined effects of the oxygen plus gaseous, condensed, or particulate impurities introduced via the gas stream. These impurities can either
increase the rate of passive oxidation (i.e., modify the transport rate of
oxygen through the protective scale), cause active oxidation, or produce
compositions which chemically attack the ceramic. The following mechanisms are of concern:

Change in the chemistry of the SiO_2 layer, increasing the oxygen diffusion
    rate
Bubble formation, disrupting the protective SiO_2 layer and allowing in-
    creased oxygen access
Decreased viscosity of the protective surface layer, which is then swept off
    the surface by the high-velocity gas flow
Formation of a molten composition at the ceramic surface that is a solvent
    for the ceramic
Localized reducing conditions, decreasing the oxygen partial pressure at
    the ceramic surface to a level at which active oxidation can occur
Formation of new surface flaws, such as pits or degraded microstructure,
    which decrease the load-bearing capability of the component

Singhal [17] has evaluated the dynamic corrosion-erosion behavior of
Si_3N_4 and SiC in a pressurized turbine test passage operating at 1100°C,

0.9 MPa pressure, and 152 m/sec gas velocity using Exxon No. 2 diesel fuel. After 250 hr of exposure, the surfaces of test specimens were smooth and free of adherent surface deposits. Average surface erosion was only 2.3 μm for the SiC and 3 μm for the Si_3N_4. No strength degradation occurred. Other tests in the same turbine test passage with 4 ppm barium present in the fuel produced drastically different results: massive surface deposits containing barium silicates with iron, magnesium, nickel, chromium, and other trace impurities. This example illustrates that small quantities of some impurities can have a pronounced effect on the corrosion behavior. This is especially important in applications where heavy residual fuels or coal-derived fuels are being considered.

Continuation of turbine passage testing without barium fuel additions has been reported by Miller et al. [18] for longer time and higher temperature. The results are summarized in Fig. 4.16. The strength of the hot-pressed Si_3N_4 was significantly degraded due to the formation of surface pits. The hot-pressed SiC showed no degradation in high-temperature strength after 4000 hr of exposure. The slight decrease in room-temperature strength was attributed to cracking of the SiO_2 surface layer during cooling.

McKee and Chatterji [19] have evaluated the stability of 97 to 99% dense sintered SiC at 900°C in a gas stream flowing at 250 m/min at 1 atm total pressure. Gases evaluated were N_2, O_2, air, O_2-N_2 mixture, 0.2% SO_2-O_2, 2% SO_2-N_2, 2% SO_2-5% CH_4-balance N_2, 10% H_2S-H_2, and H_2. Gas-molten salt environments were also evaluated using constant gas compositions and the following salt mixtures: Na_2SO_4, Na_2SO_4 + C (graphite), Na_2SO_4 + Na_2CO_3, Na_2SO_4 + Na_2O, Na_2SO_4 + $NaNO_3$, and Na_2SO_4 + Na_2S. Corrosion behavior was studied by continuous monitoring of the specimen weight in the test environment using standard thermobalance techniques, followed by examination of polished sections of the specimen cross section. The results are summarized in Fig. 4.17.

McKee and Chatterji concluded that SiC at 900°C is inert in H_2, H_2S, and high-purity N_2; that passive oxidation provides protection under normal gas turbine operating conditions or when thin condensed layers of Na_2SO_4 are present in an oxidizing atmosphere; and that corrosion occurs in the presence of Na_2O, a carbonaceous condensed phase, or a thick Na_2SO_4 surface layer.

Richerson and Yonushonis [9] have evaluated the effect on the strength of Si_3N_4 materials of 50 hr of cyclic oxidation-corrosion at specimen temperatures up to 1200°C using a combustor rig burning typical aircraft fuels with 5 ppm sea salt additions. Under conditions where Na_2SO_4 was present in the condensed form, corrosion resulting in slight material recession occurred. This was accompanied by buildup of a glassy surface layer containing many bubbles which appeared to be nucleating at the Si_3N_4 surface (Fig. 4.18). The strength of Si_3N_4 hot pressed with MgO was degraded by 30%. The strength of 2.5-g/cm^3 density reaction-bonded Si_3N_4 was degraded by

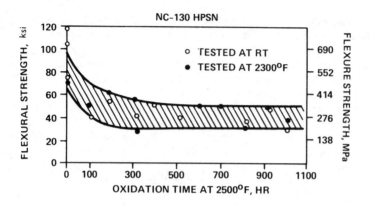

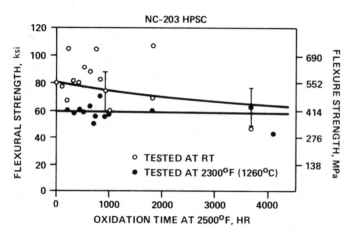

Figure 4.16 Results of exposure of hot-pressed Si_3N_4 (HPSN) and SiC (HPSC) in a turbine passage. (From Ref. 18.)

as much as 45%. Additional photomicrographs and energy-dispersive x-ray analyses further illustrating oxidation-corrosion are included in Chap. 12 (Fig. 12.24 and 12.25).

These examples show that both Si_3N_4 and SiC are susceptible to corrosion at high temperatures, especially if impurities are present. However, it should be pointed out that the rates of corrosion are still much lower than for metal gas turbine alloys at much lower temperatures.

Corrosion in Coal Combustion Environments   The oil shortage of 1973 led to a dramatic increase in evaluation of alternative energy sources and means

of achieving increased efficiency of current sources. Because of the enor-
mous coal deposits in the United States, much effort has been directed toward the development of coal-derived fuels. Concurrently, effort has been
directed toward achieving higher operating temperatures, primarily through
the use of ceramics, to increase thermal efficiency in heat engines.

Petroleum distillates used for electrical power generation are relatively
pure and can be burned without severely corroding the turbomachinery or
polluting the air. On the other hand, coal is very impure and leaves

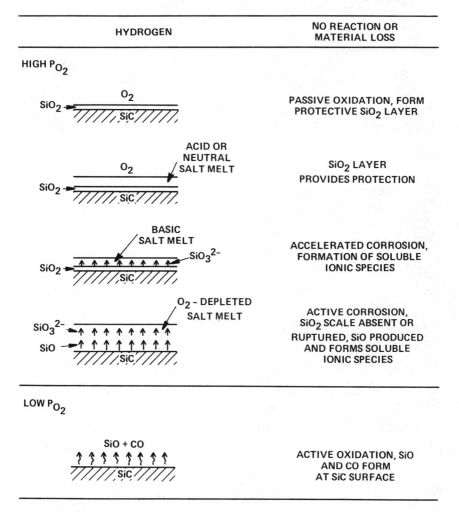

Figure 4.17 Behavior of SiC in gas-molten salt environments. (From
Ref. 19.)

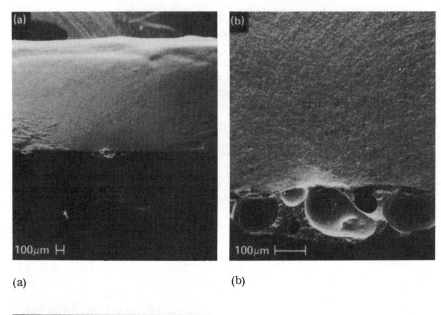

(a)   (b)

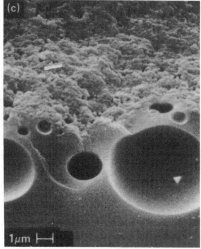

(c)

Figure 4.18   Reaction-bonded Si_3N_4 after exposure in a combustion rig with 5 ppm sea salt addition for 25 cycles of 1.5 hr at 900°C, 0.5 hr at 1120°C, and a 5-min air-blast quench. (a), (b), and (c) show the fracture surface at increasing magnification and illustrate the glassy buildup in the region of combustion gas impingement. (From Ref. 9.)

Table 4.3 Composition of Typical Coals

Location:	Illinois	Wyoming
Rank:	High-volatile bituminous B	Subbituminous B
Analysis (% by weight)		
Moisture	5.8	15.3
Volatile matter	36.2	33.5
Fixed carbon	46.3	45.2
Ash	11.7	5.7
Sulfur	2.7	2.3

Source: Compiled from Steam, Babcock and Wilcox, 38th ed., 1972, pp. 5-15; and Combustion Engineering, Combustion Engineering, Inc., 1966, pp. 13-15.

uncombusted ash or slag when burned. Table 4.3 lists the approximate composition of coal from seams in Illinois and Wyoming. Table 4.4 lists the ash or slag composition of a variety of coals.

Although the coal and ash compositions vary substantially, they all contain 5 to 15% ash and contain high concentrations of impurities such as sulfur, Fe_2O_3, CaO, and Na_2O.

The ash becomes soft and sticky for most coals at temperatures ranging from 1260°C (2300°F) to 1427°C (2600°F)[*] and becomes more fluid as the temperature is further increased. Since coal is normally burned at temperatures in excess of 1650°C (3002°F), relatively fluid slag droplets form which can corrode high-temperature surfaces and stick to lower-temperature surfaces. The buildup of slag or ash on any of the surfaces is referred to as fouling. A thin buildup can protect the surface from corrosion and erosion and in some cases can even result in a local temperature reduction. All three of these factors can increase the life of a component, especially a metal. However, a thick buildup reduces the airflow through the engine and decreases efficiency.

Fouling is an inherent problem in the direct burning of coal. A variety of approaches have been or are being studied to resolve this problem:

---

[*]Normally reported in terms of the temperature at which the viscosity is 250 P. For comparison, glycerin has a viscosity of about 250 P at -8°C (17.6°F) and about 0.25 P at 25°C (77°F).

Table 4.4 Variations in Slag Composition for Typical Coals

Location:	Zap, N.D.	Ehrenfield, Pa.	Victoria, Ill.	Hanna, Wy.
Seam:	Zap	L. Freeport	Illinois 6	80
Rank:	Lignite	Medium-volatile bituminous	High-volatile bituminous	Subbituminous
Ash composition (% wt)				
SiO_2	20–23	37	50	29
Al_2O_3	9–14	23	22	19
Fe_2O_3	6–7	34	11	10
TiO_2	0.5	0.8	0.9	0.8
CaO	18–20	0.8	9.0	18.7
MgO	6–7	0.4	1.1	2.9
Na_2O	8–11	0.2	0.35	0.2
K_2O	0.3	1.3	2.2	0.7
SO_3	21	1.6	1.6	17.78
% Ash	7–11	15.1	14.3	6.6
% Sulfur	0.6–0.8	4.04	2.6	1.2

1. Intermittent removal of buildup by thermal shock, melt-off, or passing abrasive material (such as nutshells) through the system
2. Separation of the molten slag before it enters the heat engine
3. Combustion at a temperature below the softening temperature of the ash (such as is done in a fluidized bed combustion system), so that the ash can be removed as solid particles
4. Burning the coal in an external combustor and transferring the heat to the engine working fluid with a heat exchanger (such as is done in a steam turbine system and in a closed-cycle gas turbine system)
5. Reducing the ash and impurities by a coal liquefaction process
6. Nearly eliminating the ash and impurities by a coal gasification process

It appears that fouling will prevent direct burning of coal for open-cycle gas turbine engines. Use of liquefied coal is also likely to be limited by the ash content and the resulting fouling. Thus direct burning of coal will be restricted primarily to the current steam turbine systems and possibly to

closed-cycle gas turbines, fluidized bed combustors, and magnetohydrodynamic (MHD) generation.

Ceramic linings for coal gasification systems have been evaluated by
the National Bureau of Standards. One of the evaluations was conducted at
the Conoco Coal Development Company's Lignite Gasification Pilot Plant in
Rapid City, South Dakota. The reactor vessel was 21.3 m (70 ft) high with
an inside diameter of 168 cm (66 in.) lined with 45.7 cm (18 in.) of low-
density insulating castable refractory plus 15.2 cm (6 in.) of dense abrasion-
resistant castable refractory. The dense castable consisted of approximately
37% SiO_2, 57% Al_2O_3, and 6% CaO. The gasifier operated for 5 years at
843°C (1550°F) and 1.04 MPa (150 psi) without major problems with the refractory lining. No reactions attributed to the coal were reported. The
major chemical reaction involved the refractory and the steam used in the
gasification process. Calcium aluminates in the ceramic refractory lining
reacted hydrothermally with silica under the influence of the steam to yield
calcium aluminum silicate. This acted as a strong bonding phase for the
refractory, significantly increasing the compressive strength and abrasion
resistance, both of which were beneficial.

A variety of coal gasification approaches are currently being evaluated
by the U.S. Department of Energy. Some involve temperatures in the 1500°C
(2732°F) range, where molten slag is present [20]. Studies are not far
enough along to provide life-prediction data for ceramic refractory linings.
However, work to date has shown that Al_2O_3-CrO_2 spinel compositions have
excellent corrosion resistance.

Magnetohydrodynamic (MHD) generators represent a still higher temperature requirement for ceramic materials. Operation temperatures are
in the range 1600 to 2500°C (2912 to 4532°F). The MHD generator produces
electricity by passing a moving conductive gas through a magnetic field.
Conductivity in the gas is achieved at high temperature by seeding the gas
with a salt such as K_2SO_4 or K_2CO_3 which has a low ionization potential.
This can result in an electron density of 10^{12} to 10^{13} free electrons per
cubic centimeter (compared to 10^{22} to 10^{23} for a metal conductor). In an
open-cycle MHD generator, the moving high-temperature gas can be supplied by coal or other fossil fuel combustion. The electricity is removed
from the MHD channel by ceramic electrodes. These electrodes must be
capable of transferring current densities of 1 to 5 A/cm^2 and have a life of
about 10,000 hr. They must be resistant to corrosion and erosion of the
seeded combustion gases, be resistant to electrical discharge arcs, have
good thermionic emission, and have good thermal shock resistance. The
most promising high-melting-temperature ceramic electrode materials
appear to be rare earth oxide-doped ZrO_2 and doped ThO_2 and Y_2O_3.

Ceramic insulator materials are also required in MHD generators to
separate the electrodes. High-temperature oxides (such as Al_2O_3, MgO,
and BeO), mixed oxides, and concretes are the primary candidates. The
major problem has been reaction with the alkali seeding salts, resulting in
electrical leakage and arcing.

For further information on MHD materials, the reader might refer to a review by Fehrenbacher and Tallan [21], from which the foregoing discussion was abstracted.

4.4 EROSION

Erosion resistance of a material is determined primarily by the hardness of the material compared to the hardness of other materials with which it comes in contact. Hardness is largely a function of atomic bond strength. In general, covalent and multivalent ionic ceramics have high hardness and thus have good erosion resistance. Table 4.5 compares the relative erosion resistance of some ceramics with their hardness values.

Erosion can occur by sliding motion between two surfaces or particles between the surfaces, by particulate impact, or by adhesion. All three mechanisms result in wear of metals. Only the first two are prominent in ceramics.

Sliding motion is encountered in virtually all applications of moving parts. In many cases erosion can be minimized by use of a liquid or solid lubricant. Graphite and hexagonal boron nitride are important solid lubricants. Sliding erosion can also be reduced by improving the surface finish. In many applications the degree of surface finish determines success or failure. Such is the case with ceramic bearings. Inadequate surface finish leads to spalling and accelerated destruction of both the bearing the retainer. Such is also the case with ceramic seals, especially those forming a seal around a rotating shaft. In one application a shipment of Al_2O_3 seals had

Table 4.5 Erosion Resistance Versus Hardness

Material, in increasing order of erosion resistance	Knoop hardness (kg/mm^2)
MgO	370
SiO_2	820
ZrO_2	1160
Al_2O_3	2000
Si_3N_4	2200
SiC	2700
B_4C	3500
Diamond	7000–8000

inadequate surface finish on the inner diameter. The increased friction with the rotating shaft generated enough heat to cause the metal shaft to expand thermally into the ceramic until the interference was great enough to freeze the shaft and break the ceramic seal.

Impact erosion most often involves high-velocity contact of airborne or fluid-borne particles against a surface. Rain erosion of ceramic electro-optical and electromagnetic window materials for aerospace applications is currently an important area under study. At the high relative velocities involved, rain drops hitting the surface produce substantial local Hertzian stress fields in the ceramic. This can result in surface cracks and chips that spall off after repeated impact. Fresh underlying material is exposed to impact.

Impact erosion is also being studied with respect to the potential use of ceramics for gas turbine engine components. Metcalfe [22] has reported the results of erosion tests on hot-pressed and reaction-bonded Si_3N_4 in erosion rig tests and in a small gas turbine engine. The erosion rig tests were conducted at room temperature with 80 mg of silica dust per cubic foot of air flowing 195 to 210 m/sec (650 to 700 ft/sec), using 410 stainless steel as a standard. The hot-pressed Si_3N_4 showed no erosion after 40 min at an impingement angle of 30° and only 0.13×10^{-3} cm^3 at an impingement angle of 90°. Under the same conditions reaction-bonded Si_3N_4 lost 2.2×10^{-3} and 5.6×10^{-3} cm^3, respectively, and stainless steel lost 3.7×10^{-3} and 1.9×10^{-3} cm^3. A 10-hr engine test was then conducted at a turbine inlet temperature of 925°C (1697°F) with 172.5 g (0.38 lb) of 140-mesh silica added per hour. Superalloy metal stator vanes were severely eroded, but no erosion was visible on hot-pressed Si_3N_4 vanes.

These tests illustrate some of the known characteristics of impact erosion:

1. For ceramics, the rate of erosion increases with increasing impingement angle.
2. For metals, the rate of erosion decreases as the impingement angle increases.
3. Porous ceramics, such as reaction-bonded Si_3N_4, are more susceptible to erosion than are nonporous ceramics such as hot-pressed Si_3N_4.

In addition, erosion rate is dependent on particle size, relative velocity, and the relative hardness, strength, and elastic modulus of the particle and the surface.

Combined effects should be considered when evaluating a material for wear or erosion resistance. For instance, a combination of oxidation or corrosion with erosion can be far more serious than either effect alone. Oxidized or corroded surfaces are typically much softer than the original surface and can more easily be removed by erosion, exposing a fresh surface to further oxidation or corrosion.

Erosion can affect the reliability of an application in a variety of ways. It can result in dimensional change, causing loss of critical tolerances or aerodynamic or leakage losses that reduce or restrict the performance of the component. It can also produce surface flaws or reduce cross sections, decreasing the load-bearing capability of the component. Or it can produce a roughened surface or abrasive particles that accelerate further erosion or that increase friction and limit relative motion and result in abnormally high localized stresses and fracture or chipping. Finally, erosion can remove a protective layer or produce a hole in a protective component and thus lead to failure of the system.

4.5 IMPACT

Impact can be defined in general terms as the application of a structural load to a material at a very high loading rate and usually to a localized area. Metals have some capability to survive impact. They have high fracture toughness and have ductility that can redistribute the stresses. Ceramics, because of their brittle nature and low fracture toughness, cannot redistribute stresses due to impact and are thus critically susceptible to impact damage. Therefore, impact resistance in ceramics must be achieved primarily through component design or through improvements in the fracture toughness of the materials.

One design approach to improve the impact resistance of ceramics is to shrink-fit the ceramic in a metal retainer. This places the ceramic in compression; the compressive prestress must then be exceeded by tensile loading before a crack will initiate and propagate. This approach has been used successfully for lining extrusion dies with ceramics.

Another design approach is to mount the ceramic such that impact will occur only at low angles.

The other alternative to improve impact resistance is to increase the fracture toughness of the ceramic. Five primary approaches have been pursued:

Fiber reinforcement
Second-phase dispersion
Transformation toughening
Formation of surface compression
Formation of a surface energy-absorbing layer

Fiber Reinforcement

Fabrication of composites by fiber reinforcement has been widely applied to achieve improved material toughness. Perhaps the most common example is the use of glass fibers to reinforce plastics for construction of boats, swimming pool liners, shower enclosures, sports equipment, and a wide variety of other products. Similarly, graphite fibers are used to reinforce graphite or pyrolytic carbon to produce the carbon-carbon composites used for aerospace applications.

Fiberglass and graphite-reinforced composites are relatively well developed. Other ceramic fibers are just beginning to be evaluated, especially Al_2O_3-based fibers and SiC. The Al_2O_3-based fibers have potential for reinforcing aluminum alloys to achieve increased temperature capability and improved stress rupture life. The SiC fibers are being considered for reinforcing titanium alloys and high-temperature polycrystalline ceramics such as SiC and Si_3N_4.

Perhaps the least effort has been conducted on reinforcing ceramics to improve impact resistance. The author has successfully hot-pressed a highly wear resistant Al_2O_3-based composition containing ductile stainless steel reinforcement. The resulting composite had substantially improved impact resistance and retained its usefulness and integrity even after cracks were present in the ceramic.

Brennan [23] has reinforced hot-pressed Si_3N_4 with tantalum wire and increased impact resistance in a Charpy-type test from 0.68 J (0.5 ft-lb) for unreinforced material to 21.7 J (16 ft-lb) for reinforced material. More recently, Brennan and co-workers have reinforced borosilicate glass and lithium aluminum silicate glass ceramics with SiC fibers and achieved similar improvements in impact resistance. A tungsten carbide spike was driven through one reinforced composition without secondary cracking, in much the same fashion that a nail can be driven through wood.

Second-Phase Reinforcement

Controlled dispersions of a second phase can also improve material toughness and possibly the impact resistance. This has been well demonstrated for the WC and TiC cermets containing small quantities of metals such as Co or Ni. This approach has also been demonstrated for impregnation of porous graphite with metals.

Second-phase particulate dispersions also have potential. Just as thorium oxide particles dispersed in metal superalloys provide beneficial property modifications, controlled dispersions in ceramics can be beneficial. For instance, fine particles having a different thermal expansion than the matrix can result in very fine microcracks in the microstructure. Although this will probably decrease the strength of the matrix material, it can increase the strain tolerance of the overall composite and provide potential benefits in impact resistance and thermal shock resistance. Rather than a single catastrophic crack forming, the tiny microcracks in the region of loading all deflect slightly and essentially redistribute or absorb the load.

Transformation Toughening

Transformation toughening refers to increases in fracture toughness achieved by dispersion of particles of a material that undergoes a displacive transformation in a matrix material that does not go through the same transformation. During the cooling cycle, after initial densification of the material, the dispersed particles undergo transformation which is accompanied by a volume change. The surrounding matrix material is either cracked or locally stressed by this volume change, resulting in significant increase in the fracture toughness.

Most of the work on transformation toughening has been conducted with ZrO_2 using compositions that are only partially stabilized [24, 25]. To review, ZrO_2 undergoes a 3.25% volume expansion during cooling below approximately 1000°C (1832°F) due to transformation from the tetragonal phase to the monoclinic phase. In pure polycrystalline ZrO_2, this results in catastrophic failure of the part. Addition of CaO, MgO, or Y_2O_3 to the ZrO_2 results in a cubic crystal structure that is stable over the complete temperature range and does not undergo a phase transformation. This is referred to as stabilized zirconia.

Stabilized ZrO_2 has low fracture toughness and poor resistance to impact. By not adding enough MgO, CaO, or Y_2O_3 to stabilize the ZrO_2 completely and by careful control of particle sizing and processing, mixtures of the stabilized cubic phase and the unstable monoclinic phase that have very high fracture toughness are achieved. This type of material is referred to as partially stabilized zirconia. This can be thought of as an alloy family, since many variations and combinations of properties are possible.

Partially stabilized ZrO_2 has recently been used successfully for extrusion dies for brass rod. Over 6000 extrusions were achieved for 1.9-cm (0.75-in.)-diameter bar extruded from 30.5 cm (12 in.)-diameter 76-cm (30-in.)-long billets at approximately 900°C (1650°F).[*] Metal dies typically required rework due to bore wear after only 10 to 50 extrusions.

An increase in fracture toughness has also been achieved by addition of unstabilized ZrO_2 to Al_2O_3 and Si_3N_4 [26].

Formation of Surface Compression

This approach is widely used, especially for strengthening of glass. Compressive surface layers are usually achieved in glass by either ion exchange or quenching. The ion exchange approach involves heating the glass and exposing it to cations that are larger than those initially in the glass. The larger ions, such as K^+, replace the smaller ions, such as Na^+, near the

---

[*] Anaconda Brass Company, using Zircoa partially stabilized ZrO_2 dies.

surface but not in the interior. When the glass subsequently cools, the larger ions near the surface force the surface into compression.

The quench approach is similar. The glass is heated above its softening temperature and then cooled very quickly. The surface solidifies first and then is pulled into compression as the interior cools more slowly.

Both quenching and ion exchange have been used extensively for preparation of safety glass and other glass products. The same techniques have been applied to polycrystalline ceramics [27] experimentally, but are not in general commercial use. However, they do have potential for increased impact resistance and should be considered.

Another method of achieving surface compression is by the use of laminates having a controlled thermal expansion mismatch.

Formation of a Surface Energy-Absorbing Layer

Very little work has been conducted in this area, but it is worthy of consideration for some applications. Palm [28] showed that the impact resistance of dense SiC could be improved by a surface layer of lower-density SiC. Brennan and Hulse [29] showed similar results for dense Si_3N_4 with a lower-density Si_3N_4 layer. In both cases, the low-density surface material acted as an energy-absorbing layer during impact.

Mechanisms of Impact

Some of the factors that affect impact resistance and some suggested approaches for improving impact resistance have been briefly discussed. A detailed treatment of the mechanisms of impact damage in ceramics is beyond the intended scope of this text. To delve further into this technology, Refs. 30 and 31 are recommended as a good starting point.

4.6 THERMAL SHOCK

Thermal shock refers to the thermal stresses that occur in a component as a result of exposure to a temperature difference between the surface and interior or between various regions of the component. For shapes such as an infinite slab, a long cylinder (solid or hollow), and a sphere (solid or hollow), the peak stress typically occurs at the surface during cooling according to the equation

$$\sigma_{th} = \frac{E\alpha \, \Delta T}{1 - \nu} \tag{4.4}$$

Table 4.6 Thermal Shock Resistance Parameters

Parameter designation	Parameter type	Parameter[a]	Physical interpretation/heat transfer conditions	Typical units
R	Resistance to fracture initiation	$\dfrac{\sigma(1-\nu)}{\alpha E}$	Maximum ΔT allowable for steady heat flow	°C
R'	Resistance to fracture initiation	$\dfrac{\sigma(1-\nu)k}{\alpha E}$	Maximum heat flux for steady flow	cal/cm·sec
R"	Resistance to fracture initiation	$\dfrac{\sigma(1-\nu)\alpha_{TH}}{\alpha E}$	Maximum allowable rate of surface heating	cm²·°C/sec
R'''	Resistance to propagation damage	$\dfrac{E}{\sigma^2(1-\nu)}$	Minimum in elastic energy at fracture available for crack propagation	$(\text{psi})^{-1}$
R''''	Resistance to propagation damage	$\dfrac{\gamma E}{\sigma^2(1-\nu)}$	Minimum in extent of crack propagation on initiation of thermal stress fracture	cm
R_{st}	Resistance to further crack propagation	$\left(\dfrac{\gamma}{\alpha^2 E}\right)^{1/2}$	Minimum ΔT allowed for propagating long cracks	$°C/m^{1/2}$

[a] σ, tensile strength; ν, Poisson's ratio; α, coefficient of thermal expansion; E, Young's modulus of elasticity; k, thermal conductivity; α_{TH}, thermal diffusivity; γ, fracture surface energy.
Source: Ref. 32.

where σ_{th} is the thermal stress, E the elastic modulus, α the coefficient of thermal expansion, v is Poisson's ratio, and ΔT is the temperature difference.

Equation (4.4) indicates that the thermal stress increases as the elastic modulus and thermal expansion coefficient of the material increases and as the imposed ΔT increases. From a materials point of view, the ΔT can be decreased by increasing the thermal conductivity of the material. From a design point of view, the ΔT in the material can be decreased by configuration modification and possibly by modification of the heat transfer conditions.

Hasselman [32] has defined thermal stress resistance parameters based on equation (4.4) and other equations for various heat transfer conditions and conditions of crack initiation versus crack growth. These are summarized in Table 4.6. Note that the effects of E, σ, and v are opposite for crack initiation versus crack propagation. Low E and v with high σ provide high resistance to propagation of existing cracks. This provides both a paradox and a pronounced alternative to designing for thermal shock conditions.

When selecting a ceramic material for an application where thermal shock is expected to be a problem, calculation of the appropriate thermal shock parameter for the various candidate materials can sometimes be useful. Table 4.7 shows an example of calculations of the parameter R for several materials. Of this group, the LAS is by far the most thermal shock resistant. Its thermal shock resistance is entirely due to its low coefficient of thermal expansion. Fused silica glass also has a low coefficient of thermal expansion and, similarly, excellent thermal shock resistance. Silicon

Table 4.7 Calculated Values of the Thermal Shock Parameter R for Various Ceramic Materials Using Typical Property Data

Material	Strength,[a] σ (psi)	Poisson's ratio, v	Thermal expansion, α (in./in.·°C)	Elastic modulus, E (psi)	$R = \dfrac{\sigma(1 - v)}{\alpha E}$ (°C)
Al_2O_3	50,000	0.22	7.4×10^{-6}	55×10^6	96
SiC	60,000	0.17	3.8×10^{-6}	58×10^6	230
RSSN[b]	45,000	0.24	2.4×10^{-6}	25×10^6	570
HPSN[b]	100,000	0.27	2.5×10^{-6}	45×10^6	650
LAS[b]	20,000	0.27	-0.3×10^{-6}	10×10^6	4860

[a] Flexure strength used rather than tensile strength.
[b] RSSN, reaction-sintered silicon nitride; HPSN, hot-pressed silicon nitride; LAS, lithium aluminum silicate (β-spodumene).

nitride has a much higher strength than either lithium aluminum silicate
(LAS) or fused silica, but much lower thermal shock resistance because of
its higher thermal expansion coefficient and elastic modulus. Silicon carbide has still higher α and E and correspondingly less calculated thermal
shock resistance. Al_2O_3, other oxides, and strongly ionic bonded ceramics
have even higher α and even lower thermal shock resistance.

The R parameter has only been used as an illustrative example. Other
factors, such as thermal conductivity and fracture toughness, have substantial effects. For instance, for some configurations and under some heat
transfer conditions, SiC appears to be more thermal shock resistant than
Si_3N_4. The higher thermal conductivity of the SiC redistributes the heat
and decreases stress-causing gradients. Another example is the partially
stabilized transformation-toughened ZrO_2. Even though it has a very high
coefficient of thermal expansion ($\sim$12 $\times$ 10^{-6} per °C) and only moderate
strength, it is extremely thermal shock resistant due to the high fracture
toughness. Polycrystalline aluminum titanate has shown similar thermal
shock resistance but due to a slightly different mechanism. Aluminum
titanate has extremely anisotropic thermal expansion properties in single
crystals, and polycrystalline material spontaneously microcracks during
cooling after fabrication. These fine intergranular cracks limit the strength
of the material, but provide an effective mechanism for absorbing strain
energy during thermal shock and preventing catastrophic crack propagation.

PROBLEMS

4.1 What are the primary mechanisms of creep in single-crystal and polycrystalline ceramics? What are the effects of stress and temperature?

4.2 Differentiate between creep and slow crack growth. Describe how
slow crack growth can result from creep in a polycrystalline ceramic.
Explain how slow crack growth can occur independent of creep. What effects do creep and slow crack growth have on mechanical properties under
a sustained load?

ANSWERS TO PROBLEMS

4.1 Creep in single crystals is along slip planes and results from the
movement of dislocations. Energy is required to create and move dislocations. Increasing the stress or the temperature increases dislocation movement and increases creep along slip planes.

Polycrystalline materials with large grain size can creep due to dislocation movement and slip. However, the primary mechanisms are generally
diffusion and grain boundary sliding, both of which are strongly dependent
on temperature.

4.2 Creep is a mechanism of plastic deformation under a sustained load. Slow crack growth is a mechanism for increase in the size of a flaw within the material. Creep by slip or diffusion can occur without producing a strength-limiting flaw. Creep by grain boundary sliding can be accompanied by slow crack growth. This can occur by cavitation where three grains intersect as the grains are pulled in different directions. It can also occur by grain boundary separation where two grains contact. As the grains continue to creep in different directions, the crack increases in size and eventually links up with other grain boundary cracks.

Slow crack growth can also operate independent of creep, as in the case of stress corrosion. In this case, enhanced corrosion occurs at a crack tip due to the concentrated stress, resulting in crack growth.

Both creep and slow crack growth can result in failure under sustained loading where the load is substantially smaller than is required for fast fracture.

REFERENCES

1. W. D. Kingery, H. K. Bowen, and D. R. Uhlmann, Introduction to Ceramics, 2nd ed., John Wiley & Sons, Inc., New York, 1976, Chap. 14.
2. G. Y. Chin, L. G. Van Uitert, M. L. Green, G. J. Zydzik, and T. Y. Komentani, Strengthening of alkali halides by bivalent ion additions, J. Amer. Ceram. Soc., 56, 369 (1973).
3. M. Ashby, A first report on deformation mechanism maps, Acta Met., 20, 887 (1972).
4. A. Joffe, M. W. Kupitschewa, and M. A. Lewitzky, Z. Phys. 22, 286 (1924).
5. M. S. Seltzer, High temperature creep of silicon-base compounds, Bull. Amer. Ceram. Soc., 56[4], 418 (1977).
6. D. C. Larsen and G. C. Walther, Property Screening and Evaluation of Ceramic Turbine Engine Materials, Interim Report No. 6, AFML Contract F33615-75-C-5196, 1978.
7. C. D. Greskovich and J. A. Palm, "Development of high performance sintered Si_3N_4," in Highway Vehicle Systems Contractors Coordination Meeting 17th Summary Report, NTIS Conf. 791082, Dist. Category UC-96, 1979, pp. 254-262.
8. S. M. Wiederhorn and L. H. Bolz, Stress corrosion and static fatigue of glass, J. Amer. Ceram. Soc., 53[10], 543-548 (1970).
9. D. W. Richerson and T. M. Yonushonic, Environmental effects on the strength of silicon nitride materials, in DARPA/NAVSEA Ceramic Gas Turbine Demonstration Engine Program Review, MCIC Report MCIC-78-36, 1978, pp. 247-271.
10. D. J. Godfrey and K. C. Pitman, Some mechanical properties of silicon nitride ceramics: strength, hardness and environmental effects,

in Ceramics for High Performance Applications (J. Burke, A. Gorum, and R. Katz, eds.), Brook Hill Publishing Co., Chestnut Hill, Mass., 1974, pp. 425-455. (Available from MCIC, Battelle Columbus Labs., Columbus, Ohio.)

11. K. M. Johansen, D. W. Richerson, and J. J. Schuldies, Ceramic Components for Turbine Engines, Phase II Technical Report 21-2794(08), prepared under contract F33615-77-C-5171, 1980.

12. M. S. Crowley, Hydrogen-silica reactions in refractories, Bull. Am. Ceram. Soc. 46, 679 (1967).

13. L. A. Lay, The Resistance of Ceramics to Chemical Attack, NPL Report CHEM 96, National Physical Laboratory, Jeddington, Middlesex, U.K., Jan. 1979.

14. J. J. Svec, Ceram. Ind., June 1978, p. 20.

15. A. R. Cooper, Jr., B. N. Samaddar, Y. Oishi, and W. D. Kingery, Dissolution in ceramic systems, J. Am. Ceram. Soc. 47, 37 (1964); 47, 249 (1964); 48, 88 (1965).

16. S. C. Carniglia, W. H. Boyer, and J. E. Neely, MgO refractories in the basic oxygen steelmaking furnace, in Ceramics in Severe Environments (W. Wurth Kriegel and H. Palmour III, eds.), Plenum Publishing Corp., New York, 1971, pp. 57-88.

17. S. C. Singhal, "Corrosion behavior of silicon nitride and silicon carbide in turbine atmospheres," Proceedings of the 1972 Tri-Service Conference on Corrosion, MCIC 73-19, 1973, pp. 245-250.

18. D. G. Miller, C. A. Andersson, S. C. Singhal, F. F. Lange, E. S. Diaz, E. S. Kossowsky, and R. J. Bratton, Brittle Materials Design High Temperature Gas Turbine—Materials Technology, AMMRC CTR-76-32, Vol. 4, 1976.

19. D. W. McKee and D. Chatterji, Corrosion of silicon carbide in gases and alkaline melts, J. Am. Ceram. Soc. 59(9-10), 441-444 (1976).

20. W. A. Ellingson, Materials Technology for Coal-Conversion Processes, Argonne National Laboratory Report ANL-78-54, 1978.

21. L. L. Fehrenbacher and N. M. Tallan, Electrode and insulation materials in magnetohydrodynamic generators, in Ceramics in Severe Environments (W. Wurth Kriegel and H. Palmour III, eds.), Plenum Publishing Corp., New York, 1971, pp. 503-520.

22. A. G. Metcalfe, Application of ceramics to radial flow gas turbines at Solar, in Ceramics for High Performance Applications (J. J. Burke, A. E. Gorum, and R. N. Katz, eds.), Brook Hill Publishing Co., Chestnut Hill, Mass., 1974, pp. 739-747. (Available from MCIC, Battelle Columbus Labs., Columbus, Ohio.)

23. J. J. Brennan, Increasing the impact strength of Si_3N_4 through fibre reinforcement, in Special Ceramics, Vol. 6 (P. Popper, ed.), British Ceramic Research Assoc., Manchester, U.K., 1975, pp. 123-124.

24. D. L. Porter and A. H. Heuer, Mechanisms of toughening partially stabilized zirconia (PSZ), J. Am. Ceram. Soc. 60(3-4), 183-184 (1977).

25. T. K. Gupta, Role of stress-induced phase transformation in enhancing strength and toughness of zirconia ceramics, in Fracture Mechanics of Ceramics, Vol. 4 (R. C. Bradt, D. P. H. Hasselman, and F. F. Lange, eds.), Plenum Publishing Corp., New York, 1978, pp. 877-889.

26. N. Claussen and J. Jahn, Mechanical properties of sintered and hot-pressed Si_3N_4-ZrO_2 composites, J. Am. Ceram. Soc. 61, 94-95 (1978).

27. H. P. Kirchner, Strengthening of Ceramics, Marcel Dekker, Inc., New York, 1979.

28. J. A. Palm, Improved Toughness of Silicon Carbide, Final Report on NASA Contract NAS3-17832, Jan. 1976.

29. J. J. Brennan and C. O. Hulse, Development of Si_3N_4 and SiC of Improved Toughness, Final Report, NASA-CR-135306, Oct. 1977 (N78-17216).

30. A. G. Evans and T. R. Wilshaw, Dynamic solid particle damage in brittle materials: an appraisal, J. Mater. Sci. 12, 97-116 (1977).

31. W. F. Adler, Assessment of the State of Knowledge Pertaining to Solid Particle Erosion, Final Report, ARO Contract DAAG 29-77-C-0039, CR79-680, June 1979.

32. D. P. H. Hasselman, Thermal stress resistance parameters for brittle refractory ceramics: a compendium, Bull. Am. Ceram. Soc. 49, 1033-1037 (1970).

II

PROCESSING OF CERAMICS

The relationships among atomic bonding, crystal structure, and properties for ceramics, metals, and polymers were discussed in Part I. It was shown that the theoretical strength is controlled by the strength of bonding, but that in actual ceramic components the theoretical strength is not achieved due to flaws in the fabricated material. The objectives of Part II are to review the fabrication processes used for manufacturing ceramic components, determine where in these processes strength-limiting flaws are likely to occur, and provide the reader with approaches for detecting these flaws and working with the ceramic fabricator to eliminate them.

Most ceramic fabrication processes begin with finely ground powder. Chapter 5 describes the criteria for selection of the starting powder, methods of achieving the proper particle size distribution, and requirements for pretreating the powder before it can be formed into the desired component.

Chapter 6 describes the processes used to form the ceramic powders into the component shapes. Uniaxial and isostatic pressing, slip casting, extrusion, injection molding, tape forming, and green machining are included.

The shapes resulting from the forming processes described in Chap. 6 consist essentially of powder compacts that must be densified by high-temperature processing before they will have adequate strength and other properties. The mechanisms and processes for densification are explored in Chap. 7. Some processes combine forming and densification in a single step. These include hot pressing, chemical vapor deposition, liquid particle spray, and cementicious bonding. These are also discussed in Chap. 7.

Ceramic components requiring close tolerances must be machined after densification. This machining step can be as expensive as all the other process steps together and thus must be thoroughly understood by the engineer. Chapter 8 reviews the mechanisms of material removal during machining, the effects on the strength of the ceramic, and guidelines for selection of a machining method and abrasive material.

Chapter 9 discusses quality control and quality assurance methods for ceramic components. Use of specifications and in-process certification will be described. Nondestructive inspection techniques for both internal and surface flaws are reviewed, including radiography, ultrasonics, image enhancement, penetrants, and some relatively new techniques that look promising.

Powder Processing

The nature of the raw material has a major effect on the final properties of a ceramic component. Purity, particle size distribution, reactivity, polymorphic form, availability, and cost must all be considered and carefully controlled. In this chapter we discuss the types and sources of raw materials and the processing required to prepare them into the appropriate purity, particle size distribution, and other conditions necessary to achieve optimum results in later processing steps.

5.1 RAW MATERIALS

Traditional Ceramics

Ceramics have been produced for centuries. The earliest ceramic articles were made from naturally occurring raw materials. Early civilizations found that clay minerals became plastic when water was added and could be molded into shapes. The shape could then be dried in the sun and hardened in a high-temperature fire. The word "ceramic" comes from the Greek word keramos, which translates roughly as "burnt stuff."

Many of the raw materials used by the ancient civilizations are still used today and form the basis of a sizable segment of the ceramic industry [1-3]. These ceramic products are often referred to as traditional ceramics. Important applications of traditional ceramics are listed in Table 5.1. Some of the naturally occurring minerals and their sources and uses are summarized in the following paragraphs.

The clay minerals are hydrated aluminosilicates which have layer structures. There are a variety of clay minerals, including kaolinite $[Al_2(Si_2O_5)(OH)_4]$, halloysite $[Al_2(Si_2O_5)(OH)_4 \cdot 2H_2O]$, pyrophyllite $[Al_2(Si_2O_5)_2(OH)_2]$, and montmorillonite $[Al_{1.67}(Na,Mg)_{0.33}(Si_2O_5)_2(OH)_2]$. All are secondary in origin, having formed by weathering of igneous rocks under the influence of water, dissolved CO_2, and organic acids. The largest deposits were formed when feldspar $(KAlSi_3O_8)$ was eroded from rocks such as granite and deposited in lake beds and then altered to a clay.

Table 5.1 Traditional Ceramics

Whitewares	Dishes, plumbing, enamels, tiles
Heavy clay products	Sewer pipe, brick, pottery, sewage treatment, and water purification components
Refractories	Brick, castables, cements, crucibles, molds
Construction	Brick, block, plaster, concrete, tile, glass, fiberglass
Abrasive products	Grinding wheels, abrasives, milling media, sandblast nozzles, sandpaper
Glass	Too numerous to list

The importance of clays in the evolution of traditional ceramic processing cannot be overemphasized. The plasticity developed when water is added provides the bond and workability so important in the fabrication of pottery, dinnerware, brick, tile, and pipe.

Silica (SiO_2) is a major ingredient in glass, glazes, enamels, refractories, abrasives, and whiteware. Its major sources are in the polymorphic form quartz, which is the primary constituent of sand, sandstone, and quartzite.

Feldspar is also used in glass, pottery, enamel, and other ceramic products. Feldspar minerals range in composition from $KAISi_3O_8$ to $NaAISi_3O_8$ to $CaAl_2Si_2O_8$ and act as a flux (reduces the melting temperature) in a composition. Nepheline syenite ($Na_2Al_2Si_2O_8$) is used in a similar fashion.

Other naturally occurring minerals used directly in ceramic compositions include talc [$(Mg_3(Si_2O_5)_2(OH_2]$, asbestos [$(Mg_3Si_2O_5)(OH)_4$], wollastonite ($CaSiO_3$), and sillimanite (Al_2SiO_5).

Modern Ceramics

During the past 50 years scientists and engineers have acquired a much better understanding of ceramic materials and their processing and have found that naturally occurring minerals could be refined or new compositions synthesized to achieve unique properties. These refined or new ceramics are often referred to as modern ceramics. They typically are of highly controlled composition and structure and have been engineered to fill the needs of applications too demanding for traditional ceramics. Modern ceramics include the oxide ceramics (such as Al_2O_3, ZrO_2, ThO_2, BeO, MgO, and

$MgAl_2O_4$), magnetic ceramics (such as $PbFe_{12}O_{19}$, $ZnFe_2O_4$, and $Y_6Fe_{10}O_{24}$), ferroelectric ceramics (such as $BaTiO_3$), nuclear fuels (such as UO_2 and UN), and nitrides, carbides, and borides (such as Si_3N_4, SiC, B_4C, and TiB_2) [4-6]. Table 5.2 summarizes many of the modern ceramic applications. Emphasis in this book will be on the modern ceramics as they are the ones with which an engineer is most likely to become involved. The following sections describe how some of the modern ceramic starting powders are refined or synthesized.

Aluminum Oxide Powder

Aluminum oxide (Al_2O_3) occurs naturally as the mineral corundum, which is better known to most of us when it is in gem-quality crystals called ruby and sapphire. Ruby and sapphire are precious gems because of their chemical inertness and hardness. Al_2O_2 powder is produced in large quantity from the mineral bauxite by the Bayer process. Bauxite is primarily colloidal aluminum hydroxide intimately mixed with iron hydroxide and other impurities. The Bayer process involves the selective leaching of the alumina by caustic soda and precipitation of the purified aluminum hydroxide. The resulting fine-particle-size aluminum hydroxide can then be thermally converted to Al_2O_3 powder, which is used to manufacture polycrystalline Al_2O_3-based ceramics.

Al_2O_3 powder is used in the manufacture of porcelain, alumina laboratory ware, crucibles and metal casting molds, high-temperature cements,

Table 5.2 Modern Ceramics

Electronics	Heating elements, dielectrics, substrates, semiconductors, insulators, transducers, lasers, hermetic seals, igniters
Aerospace and automotive	Reentry, radomes, turbine components, heat exchangers, emission control
Medical	Prosthetics, controls
High-temperature structural	Kiln furniture, braze fixtures, advanced refractories
Nuclear	Fuels, controls
Technical	Laboratory ware
Miscellaneous	Cutting tools, wear-resistant components, armor, magnets, glass-ceramics, single crystals, fiber optics

wear-resistant parts (sleeves, tiles, seals, etc.), sandblast nozzles, armor, medical components, abrasives, refractories, and a variety of other components. Many hundreds of tons of alumina powder and alumina-based articles are produced each year. It has even been used to make extrusion dies for corn chips and mixing valves for water faucets.

Magnesium Oxide Powder   Magnesium oxide occurs naturally as the mineral periclase, but not in adequate quantity or purity for commercial requirements. Most MgO powder is produced from $MgCO_3$ or from seawater. It is extracted from seawater as the hydroxide, which is easily converted to the oxide by heating at the appropriate temperature.

MgO powder is used extensively for high-temperature electrical insulation and in refractory brick.

Silicon Carbide Powder   SiC has been found occurring naturally only as small green hexagonal plates in meteoric iron. This same hexagonal polymorph (α-SiC) has been synthesized commercially in large quantities by the Acheson process. This fascinating process appears crude, but is cost effective and simultaneously produces lower-grade SiC for abrasives and high-grade SiC for electrical applications. The Acheson process consists essentially of mixing SiO_2 sand with coke in a large elongated mound and placing large carbon electrodes in opposite ends. An electric current is then passed between the electrodes, resistance-heating the coke in the mound to about 2200°C. In this temperature range the coke reacts with the SiO_2 to produce SiC plus CO gas. Heating is continued until the reaction is completed in the interior of the mound. After cooling, the mound is broken up and sorted. The core contains intergrown green hexagonal SiC crystals which are low in impurities and suitable for electronic applications. Around the core is a zone of lower purity that is used for abrasives. The outer layer of the mound is a mixture of SiC and unreacted SiO_2 and carbon which is added to silica sand and coke for the next batch.

SiC can be prepared from almost any source of silicon and carbon. For instance, it has been prepared in the laboratory from a mixture of silicon metal powder and sugar. It has also been prepared from rice hulls. It can also be prepared from silicon tetrachloride ($SiCl_4$) and some silanes. In this case, the cubic β-SiC polymorph normally forms.

SiC is used for high-temperature kiln furniture, electrical-resistance heating elements, grinding wheels and abrasives, wear-resistance applications, incinerator lining, and is currently being evaluated for highly stressed components in heat engines.

Silicon Nitride Powder   Silicon nitride does not occur naturally. It has been synthesized by several different process. Most of the powder available commercially has been made by the reaction of silicon metal powder with

nitrogen at temperatures in the range 1250 to 1400°C and consists of a mixture of α-Si_3N_4 and β-Si_3N_4 polymorphs. This Si_3N_4 is not a ready-to-use
powder when it is removed from the furnace. It is loosely bonded and must
be crushed and sized first. The resulting powder is not of high purity, but
contains impurities such as Fe, Ca, and Al, which were originally present
in the silicon, plus impurities picked up during crushing and grinding.

Higher-purity Si_3N_4 powder has been made by reduction of SiO_2 with
carbon in the appropriate nitrogen environment and reaction of $SiCl_4$ or
silanes with ammonia. Both of these methods produce very fine particle
size powder that does not require further grinding for use. In fact, some
of these powders are so fine that they require coarsening by heat treating
(calcining) before they are suitable for shape-forming operations.

High-purity Si_3N_4 powder has recently been made in small quantities
by laser reaction [7]. A mixture of gaseous silane (SiH_4) and ammonia is
exposed to the coherent light output of a CO_2 laser. The silane has high
absorption for the wavelengths involved, resulting in the heat required for
reaction. The resulting Si_3N_4 is in spherical particles of a uniform size
for the given gas flow and laser power conditions. Particles typically in
the range 200 to 1000 Å can be produced.

Raw Materials Selection Criteria

The selection criteria for ceramic starting powders are dependent on the
properties required in the finished component. Purity, particle size distribution, reactivity, and polymorphic form can all affect the final properties
and thus must be considered from the outset.

Purity   Purity strongly influences high-temperature properties such as
strength, stress rupture life, and oxidation resistance, as discussed in
Chaps. 3 and 4. The effect of the impurity is dependent on the chemistry
of both the matrix material and the impurity, the distribution of the impurity, and the service conditions of the component (time, temperature,
stress, environment). For instance, Ca severely decreases the creep
resistance of Si_3N_4 hot-pressed with MgO as a densification (sintering) aid,
but appears to have little effect on Si_3N_4 hot pressed with Y_2O_3 as the densification aid [8,9]. In the former case, the Ca is concentrated at the grain
boundaries and depresses the softening temperature of the grain boundary
glass phase. In the latter case, the Ca is apparently absorbed into solid
solution by the crystalline structure and does not significantly reduce the
refractoriness of the system.

Impurities present as inclusions do not appreciably affect properties
such as creep or oxidation, but do act as flaws that can concentrate stress
and decrease component tensile strength. The effect on strength is dependent on the size of the inclusion compared to the grain size of the ceramic
and on the relative thermal expansion and elastic properties of the matrix

and inclusion. Tungsten carbide inclusions in Si_3N_4 have little effect on the
strength; Fe and Si have a large effect.

Particle Size and Reactivity The effects of impurities are important for
mechanical properties, but may be even more important for electrical,
magnetic, and optical properties. Electrical, magnetic, and optical properties are usually carefully tailored for a specific application, often by
closely controlled addition of a dopant. Slight variations in the concentration
or distribution of the dopant severely alters the properties. Similarly, the
presence of unwanted impurities can poison the effectiveness of the dopant
and cause the device to operate improperly.

Particle size distribution is important, depending on which consolidation
or shaping technique is to be used. In most cases the objective of the consolidation step is to achieve maximum particle packing and uniformity, so
that minimum shrinkage and retained porosity will result during densification. A single particle size does not produce good packing. Optimum
packing for particles all the same size results in over 30% void space.
Adding particles of a size equivalent to the largest voids reduces the void
content to 26%. Adding a third, still smaller particle size can reduce the
pore volume to 23%. Therefore, to achieve maximum particle packing, a
range of particle sizes is required.

A controlled optimum particle size distribution is required to achieve
maximum, reproducible strength. The strength is controlled by flaws in
the material. A single particle which is significantly larger than the other
particles in the distribution can become the critical flaw that limits the
strength of the final component. Similarly, a large void resulting from a
nonhomogeneous particle size distribution or from particles too close to
the same size may not be eliminated during sintering and may become the
strength-limiting flaw.

Small particle size is important if high strength is the primary objective. However, there are many applications where strength is not the primary criterion. Refractories are a good example. Most refractories
contain either large particles or high porosity as an important constituent
in achieving the desired properties.

Another important aspect of starting powder is reactivity. The primary
driving force for densification of a compacted powder at high temperature
is the change in surface free energy. Very small particles with high surface area have high surface free energy and thus have a strong thermody-
namic drive to decrease their surface area by bonding together. Very
small particles, approximately 1 μm or less, can be compacted into a
porous shape and sintered at high temperature to near-theoretical density.
Transparent polycrystalline alumina for sodium vapor lamp envelopes is a
good example. To achieve transparency, virtually all the initial pores
must be removed during sintering. Highly reactive starting powder with an

average particle size of about 0.3 μm is used as the raw material. Another example is sintered Si_3N_4. Starting powder of approximately 2 μm average particle size only sinters to about 90% of theoretical density. Submicron powder with a surface area roughly greater than 10 m^2/g sinters to greater than 95% of theoretical density.

Particle size distribution and reactivity are also important in determining the temperature and the time at the temperature necessary to achieve sintering. Typically, the finer the powder and the greater its surface area, the lower are the temperature and time at temperature for densification. This can have an important effect on strength. Long times at temperature result in increased grain growth and lower strength. To optimize strength, a powder that can be densified quickly with minimal grain growth is desired.

Polymorphic Form   Many ceramics occur in different polymorphic forms, as discussed in Chap. 1. For most applications one polymorph is preferred. For instance, α-Si_3N_4 is superior to β-Si_3N_4 as a starting powder for hot pressing. A similar situation exists for SiC. The stable form at high temperature is the hexagonal α-SiC polymorph. α-SiC starting powder can be densified by either hot pressing or pressureless sintering over a relatively broad temperature range without problems associated with polymorphic transformation. The lower-temperature form, cubic β-SiC, can also be densified, but in a much more limited temperature range.

5.2 POWDER SIZING

Section 5.1 explained the importance of particle size distribution in achieving the optimum properties in the final component. Raw materials are not usually available with the optimum particle size distribution. The ceramic fabricator must further process the raw materials to the specifications. The following techniques are used:

Screening	Hammer milling
Air classification	Precipitation
Élutriation	Freeze drying
Ball milling	Laser
Attrition milling	Plasma
Vibratory milling	Calcining
Fluid energy milling	

The following paragraphs describe briefly each of these approaches and discuss their limitations and how these limitations may alter the properties of the material.

Screening

Screening is a positive method of particle sizing. The powder is poured
onto a single screen having the selected size openings or on a series of
screens, each subsequently with smaller openings. The particles are separated into size ranges; the particles larger than the screen openings remain
on the screen and smaller particles pass through until they reach a screen
with holes too small to pass through.

   Screen sizes are classified according to the number of openings per
linear inch and are referred to as mesh sizes. A 16 mesh screen has 16
equally spaced openings per linear inch; a 325 mesh screen has 325. Table
5.3 compares the mesh size of standard screens with the actual size of the
openings.

   Raw and processed ceramic materials are often supplied according to
screen size. For instance, a −325-mesh powder has all passed through a
325-mesh screen and should contain no particles larger than 44 μm (0.0018
in.). A powder designated −100 mesh +150 mesh consists of a narrow particle size range that was small enough to pass through a 100-mesh screen
but too large to pass through a 150-mesh screen. Powders containing a
broad particle size range can also be classified according to screen size,
for example:

Size range	Weight (%)
−80 + 100	5
−100 + 150	8
−150 + 220	13
−220 + 280	20
−280 + 325	18
−325	36

   Screening can be conducted dry or with the particles suspended in a
slurry. Dry screening is used most frequently for larger particles and is
a fast and effective approach. It is used in the mining industry and in many
phases of the ceramic industry, especially in the sizing of abrasives. For
free-flowing particles, dry screening can normally be effective down to
about 325 mesh. Below this the particles are so fine that they either tend
to agglomerate or clog the screen. Some automatic screen systems use
airflow or vibration to aid in screening powders that have a significant portion of particles in the range 325 mesh or smaller.

Table 5.3 ASTM Standard Screen Sizes

"Mesh" sieve designation	Sieve opening	
	mm	in.
4	4.76	0.187
6	3.36	0.132
10	2.00	0.0787
12	1.68	0.0661
16	1.19	0.0469
20	0.84	0.0331
40	0.42	0.0165
80	0.177	0.0070
120	0.125	0.0049
170	0.088	0.0035
200	0.074	0.0029
230	0.063	0.0025
270	0.053	0.0021
325	0.044	0.0017
400	0.037	0.0015

Source: ASTM E11, Annual Book of ASTM Standards, American Society for Testing and Materials, Philadelphia, 1970.

Suspending the particles in a dilute water or other liquid suspension (slurry) also aids in screening fine particles. Slurries can normally be screened easily through at least 500 mesh as long as the solids content in the slurry is low and fluidity is high. For very fine powder, this is a useful method of assuring that no particles larger than an acceptable limit (determined by the screen size selected) are left in the powder. Since isolated large particles in a powder of fine particle size distribution often become the strength-limiting flaw in the final component, wet screening can be used as an in-process quality control step.

Screening does have limitations. As with any process, it is accurate only as long as the equipment is properly maintained. Distorted or broken

screens pass larger particles than specified. Screens of 325 mesh and
finer are frequently damaged because of the fragile nature of the thin filaments required to construct the screen. The user tends to get impatient
due to the slow feed rate of some powders and tries to force the powder
through by brushing or scraping the screen. Once the screen is damaged,
particle size control and knowledge of the particle size distribution are lost.
    Another limitation of screening is related to the nature of the powder.
If the powder tends to compact or agglomerate, groups of particles will act
as a single particle and result in inaccurate screening. Similarly, packing
or agglomeration can clog the screen and prevent further screening or decrease efficiency.

Air Classification

Air classification (also referred to as air separation) is used to separate
coarse and fine fractions of dry ceramic powders. A schematic of an air
classifier is shown in Fig. 5.1. Separation is achieved by control of hori-
zontal centrifugal force and vertical air currents within the classifier. Particles enter the equipment along the centerline and are centrifugally accelerated outward. As the coarse particles move radially away from the center
into the separating zone, they lose velocity and settle into a collection cone.
The finer particles are carried upward and radially by the air currents
through selector blades. These selector blades impart an additional centrifugal force to the particles and cause additional coarse particles to settle
into the coarse collection cone. The fines are then carried by the airflow
to a separate cone for collection.
    Air classifiers have been used to separate particles in the approximate
range 40 to 400 mesh at rates exceeding 400 tons/hr. They are used extensively in the cement industry for sizing fine particles to close size and surface area specification (range -170 mesh). They are also used to remove
undesirable fines in other ceramic industries where a coarse aggregate is
required.
    Special air classifiers are available for isolating fine powders in the
range below 20 μm. Separation can be done with reasonable accuracy down
to about 10 μm. However, below 10 μm, it is difficult to obtain a cut that
is completely free of larger particles. For instance, if a powder were
desired with no particles larger than 5 μm, it could not be guaranteed by
air classification.
    Air classification is frequently linked directly to milling, crushing,
grinding, or other comminution equipment in a closed circuit. Particles
from the mill are discharged directly into the air classifier. The fines are
separated and the coarse is returned to the mill for further grinding. One
type of unit combines size reduction and classification into a single piece of
equipment. The coarse powder particles are carried by high-velocity air
through two opposing nozzles. Where the two air streams meet, particles

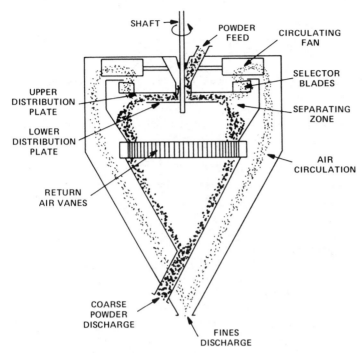

Figure 5.1 Drawing of an air classifier, showing the paths of the coarse and fine particles. (Courtesy of the Sturtevant Mill Company, Boston, Mass.)

strike each other and are shattered into smaller particles. The air carrying the particles flows vertically. Large particles pass through a centrifugal-type air classifier at the top of the unit, where additional controlled sizing is accomplished.

Air classification has its advantages and limitations. It is an efficient and high-volume approach for separating coarse particles from fine particles and producing controlled size ranges roughly from 40 to 400 mesh. However, it is limited in its efficiency and accuracy in producing controlled sizing of particles below 10 μm.

Another concern of the engineer is the presence of contamination. Sliding motion and impact of the ceramic particles against equipment surfaces results in some metallic contamination. The amount of contamination is less than for comminution equipment, but may still cause a problem for some applications. Therefore, the engineer should be aware of the air classification process and its potential for affecting material application specification.

During development, an engineer sometimes needs to have an experimental batch of powder air classified. This can be done on a contract basis with many of the equipment manufacturers at their R&D facilities. Again, though, it should be realized by the engineer that contamination can be picked up, especially since R&D equipment is used on a batch basis with a different material in each batch. Even with careful cleaning (which is not automatically done between batches), foreign particles will remain in the equipment. The best procedure is to insist on careful cleaning, run a sacrificial batch of your powder prior to your study batch, and be aware that contamination may occur that would not occur in production.

Elutriation

Elutriation is a general term that refers to particle size separation based on settling rate; i.e., large or high-specific-gravity particles settle more rapidly from a suspension than do small or low-specific-gravity particles. Air classification is a form of elutriation where the suspending medium is air. For this book the term elutriation is used to describe particle sizing by settling from a liquid suspension.

Elutriation is frequently used in the laboratory for obtaining very fine particle distributions free of large particles. The powder is mixed with water or other liquid and usually with a wetting agent and possibly a deflocculant* to yield a dilute suspension. Stirring or mixing is stopped and settling is allowed to occur for a predetermined time. The time is based on the particle size cut desired. The fluid containing the fine particles is then decanted or siphoned and the remaining fluid and residue discarded or used for some other purpose.

A major problem with elutriation is that the fines must be extracted from the fluid before they can be used. This can be done by evaporating the fluid or by filtration. Both tend to leave the fines compacted or crusted rather than as a free-flowing powder, thus requiring additional process steps before the powder can be used. Also, unless the elutriation and liquid removal are conducted in a closed system, chances for contamination are high.

Ball Milling

The desired particle size distribution usually cannot be achieved simply by screening, classifying, or elutriating the raw material. More typically, a particle size reduction (comminution) step is required. Ball milling is

---

*The wetting agent and deflocculant help to break up agglomerates.

is one of the most widely used. Ball milling consists of placing the particles to be ground (the "charge") in a closed cylindrical container with
grinding media (balls, short cylinders or rods) and rotating the cylinder
horizontally on its axis so that the media cascade. The ceramic particles
move between the much larger media and between the media and the wall of
the mill and are effectively broken into successively smaller particles.

The rate of milling is determined by the relative size, specific gravity,
and hardness of the media and the particles. High-specific-gravity media
can accomplish a specified size reduction much more quickly than can low-
specific-gravity media. The following media are commonly used and are
listed in descending order of specific gravity: WC, steel, ZrO_2, Al_2O_3,
and SiO_2.

Contamination is a problem in milling. While the particle size is being
decreased, the mill walls and media are also wearing. Milling Al_2O_3
powder with porcelain or SiO_2 media can result in about 0.1% contamination per hour. Some Si_3N_4 powder milled in a porcelain-lined mill with
porcelain cylinders picked up nearly 6% contamination in 72 hr of milling
time. The contamination in the Si_3N_4 resulted in a decrease in the high-
temperature strength by a factor of 3 and nearly an order-of-magnitude
decrease in creep resistance in the final part.

Contamination can be controlled by careful selection of the mill lining
and the media. Polyurethane and various types of rubber are excellent
wear-resistant linings and have been used successfully with dry milling
and with water as a milling fluid. However, some milling is conducted
with organic fluids that may attack rubber or polyurethane. Very hard
grinding media can reduce contamination because they wear more slowly.
WC is good for some cases because its high hardness reduces wear and its
high specific gravity minimizes milling time. If contamination from the
media is an especially critical consideration, milling can be conducted with
media made of the same composition as the powder being milled. Another
approach is to mill with steel media and remove the contamination by acid
leaching.

Milling can be conducted either dry or wet. Dry milling has the advantage that the resulting powder does not have to be separated from a liquid.
The major concern in dry milling is that the powder does not pack in the
corners of the mill and avoid milling. The powder must be kept free flowing. One method of accomplishing this is to use a dry lubricant such as a
stearate. In some cases, humidity or moisture in the powder causes packing. This has been resolved through the use of a heated mill.

Wet milling is usually very efficient if the correct ratio of fluid to powder to milling media is used. The ratio varies for different materials and
usually has to be optimized experimentally. A slurry of the consistency
of syrup or slightly thicker mills effectively.

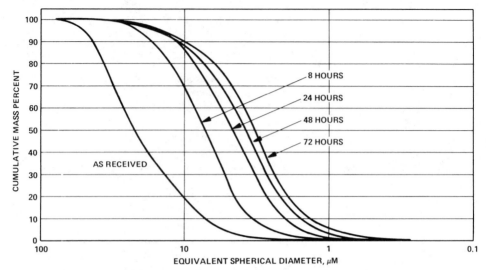

Figure 5.2 Particle size distribution of silicon powder as a function of milling time. (From Ref. 24.)

Milling produces a broad particle size distribution rather than a narrow particle size range as achieved by screening. Milling can readily reduce the average particle size to 5 μm or less. Figure 5.2 shows particle size distribution curves for silicon powder ball-milled for use in the fabrication of reaction-bonded Si_3N_4 parts. Initial particle size reduction was rapid but decreased as the powder became finer.

Besides producing the required particle size distribution, ball milling can also produce a very active powder that is easier to density in later process steps. In some cases, this is achieved by an active surface condition. In other cases it appears to be achieved by increased strain energy in the particle.

At this point, the non-ceramic engineer may wonder why he or she needs to know the details about particle sizing methods, since he or she will probably only be purchasing the final part and will not be involved in the material processing. This is probably true, but it is still the engineer's responsibility to make sure the part works in the particular application. Current engineering requirements of materials are often far more demanding than they used to be and traditional controls and techniques used by ceramic manufacturers may require upgrading. The engineer knows the application requirements better than the supplier does and will thus need extensive liaison with the supplier to achieve the material objectives. It is not unusual for an engineer to request or suggest a process change or modification to the supplier.

Attrition Milling

Figure 5.3 shows a schematic of an attrition mill. It is similar to a ball mill since it is cylindrical and contains balls or grinding media, but rather than the cylinder rotating, the balls are agitated by a series of stirring arms mounted to an axial shaft. Herbell et al. [10] and Claussen and Jahn [11] report that attrition milling is quicker than ball milling, is more efficient at achieving fine particle size, and results in less contamination. Furthermore, attrition milling can easily be conducted dry, wet, or with a vacuum or inert gas atmosphere.

Data from Herbell et al. [10] for dry attrition milling of silicon powder are summarized in Table 5.4. Although the attrition mill was lined with an iron-base alloy, no significant iron or carbon was picked up, even after 18 hr of milling. The oxygen content did increase by interaction with air, as would normally be expected for fine particles of silicon, as the surface area increased.

Even though the average particle size was substantially reduced by dry attrition milling, some particles in the range 40 μm (0.0018 in.) were still present and ultimately controlled the strength of the final component. Evidently, some particles were trapped or packed in regions of the mill where they did not receive adequate milling. This did not occur for wet milling.

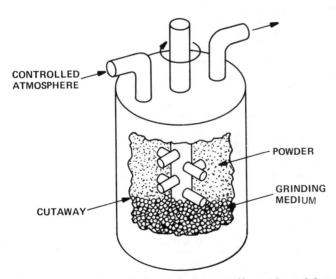

Figure 5.3 Schematic of an attrition mill. (Adapted from T. P. Herbell and T. K. Glasgow, NASA, presented at the DOE Highway Vehicle Systems Contractors Coordination Meeting, Dearborn, Mich., Oct. 17-20, 1978.)

Table 5.4 Changes during Attrition Milling of Silicon Powder

Milling time (hr)	Surface area (m^2/g)	Iron content (wt %)	Carbon content (wt %)	Oxygen content (wt %)
0	3.0	0.62	0.03	0.60
1	11.5	0.59	0.04	2.29
4	14.5	0.58	0.04	2.57
18	23.3	0.55	0.04	3.67

Source: Ref. 10.

Stanley et al. [12] have described a different configuration of attrition mill that has been used primarily for wet milling. It consists of a rotating cylindrical cage of vertical bars surrounded by a stationary cylindrical cage of vertical bars. The material to be ground is mixed with water or other fluid plus sand-size grinding media. Materials such as ZrO_2, Al_2O_3, and SiO_2 have been ground to submicron size in a few hours, compared to 30 hr for vibratory milling and much longer times for ball milling. The primary problem with this attrition milling approach is the amount of contamination and the difficulty of separating the powder from the media. For example, in one case 20 to 30% of the media was ground to -325 mesh (< 44 μm) and was not successfully separated from the 0.1-μm milled powder.

Vibratory Milling

Vibratory milling is substantially different from ball milling or attrition milling. The energy for comminution is supplied through vibration rather than tumbling or mechanical stirring. The powder is placed in the stationary chamber of the mill together with suitable grinding media and a liquid. When the mill is turned on, vibration is transmitted (usually from the bottom center of the mill) through the chamber and into the media and powder. This results in two types of movement. First, it causes a cascading or mixing action of the contents of the milling chamber. Second, it causes local impact and shear fracturing of the powder particles between adjacent grinding media.

Vibratory milling is relatively fast and efficient and yields a finer powder than is usually achieved by ball milling. The vibratory mill chamber is typically lined with polyurethane or rubber and minimizes contamination. Vibratory mills are not used exclusively for powder processing. They are used extensively for deburring and cleaning of metal parts.

Fluid Energy Milling

Fluid energy mills achieve particle size reduction by particle-particle impact in a high-velocity fluid [13]. The fluid can be compressed air, nitrogen, carbon dioxide, superheated steam, water, or any other gas or liquid
compatible with the specific equipment design. The powder is added to the
compressed fluid and accelerated to sonic or near-sonic velocity through
jets leading into the grinding chamber. The grinding chamber is designed
to maximize particle-particle impact and minimize particle-wall impact.
    Fluid energy milling can achieve controlled particle sizing with mini-
mum contamination. Most jet mills have no moving parts and can be easily
lined with polyurethane, rubber, wear-resistant steel, and even ceramics.
Examples of particle sizing capabilities are shown in Table 5.5 [14]. Output capacity can range from a few grams per hour to thousands of kilograms
per hour, depending on the size of the equipment.
    The primary problem with fluid energy milling is collecting the powder.
Large volumes of gases must be handled. Cyclones are not efficient for
micrometer-sized particles and filters clog rapidly.

Hammer

Table 5.5 Typical Grinding Data for Fluid Energy Milling

Material	Mill diameter		Grinding medium	Material feed rate		Average particle size obtained	
	cm	in.		kg/hr	lb/hr	μm	in.
Al_2O_3	20.3	8	Air	6.8	15	3	0.00012
TiO_2	76.2	30	Steam	1020	2250	<1	<0.00004
TiO_2	106.7	42	Steam	1820	4000	<1	<0.00004
MgO	20.3	8	Air	6.8	15	5	0.0002
Coal	50.8	20	Air	450	1000	5-6	~0.00025
Cryolite	76.2	30	Steam	450	1000	3	0.00012
DDT 50%	61.0	24	Air	820	1800	2-3	~0.0001
Dolomite	91.4	36	Steam	1090	2400	<44	<0.0018
Sulfur	61.0	24	Air	590	1300	3-4	~0.00014
Fe_2O_3	76.2	30	Steam	450	1000	2-3	~0.0001

Source: Ref. 14, courtesy of Sturtevant Mill Company, Boston, Mass.

Hammer Milling

Hammer milling is also an impact approach, but the impact is applied by rapidly rotating blades, rather than particle-particle attrition. High levels of energy can be achieved, but particle sizing is difficult to control and contamination from the equipment is generally high. Hammer milling is usually used as an intermediate grinding step following rough crushing with a cone crusher or rolls and preceding ball milling, vibratory milling, or other fine-particle sizing techniques.

Precipitation

Precipitation of soluble salts followed by thermal decomposition to the oxide is a widely used method of both particle sizing and purifying of oxide ceramics [15]. Analogous techniques in controlled atmospheres have also been used to produce nonoxide ceramic powders.

The Bayer process for producing Al_2O_3 from bauxite relies on controlled precipitation. The aluminum hydroxide in bauxite is dissolved with caustic soda and separated from the nonsoluble impurities in the bauxite by filtering. Aluminum trihydrate is then precipitated by changing the pH, controlling the resulting particle size through addition of seed crystals. Table 5.6 lists the characteristics of commercially available Al_2O_3 powders and illustrates the variations in particle size, surface area, and purity that can be achieved in the precipitation process [16]. The very fine reactive alumina powders first became available in 1966 and have led to dramatic improvements in the properties of Al_2O_3 components and an expansion in the number and range of applications.

Freeze Drying

Freeze drying (also known as cryochemical processing) is a relatively new process which was first reported in 1968 by Schnettler et al. [17]. It has potential for producing uniform particle and crystallite sizing of very pure, homogeneous powder.

There are four steps in freeze drying:

1. A mixture of soluble salts containing the desired ratio of metal ions is dissolved in distilled water.
2. The solution is formed into droplets usually 0.1 to 0.5 mm in diameter and rapidly frozen so that no compositional segregation can occur and so that the ice crystals that nucleate are very small.
3. The water is removed in vacuum by sublimation, with care to avoid any liquid phase, thus preventing any chance for segregation.

4. The resulting powder is then calcined (heat-treated) at a temperature
    that decomposes the crystallized salts and converts them to very fine
    crystallites of the desired oxide or compound.

Most of the reported freeze drying has been conducted with sulfates.
Each frozen droplet contains sheaths of sulfate crystals radiating from the
center to the surface. During calcining, the oxide crystallites form in an
oriented chainlike fashion along these radial sheaths. The size of the crystallites can be controlled by the time and temperature of calcining. Johnson and Schnettler [18] reported that a 10-hr 1200°C heat treatment of freeze-
dried aluminum sulfate resulted in Al_2O_3 crystallites averaging 1500 Å and
ranging from 600 to 2600 Å.

Not all compositions can be achieved by freeze drying. The two primary
limitations are (1) the nonavailability of soluble salts containing the required
metal ions and (2) the reaction of some salt combinations to form a precipitate. For instance, it has been difficult to find soluble salts that will yield
lead zirconate titanate. Similarly, a mixture of barium acetate plus ferrous
sulfate results in precipitation of insoluble $BaSO_4$.

Rigterink [19] described a laboratory apparatus and potential production
approaches for freeze drying. The laboratory apparatus in quite simple
and inexpensive and is adequate for making small batches for material
processing development. The apparatus consists of a beaker containing hexane which is being mechanically stirred or swirled rapidly. The hexane
container is surrounded by a bath of dry ice and acetone. The salt solution
is forced through a glass nozzle as uniform drops of the desired size into
the swirling hexane. The drops freeze rapidly and settle to the bottom,
where they can be removed later by sieve.

Laser

The use of laser energy for synthesizing Si_3N_4 powder was discussed earlier in this chapter. It has also been used successfully to produce controlled
particle sizes of silicon and SiC and appears to have potential for synthesizing
oxides and other ceramic powders.

Plasma

Plasma spray techniques for applying coatings by depositing plasma-melted
particles on a surface are discussed in Chap. 7. Particle sizing can be
achieved by spraying the particles into an open space or a quench medium.
Any oxide or ceramic composition that melts without decomposing can be
handled by this approach. However, plasma spraying is expensive and the
resulting powder is not optimum for most ceramic forming processes.

Table 5.6 Range in Powder Characteristics Available for Al_2O_3 Produced by Precipitation Using the Bayer Process

	Type[a]	Percent Na_2O	Average particle size (μm)	Surface area (m^2/g)	Compaction ratio	Final density
Nonreactive, high purity	A-10	0.06	>5	0.2	--	--
	C-75	0.01	2.8	0.49	2.25	2.97[b]
	A-14	0.06	--	0.6	--	--
	RC-122	0.03	2.6	0.35	2.35[c]	3.46[b]
Nonreactive, intermediate purity	A-12	0.24	--	0.5	--	--
	RC-24	0.23	2.9	0.53	2.35[c]	3.50[b]
	C-73	0.2	--	0.34	--	--
Nonreactive, low purity	A-2	0.46	3.25	--	2.21	3.22[b]
	RC-20	0.40	2.7	0.87	2.31[c]	3.37[b]
	C-70	0.50	2.5	0.66	2.28	3.18[b]
	RC-25	0.31	2.6	0.95	2.33[c]	3.52[b]

	A-5	0.35	4.7	--	2.30	3.07[b]
	C-71	0.75	2.2	0.71	2.24	3.33[b]
Reactive, high purity	XA-139	0.008	0.42	6.5	2.00	3.90[d]
	ERC-HP	0.008	0.55	7.4	2.15[e]	3.95[d]
	A-16	0.06	0.6	6.5	--	--
	RC-172	0.04	0.6	4.0	2.21[e]	3.94[d]
Reactive, low purity	A-3	0.36	0.63	9.0	1.79	3.64[b]
	RC-23	0.30	0.6	7.5	1.91[c]	3.86[b]
Reactive, nonreactive mixture	RC-152	0.05	1.5	2.6	2.30[c]	3.46[b]
	A-15	0.07	--	--	--	--

[a] C, Alcan; A, Alcoa; RC, Reynolds.
[b] For 1620°C/1 hr sintering.
[c] 20-g pellet pressed at 4000 psi, 4-hr grind.
[d] For 1510°C/2 hr sintering.
[e] 10-g pellet pressed at 5000 psi, 4-hr grind.
Source: Adapted from Ref. 16.

Other powder synthesis or sizing approaches using plasma equipment technology do exist, but very little information has been published and the specific techniques evidently are proprietary.

Calcining

Calcining is widely used for preparation of precursor ceramic powders for later processing steps and for achieving the optimum particle sizing. It consists essentially of a high-temperature heat treatment selected for each powder and optimized by prior research to achieve a specific objective. In one case, calcining is used to decompose a salt or a hydrate to the oxide, such as that described earlier for achieving an oxide from a freeze-dried sulfate solution. In another case, calcining is used to achieve the desired crystal phase and particle size, as described for preparation of Al_2O_3 from the Bayer process.

Calcining is also used for dehydration. The preparation of plaster from gypsum is a good example. Heat treatment at the right temperature partially dehydrates the gypsum ($CaSO_4 \cdot 2H_2O$) to $CaSO_4 \cdot 1/2H_2O$. This hemi-hydrate is plaster of paris and will rehydrate to form gypsum when water is added. Calcining at too high a temperature removes all the water from the gypsum and produces an anhydrous powder that will not easily rehydrate.

One of the most common uses of calcining is for coarsening, i.e., increasing the particle size of a very fine powder which has poor packing properties to a less-fine powder with improved packing properties. Usually, the coarsening consists of loose bonding of adjacent particles to form aggregates which act like larger particles during later processing steps, such as compaction.

5.3 PRECONSOLIDATION

The sized powders described in Sec. 5.2 are compacted into the desired shapes by techniques such as pressing, slip casting, and injection molding (discussed in Chap. 6) and then strongly bonded or densified (discussed in Chap. 7). To achieve a final component having uniform properties and no distortion requires a uniform particle compact. To achieve the required uniformity, the powder usually requires special treatments or processing prior to compaction. Table 5.7 summarizes some of these special preconsolidation considerations for several compaction or consolidation approaches.

The preconsolidation steps are essential to minimize severe fabrication flaws that can occur in later processing steps. For instance, a powder that is not free flowing can result in poor powder distribution in the pressing die and distortion or density variation in the final part. Similarly, improper viscosity control of a casting slurry can result in incomplete fill of the mold or a variety of other defects during slip casting. Inadequate deairing of

Table 5.7 Preconsolidation Steps for Several Consolidation Approaches

Pressing	Slip casting	Injection molding
Binder addition	Slurry preparation	Thermoplastic addition
Lubricant addition	Binder addition	Plasticizer addition
Sintering aid addition	Deflocculant addition	Wetting agent addition
Preparation of a free-flowing powder by spray drying or granulation	pH control	Lubricant addition
	Viscosity control	Sintering aid addition
	Percent solids control	Deairing
	Deairing	Granulation or pelletizing

either a slurry or an injection molding mix can result in a strength-limiting void in the final slip-cast or injection-molded part. Such fabrication flaws can reduce the strength of a material to a fraction of its normal value.

The following sections describe briefly some of the reasons and techniques of preconsolidation steps. Emphasis will be on preparation of a free-flowing powder suitable for pressing. Preconsolidation considerations for slip casting and injection molding are discussed in Chap. 6 as part of those specific processing steps.

Additives

Additives are required for different reasons, depending on the specific forming process. However, several general comments are relevant to most forming approaches:

1. Binders are added to provide enough strength in the "green" body (unfired compact) to permit handling, "green" machining, or other operations prior to densification.
2. Lubricants are added to decrease particle-particle and particle-tool friction during compaction.
3. Sintering aids are added to activate densification.
4. Deflocculants, plasticizers, wetting agents, and thermoplastics are added to yield the rheological (flow) properties necessary for the specific shape-forming process (discussed in detail in Chap. 6).

Table 5.8 Function of Additives to Ceramics

Additive	Function
Binder	Green strength
Lubricant	Mold release, interparticle sliding
Plasticizer	Rheological aids, improving flexibility of binder films, allowing plastic deformation of granules
Deflocculant	pH control, particle-surface charge control, dispersion, or coagulation
Wetting agent	Reduction of surface tension
Water retention agent	Retain water during pressure application
Antistatic agent	Charge control
Antifoam agent	Prevent foam
Foam stabilizer	Strengthen desired foam
Chelating or sequestering agent	Deactivate undesirable ions
Fungicide and bactericide	Stabilize against degradation with aging
Sintering aid	Aid in densification

Source: Ref. 20.

Table 5.8 further summarizes the function of additives [20].

A wide variety of binders are available, as shown by the partial listing in Table 5.9. Selection depends on a number of variables, including green strength needed, ease of machining, compatibility with the ceramic powder, and nature of the consolidation process. Gums, waxes, thermoplastic resins, and thermosetting resins are not soluble in water[*] and do not provide a benefit for slip casting, but are excellent for the warm mixing used to prepare a powder for injection molding. Organic binders can be burned off at low temperature and result in minimal contamination, whereas inorganic binders become a part of the composition.

[*]Emulsions can be used.

Table 5.9 Examples of Binders Used in Ceramic
           Processing

Organic	Inorganic
Polyvinyl alcohol (PVA)	Clays
Waxes	Bentonites
Celluloses	Mg-Al silicates
Dextrines	Soluble silicates
Thermoplastic resins	Organic silicates
Thermosetting resins	Colloidal silica
Chlorinated hydrocarbons	Colloidal alumina
Alginates	Aluminates
Lignins	Phosphates
Rubbers	Borophosphates
Gums	
Starches	
Flours	
Casein	
Gelatins	
Albumins	
Proteins	
Bitumens	
Acrylics	

Spray Drying

Spray drying is commonly used in ceramic processing to achieve a uniform,
free-flowing powder [21-23]. As shown in Fig. 5.4, a spray drier consists
of a conical chamber that can either be heated or has an inlet for hot air.
The powder to be spray-dried is suspended in a slurry with the appropriate
additives. Slurry preparation is most frequently done in a ball mill. The
slurry is fed into the top of the spray drier through an atomizer and is
swirled around by the hot air circulating in the conical spray drier chamber.
The water evaporates and the powder forms into round, soft agglomerates.

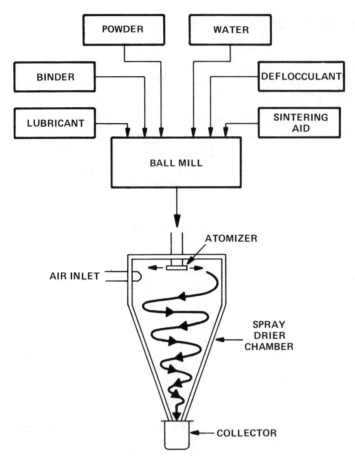

Figure 5.4 Schematic of the spray-drying process for achieving free-flowing spherical powder agglomerates containing a uniform level of additives.

These are less than 1 mm in diameter and are easily crushed during later pressing operations. Because all the particles have a spherical shape, the powder flows almost like a liquid and provides very uniform die fill for automated pressing.

Granulation

Granulation is another approach to achieving better flow properties of a powder. In this case, a slurry is not prepared. Instead, only a damp or plastic mix is prepared, usually with equipment such as a mix muller, a sigma mixer, or any of a variety of other commercially available mixer designs. For laboratory-scale batches, this mixing can be accomplished with a mortar and pestle or even by hand. The damp or plastic material is then forced through orifices of the desired size or screened. The resulting particle agglomerates are usually harder and more dense than spray-dried agglomerates and irregular in shape. They do not flow as readily, but do tend to pack to a lower volume.

The major advantage of granulation is that the powder is prepacked and takes up less volume during pressing, extrusion, or injection molding. In some cases, powders are precompacted at 703 to 1055 kg/cm^2 (10,000 to 15,000 psi) prior to granulation to assure relatively tight packing of the powder in each agglomerate. However, if this pressure is not equaled or exceeded during the shape-forming process, the agglomerates will not be crushed or fused together and the finished part will retain the identity of the agglomerates. This will consist of inhomogeneity in the part and for many applications will require that the part be rejected.

5.4 POWDER PROCESSING SUMMARY

Proper powder selection, sizing, and preconsolidation processing are vital to achieving the final desired shape and properties. Contamination and non-uniformity during powder processing will be carried throughout the rest of the processing and will usually lead to a deficient or rejectable component. This can only be avoided if quality control (QC) procedures are incorporated into each processing step. QC is not only the responsibility of the ceramic manufacturer, but also of the engineer who must use the finished parts or build them into a system. The interactions between QC and processing and the ways an engineer can be involved are discussed in Chap. 9. It might be worthwhile to preread the early sections of Chap. 9 before continuing on to Chap. 6.

REFERENCES

1. F. H. Norton, Elements of Ceramics, 2nd ed., Addison-Wesley Publishing Co., Inc., Reading, Mass., 1974.
2. F. V. Tooley, ed., Handbook of Glass Manufacture, Vols. 1 and 2, Ogden Publishing Company, New York, 1961.
3. F. H. Norton, Refractories, 4th ed., McGraw-Hill Book Company, New York, 1968.
4. J. E. Burke, ed., Progress in Ceramic Science, Vols. 1-4, Pergamon Press, Inc., Elmsford, N.Y., 1962-1966.

5. A. M. Alper, ed., High Temperature Oxides, Parts I-IV, Academic Press, Inc., New York, 1970-1971.

6. E. Ryshkewitch, Oxide Ceramics, Academic Press, Inc., New York, 1960.

7. J. S. Haggerty and W. R. Cannon, Sinterable Powders from Laser-Driven Reactions, Massachusetts Inst. of Tech., Cambridge, Mass., ONR Contract Report, Oct. 1978 (AD-A063 064).

8. D. W. Richerson and M. E. Washburn, "Hot pressed silicon nitride," U.S. Patent No. 3,836,374, Sept. 17, 1974.

9. G. E. Gazza, Effect of yttria additions on hot-pressed Si_3N_4, Bull. Am. Ceram. Soc. 54, 778-781 (1975).

10. T. P. Herbell, T. K. Glasgow, and H. C. Yeh, Effect of Attrition Milling on the Reaction Sintering of Silicon Nitride, National Aeronautics and Space Administration, Lewis Research Center, Report No. NASA-TM-78965, 1978 (N78-31236).

11. N. Claussen and J. Jahn, Mechanical properties of sintered and hot-pressed Si_3N_4-ZrO_2 composites, J. Am. Ceram. Soc. 61, 94-95 (1978).

12. D. A. Stanley, L. Y. Sadler III, and D. R. Brooks, First Proc. Int. Conf. Particle Technol., 1973.

13. C. Greskovich, in Treatise on Materials Science and Technology, Vol. 9: Ceramic Fabrication Processes (F. F. Y. Wang, ed.), Academic Press, Inc., New York, 1976, pp. 28-33.

14. Sturtevant Mill Co., Boston, Mass., Bulletin No. 091, Sturtevant Micronizer Fluid Energy Mills.

15. D. W. Johnson and P. K. Gallagher, Reactive powders from solution, in Ceramic Processing Before Firing (G. Y. Onoda, Jr., and L. L. Hench, eds.), John Wiley & Sons, Inc., New York, 1978, pp. 125-139.

16. W. M. Flock, Bayer-processed aluminas, in Ceramic Processing Before Firing (G. Y. Onoda, Jr., and L. L. Hench, eds.), John Wiley & Sons, Inc., New York, 1978, pp. 85-100.

17. F. J. Schnettler, F. R. Monforte, and W. H. Rhodes, A cryochemical method for preparing ceramic materials, in Science of Ceramics (G. H. Stewart, ed.), The British Ceramic Society, Stoke-on-Trent, U.K., 1968, pp. 79-90.

18. D. W. Johnson and F. J. Schnettler, Characterization of freeze-dried Al_2O_3 and Fe_2O_3, J. Am. Ceram. Soc. 53, 440-444 (1970).

19. M. D. Rigterink, Advances in technology of the cryochemical process, Am. Ceram. Soc. Bull. 51, 158-161 (1972).

20. A. G. Pincus and L. E. Shipley, The role of organic binders in ceramic processing, Ceram. Ind. 92, 106-109 (1969).

21. E. H. Rasmussen, Instrumentation for spray driers, Am. Ceram. Soc. Bull. 39, 732-734 (1930).

22. E. J. Motyl, West Electr. Eng. 7, 2-10 (1963).

23. H. J. Helsing, Powder Metall. Int. $\underline{2}$, 62–65 (1969).
24. D. W. Richerson and T. M. Yonushonis, Environmental effects on the strength of silicon nitride materials, DARPA/NAVSEA Ceramic Gas Turbine Demonstration Engine Program Review, MCIC Report MCIC-78-36, 1978, pp. 247–271.

6

Shape-Forming Processes

The properly sized and preconsolidated powders are now ready for forming into the required shapes. Table 6.1 summarizes the major techniques for consolidation of powders and producing shapes. In this chapter we examine the major approaches in terms of the process steps and controls involved, the types of strength-limiting flaws that may result, and the range of shapes that can be produced.

6.1 PRESSING

Pressing is accomplished by placing the powder (premixed with suitable binders and lubricants and preconsolidated so that it is free flowing) into a die and applying pressure to achieve compaction. Typical flow sheets for granulated and spray-dried powders are shown in Fig. 6.1. The following section describes some of the critical factors for individual steps in the processing and some of the common problems encountered.

Binder and Lubricant Selection

Organic materials that provide a temporary bond between the ceramic particles are often required for pressing. These binders provide lubrication during pressing and give the pressed part enough strength and toughness that it can be handled and even machined prior to densification. The amount of binder required is quite low, typically ranging from 0.5 to 5%.

Binder selection is dependent on the type of pressing that will be conducted. Some binders such as waxes and gums are very soft and quite sensitive to temperature variations. These generally do not require moisture or lubricant additions prior to pressing, but must be handled more carefully to avoid changes in granule size that might alter flow characteristics into the pressing die or result in inhomogeneous density distribution. The soft binders also have a tendency to produce flash between the die components, which can cause sticking or reduce the production rate.

178

Table 6.1   Major Techniques for Powder Consolidation and
            Shape Forming

Pressing	Plastic forming
Uniaxial pressing	Extrusion
Isostatic pressing	Injection molding
Hot pressing[a]	Transfer molding
Hot isostatic pressing[a]	Compression molding
Casting	Others
Slip casting	Tape forming
Thixotropic casting	Flame spray[a]
Soluble-mold casting	Green machining

[a] These techniques combine consolidation and densification in
one step and are discussed in Chap. 7.

Other binders can be classified as hard; i.e., they produce granules
that are hard or tough. These granules have the advantage that they are
dimensionally stable and free flowing and are therefore excellent for high-
volume production with automated presses. However, these are generally
not self-lubricating and thus require small additions of lubricant and moisture prior to pressing. They also require higher pressure to assure uniform
compacts. If the starting powder agglomerates are not completely broken
down into a continuous compact during pressing, artifacts of the approximate size of the agglomerates will persist through the remaining process
steps and may act as large flaws, which will limit the strength.

Dextrine, starches, lignins, and acrylates produce relatively hard
granules. Polyvinyl alcohol and methyl cellulose result in slightly softer
granules. Waxes, wax emulsions, and gums produce soft granules.

Binder selection must also be compatible with the chemistry of the
ceramic and the purity requirements of the application. The binder must
be removed prior to densification of the ceramic. Organic binders can be
removed by thermal decomposition. If reaction between the binder and the
ceramic occurs below the binder decomposition temperature or if the
ceramic densifies below this temperature, the final part will be contaminated or may even be cracked or bloated. If the temperature is raised too
rapidly, the binder may char rather than decompose, leaving carbon.

Lubricants are used to aid in the redistribution of particles during
pressing to obtain maximum packing, to improve powder flow into the die,
and to minimize die sticking. Paraffin oil is often blended with granulated

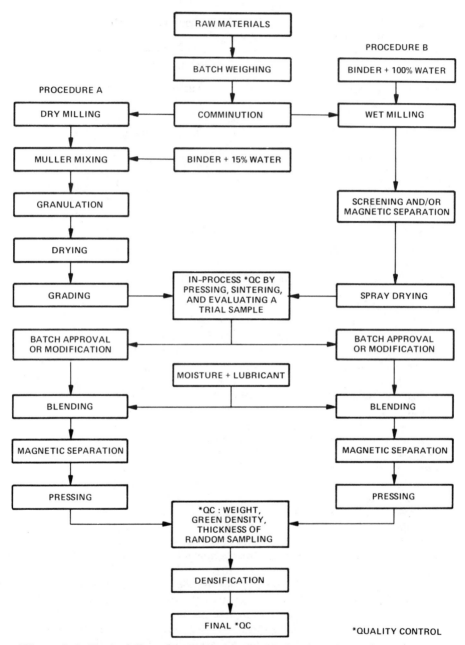

Figure 6.1 Typical flow sheets for fabrication by pressing. [Reprinted from <u>Ceramic Fabrication Processes</u> (W. D. Kingery, ed.), by permission of The MIT Press, Cambridge, Mass. © 1963, The MIT Press.]

or spray-dried powder prior to pressing. This is usually not required if waxes or wax emulsions are used as binders. Dry lubricants such as the stearates are also often used.

Uniaxial Pressing

Uniaxial pressing involves the compaction of powder into a rigid die by applying pressure along a single axial direction through a rigid punch, plunger, or piston. Thurnauer [1] and Whittemore [2] provide introductory descriptions of the approaches and some of the critical considerations of uniaxial pressing, especially automated pressing. The following discussion is derived largely from these two references.

Dry Pressing   Most automated pressing is conducted with granulated or spray-dried powder containing 0 to 4% moisture. This is referred to as dry, semidry, or dust pressing. Compaction occurs by crushing of the granules and mechanical redistribution of the particles into a close-packed array. The lubricant and binder usually aid in this redistribution and the binder provides cohesion. High pressures are normally used for dry pressing to assure breakdown of the granules and uniform compaction.

High production rates and close tolerances can be achieved with dry pressing. Millions of capacitor dielectrics approximately 0.050 cm (0.020 in.) thick are made to close tolerances and with tightly specified electrical properties. Millions of electrical substrates, packages (thin-walled insulator "boxes" for isolating miniature electronic circuits), and other parts for a wide variety of applications are made by dry pressing. Dimensional tolerances to ±1% are normally achieved in routine applications, and closer tolerances have been achieved in special cases.

Wet Pressing   Wet pressing involves a feed powder containing 10 to 15% moisture and is often used with clay-containing compositions. This feed powder deforms plastically during pressing and conforms to the contour of the die cavity. The pressed shape usually contains flash (thin sheets of material at edges where the material extruded between the die parts) and can deform after pressing if not handled carefully. For these reasons, wet pressing is not well suited to automation. Also, dimensional tolerances are usually only held to ±2%.

Types of Presses   Most presses are either mechanical or hydraulic. Mechanical presses typically have a higher production rate. Figure 6.2 shows schematically the pressing sequence of a typical single-stroke mechanical press. The punches preposition in the die body to form a cavity predetermined (based on the compaction ratio of the powder) to contain the correct volume to achieve the required green dimensions after compactions. The feed shoe then moves into position and fills the cavity with free-flowing

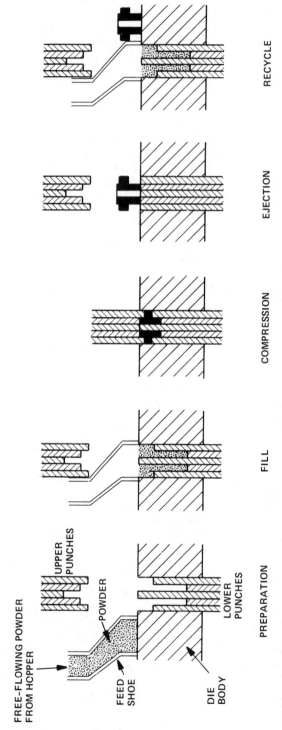

Figure 6.2 Typical single-stroke automated mechanical press cycle. (Adapted from Ref. 1.)

powder containing suitable binders, moisture, and lubricant. The feed shoe retracts, smoothing the powder surface as it passes, and the upper punches move down to precompress the powder. The upper and lower punches then simultaneously compress the powder as they independently move to preset positions. The upper punches retract and the lower punches eject the compact from the die body. The feed shoe then moves into position and pushes the compact away from the punches as the punches reset to accept the correct powder fill. This cycle repeats typically 6 to 100 times per minute, depending on the press and the shape being fabricated. Presses of this type generally have a capacity from 1 to 20 tons, but some operate up to 100 tons.

Another type of mechanical press is the rotary press. Numerous dies are placed on a rotary table. The die punches pass over cams as the table rotates, resulting in a fill, compress, and eject cycle similar to the one described for a single-stroke press. Production rates in the range of 2000 parts per minute can be achieved with a rotary press. Pressure capability is in the range 1 to 100 tons.

Another type of mechanical press is the toggle press. It is commonly used for pressing refractory brick and is capable of exerting pressure up to about 800 tons. The toggle press closes to a set volume so that the final density is controlled largely by the characteristics of the feed.

Hydraulic presses transmit pressure via a fluid against a piston. They are usually operated to a set pressure, so that the size and characteristics of the pressed component are determined by the nature of the feed, the amount of die fill, and the pressure applied. Hydraulic presses can be very large. Whittemore [2] mentions presses that can exert 5000 tons of pressure. Hydraulic presses have a much lower cycle rate than mechanical presses.

Uniaxial Pressing Problems   The following are some of the problems that can be encountered with uniaxial pressing:

Improper density or size
Die wear
Cracking
Density variation

The first two are easy to detect by simple measurements on the green compact immediately after pressing. Improper density or size are often associated with off-specification powder batches and are therefore relatively easy to resolve. Die wear shows up as progressive change in dimensions. It should also be routinely handled by the process specification and quality control (see Chap. 9). However, since not all fabrication operations have suitable quality control, off-dimension parts can be delivered, and it then becomes the responsibility of the engineer using the parts to track down the source of the problem and make sure that the problem is resolved.

The source of cracking may be more difficult to locate. It may be due to improper die design, air entrapment, rebound during ejection from the die, die wear, or other causes. Furthermore, cracks may not be readily visible by routine inspection and may not even be detected by dye penetrant or x-ray radiography inspection. Their presence may not become known until the part fails in service. Again, the responsibility for locating a solution to the problem falls in the hands of the engineer in charge of the use of the component. In this case, the engineer must either work closely with the manufacturer to identify the source of cracking or must locate another source of the component.

Perhaps the most important problem to be overcome in uniaxial pressing is nonuniform density. Density variation in the green compact causes warpage, distortion, or cracking during firing. One source of density variation is the friction between the powder and the die wall and between powder particles [3-5]. As shown in Fig. 6.3, a uniaxial pressure applied from one end of a die full of powder will be dissipated by friction so that a substantial portion of the powder will experience much lower than the applied pressure. These areas will compact to a lower density than the areas exposed to higher pressure. The pressure difference increases as the length-to-diameter ratio increases. During firing, the lower-density areas will either not densify completely or will shrink more than surrounding areas. Both will result in flaws that can cause rejection of the part.

Use of suitable binders and lubricants can reduce both die wall and particle-particle friction and thus reduce density variation in the compact. Applying pressure from both ends of the die also helps. But perhaps the most fruitful approach is to work with an experienced die designer who knows the limitations of shape compaction in terms of width-to-thickness variation, shapes and tolerances achievable, and other factors relevant to the specific configuration and application.

Isostatic Pressing

Isostatic pressing (also called hydrostatic pressing or molding) involves application of pressure equally to the powder from all sides. This substantially reduces the problems of nonuniformity due to die wall and powder friction and permits uniform compaction of larger volumes of powder, including shapes with a large length-to-diameter ratio.

A schematic of an isostatic pressing apparatus is illustrated in Fig. 6.4. It consists of a thick-walled steel pressure vessel with either a breach-lock or pressure-seal cover. The powder is enclosed in a liquid-tight rubber mold and immersed in the fluid in the pressure vessel. Glycerine, hydraulic oil, water (with a suitable rust inhibitor), or other essentially noncompressible fluid is used. The fluid is pressurized, transmitting the pressure uniformly to all surfaces of the mold. The rubber deforms as the powder compacts, but springs back after the pressure releases and allows easy removal of the pressed part.

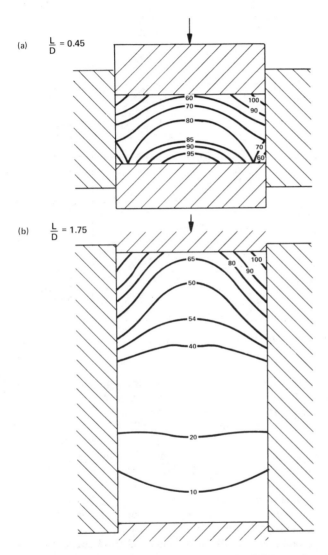

Figure 6.3 Pressure variations in uniaxial pressing due to die wall friction and particle-particle friction, which lead to nonuniform density of the
pressed compact. [Adapted from Ceramic Fabrication Processes (W. D.
Kingery, ed.), by permission of The MIT Press, Cambridge, Mass. ©
1963, The MIT Press.]

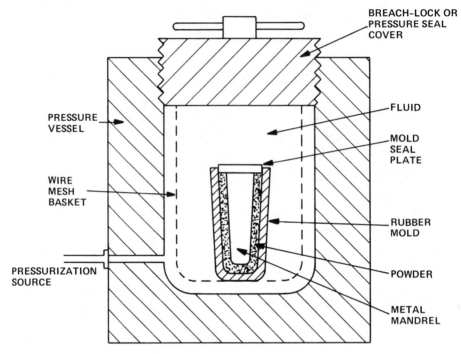

Figure 6.4 Schematic of an isostatic pressing system.

Laboratory isostatic presses have been built with pressure capabilities ranging from 35 to 1380 MPa (5000 to 200,000 psi). However, production units usually operate at 207 MPa (30,000 psi) or less.

A major concern in isostatic pressing is uniform fill of the mold. This is usually achieved by use of vibration plus free-flowing spray-dried or granulated powder. Since higher pressures are usually achieved by isostatic pressing than by uniaxial pressing and since these pressures are applied uniformly, a greater degree of compaction is achieved. This usually results in improved densification characteristics during the subsequent sintering step of processing and a more uniform, defect-free component.

The primary disadvantages of isostatic pressing are a limited production rate and difficulty in achieving close tolerances and good surface finish. Developments in the spark-plug insulator industry demonstrated that these limitations can be overcome. This was achieved by the use of a thick-walled single- or multicomposition rubber mold with controlled fluid passages for pressurization. The rubber material was flexible enough to transmit pressure but not to deform irregularly. The mold could be filled, pressed, and emptied automatically at a rate of 1000 to 1500 cycles per hour [6].

Applications of Pressing

Uniaxial pressing is widely used for compaction of small shapes, especially
of insulating, dielectric, and magnetic ceramics for electrical devices.
These include simple shapes, such as bushings, spacers, substrates, and
capacitor dielectrics, and more complex shapes, such as the bases or
sockets for tubes, switches, and transistors. Uniaxial pressing is also
used for the fabrication of tiles, bricks, grinding wheels, wear-resistant
plates, crucibles, and an endless variety of parts.

Isostatic pressing, typically in conjunction with green machining (which
is discussed briefly later in the chapter), is used for configurations that
cannot be uniformly pressed uniaxially or that require improved properties.
Large components such as radomes, cone classifiers, and cathode-ray-tube
envelopes have been fabricated by isostatic pressing. So also have bulky
components for the paper industry. Small components with a large length-
to-width ratio are also fabricated by isostatic pressing and machining.

Figure 6.5 shows a variety of ceramic parts that have been fabricated
by uniaxial and isostatic pressing.

6.2 CASTING

When most engineers hear the term "casting," they automatically think of
metal casting where a shape is formed by pouring molten metal into a mold.
A limited amount of casting of molten ceramics is done in the preparation
of high-density Al_2O_3 and Al_2O_3-ZrO_2 refractories and in preparation of
some abrasive materials. In the latter case, casting from a melt into
cooled metal plates produces rapid quenching, which results in very fine
crystal size that imparts high toughness to the material.

More frequently, the casting of ceramics is done by a room-temperature operation in which ceramic particles suspended in a liquid are cast
into a porous mold which removes the liquid and leaves a particulate compact in the mold. There are a number of variations to this process, depending on the viscosity of the ceramic-liquid suspension, the mold, and
the procedures used. The most common is referred to as slip casting.
The principle and controls for slip casting are similar to those of the other
particulate ceramic casting techniques. Slip casting is described in detail,
followed by a brief description of less used techniques.

Slip Casting

Most commercial slip casting is conducted with ceramic particles suspended in water and cast into porous plaster molds. Figure 6.6 shows the
critical process steps in slip casting and some of the process parameters

Figure 6.5 Ceramic parts formed by uniaxial and isostatic pressing, some
with green machining. (Courtesy of Western Gold and Platinum Company,
Belmont, Calif., Subsidiary of GTE Sylvania, Inc.)

that must be carefully controlled to optimize strength or other critical
properties.

Raw Materials   Selection of the starting powder is dependent on the requirements of the application. Most applications require a fine powder, typically
-325 mesh. Applications requiring high strength require even finer powders, averaging under 5 μm, with a substantial portion under 1 μm. Some
applications, such as kiln furniture, which must withstand cyclic thermal
shock, may require a bimodal particle size with some particles considerably larger than 325 mesh.

The chemical composition is frequently an important consideration in selecting the starting powders and additives. Impurities and second phases can have pronounced effects on high-temperature properties.

Powder Processing As discussed in Chap. 5, powder as received from the supplier does not usually meet all the specifications for the shape-forming process or application and must be processed. Processing for slip casting usually involves particle sizing to achieve a particle size distribution that will yield maximum packing and uniformity during casting. Often, particle sizing is combined in one step with addition of binders, wetting

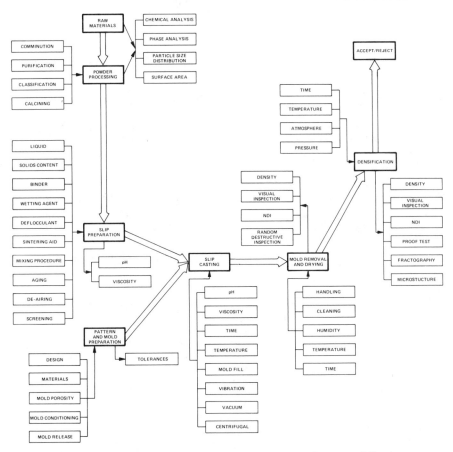

Figure 6.6 Critical process steps in slip casting and some of the process parameters that must be carefully controlled.

agents, deflocculants, and densification aids and with slip preparation. This is usually done by ball milling, but can also be done by vibratory milling or other process that provides wet milling. After milling, the slip is screened and perhaps passed through a magnetic separator to remove iron contamination. Slight adjustment might be required to achieve the desired viscosity and then the slip is ready for aging, deairing, or casting.

Slip Preparation To understand slip preparation considerations one needs to understand a little about rheology. Rheology is the study of flow characteristics of matter, e.g., suspensions of solid particles in a liquid, and is described quantitatively in terms of viscosity η. For low concentrations of spherical particles where no interaction occurs between either the particles or the particles and the liquid, the Einstein relationship applies:

$$\frac{\eta}{\eta_0} = 1 + 2.5V \tag{6.1}$$

where η is the viscosity of the suspension, η_0 the viscosity of the suspending fluid, and V the volume fraction of solid particles. This idealized relationship implies that the resulting viscosity is essentially controlled by the volume fraction of solids. In actual systems, the volume fraction does have a major effect, but so also do particle size, particle surface charges, and particle shape. The following discussion on the nature of these effects is largely derived from Michaels [7].

Decreasing particle size typically results in an increase in viscosity because of increased particle-particle interaction. For a given volume fraction of solids, the smaller the particle size, the closer the particles will be to each other and the greater the chance for interaction. Michaels used the following example assuming 40 vol % spherical particles:

Particle diameter (μm)	Mean distance between particle surfaces (Å)
10	9200
1	920
0.1	92
0.05	20
0.01	9.2

Van der Waals forces of attraction at surfaces are generally strong enough to affect another surface within a range of about 20 Å. It is apparent from the example above that, even in a relatively dilute suspension, considerable particle-particle attraction can occur and that the effect will be increased as the mean particle size decreases or the percent of submicron particles increases. This particle-particle attraction will increase the viscosity by increasing the force required to move one particle past another.

Particle shape is also important. Nonspherical particles, especially rods or plates, approach each other much more closely in a suspension than do spherical particles of an equivalent volume. This consideration is especially relevant in suspensions of clay particles. Clay was traditionally added to a casting slip to provide controlled suspension characteristics. Clays such as kaolinite are made up of thin hexagonal plate-shaped crystals. Michaels [7] used the following example to illustrate the shape effect:

Edge length of plate (μm)	Thickness of plate (μm)	Volume fraction of solid at which mean distance between particle surfaces is 20 Å
10	1	30
1	0.1	30
0.1	0.01	28
0.01	0.001	17

This can be compared to a mean spacing of 20 Å with 0.01-μm-diameter spherical particles at a volume fraction of 40% in the previous example.

For high-solid-content suspensions, particle-particle attraction results in the formation of agglomerates. In some cases these agglomerates can act essentially like roughly spherical particles and result in a decrease in viscosity. In other cases, especially for very high solids content, the agglomerates can interact with each other and further increase the viscosity.

Dispersion and flocculation (agglomeration) of ceramic particles in a fluid are strongly affected by the electrical potential at the particle surface, adsorbed ions, and the distribution of ions in the fluid adjacent to the particle. Thus the chemical and electronic structure of the solid, the pH of the fluid, and the presence of impurities are all critical considerations in the preparation of a slip for casting. Kaolinite clay has been studied extensively and is a good example.

At pH 6 or higher, where low concentrations of sodium or lithium cations are present, kaolinite is well dispersed in water. Under these conditions, each particle has a slight negative charge and the particles repel each other. However, if aluminum or iron salts are present in low concentration ($\sim 10^{-5}$ molar), the net charge on each particle is decreased and flocculation occurs. On the other hand, if the pH is below 6 and a $\sim 10^{-3}$ molar concentration of aluminum or ferric halides is present, the kaolinite will be dispersed. This is because the charge has been reversed under these conditions and the particles again repel each other because they have adequate levels of like charge. A similar situation exists when the pH is below 2 and monovalent anions such as chloride, nitrate, or acetate are present.

Low concentrations (0.005 to 0.3%) of certain organic and inorganic compounds have a strong dispersing effect on kaolinite suspensions. Some of these include sodium silicate, sodium hexametaphosphate (Calgon), sodium oxalate, sodium citrate, and sodium carbonate. These tend to ion exchange with ions such as calcium and aluminum, which prevent surface charge buildup and leave sodium, which allows a residual charge and causes repulsion between particles. Approximately 0.1% addition of sodium silicate reduces the viscosity by a factor of about 1000 [8].

Casting slips can also be formed with ceramics other than clays [9]. Success in casting depends largely on the ability to disperse these nonclay particles and achieve high solids concentrations. For most oxides, dispersion can be controlled by pH using the polar properties of water and the ion concentrations of acids or bases to achieve charged zones around the particles so that they repel each other. For instance, an Al_2O_3 slip with a specific gravity of 2.8 g/cm^3 had a viscosity of 65 cP at a pH of 4.5, but 3000 cP at a pH of 6.5 [10]. The slip at a pH of 4.5 was well dispersed and had good casting properties. It also was not extremely sensitive to changes in the solids content. Reducing the specific gravity to 2.6 g/cm^3 resulted in only a factor of 2 decrease in viscosity. For comparison, decreasing the specific gravity of the pH 6.5 slip to 2.6 g/cm^3 resulted in a tenfold decrease in viscosity.

Proper dispersion of the slip is perhaps the most important parameter in slip casting and warrants a more thorough discussion than some of the other parameters.

The actual physical preparation of the slip can be done by a variety of techniques. Perhaps the most common is wet ball milling or mixing. The ingredients, including the powder, binders, wetting agents, sintering aids, and dispersing agents, are added to the mill with the proper proportion of the selected casting liquid and milled to achieve thorough mixing, wetting, and (usually) particle size reduction. The slip is then allowed to age until its characteristics are relatively constant. It is then ready for final viscosity checking (and adjustment, if necessary), deairing, and casting.

Mold Preparation   The mold for slip casting must have controlled porosity so that it can remove the fluid from the slip by capillary action. The mold must also be low in cost. The most common mold material is plaster [11].

Plaster molds are prepared by mixing water with plaster of paris powder, pouring the mix into a pattern mold, and allowing the plaster to set. This produces a smooth-surface mold duplicating the contours of the pattern. For a complex shape, the mold is made in segments, each of which is sized so that it can be removed after slip casting without damaging the delicate casting.

Plaster of paris (hemihydrate) is partially dehydrated gypsum:

$$CaSO_4 \cdot 2H_2O \xrightarrow{180°C} H_2O + CaSO_4 \cdot 1/2H_2O \qquad (6.2)$$

   Gypsum   Hemihydrate

The reaction is reversible; addition of water to the hemihydrate results in precipitation of very fine needle-shaped crystals of gypsum which intertwine to form the plaster mold. The reaction is satisfied chemically by addition of 18% water, but considerably more water is necessary to provide a mixture with adequate fluidity for mold making. This extra water fills positions between the gypsum crystals during precipitation and results in very fine capillary porosity after the finished plaster mold has been dried. It is this porosity that draws the water out of the slip during slip casting. The amount of porosity can be controlled by the amount of excess water added during fabrication of the plaster mold. For normal slip casting, 70 to 80 wt % water is used.

The setting rate of plaster can be widely varied by impurities.

Casting   Once the mold has been fabricated and properly dried and an optimum slip has been prepared, casting can be conducted. Many options are available, depending on the complexity of the component and other factors:

Simple casting into a one-piece mold
Simple casting into a multipiece mold
Drain casting
Solid casting
Vacuum casting
Centrifugal casting
Casting with nonabsorbing pins or mandrels inserted into the mold

Figure 6.7 illustrates schematically drain casting. The slip is poured into the mold and water is sucked out where the slip is in contact with the mold, leaving a close-packed deposition of particles growing into the slip from the mold walls. The slip is left in the mold until the desired thickness is built up, at which time the remaining slip is drained from the mold.

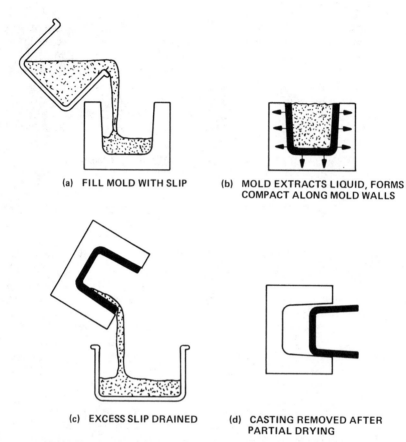

(a) FILL MOLD WITH SLIP   (b) MOLD EXTRACTS LIQUID, FORMS
                                       COMPACT ALONG MOLD WALLS

(c) EXCESS SLIP DRAINED   (d) CASTING REMOVED AFTER
                                       PARTIAL DRYING

Figure 6.7 Schematic illustrating the drain-casting process.

Drain casting is the most common slip-casting approach. It is used for art
casting (figurines), sinks and other sanitary ware, crucibles, and a variety
of other products.

   Solid casting is identical to drain casting except that slip is continually
added until a solid casting has been achieved.

   Vacuum casting can be conducted either with the drain or solid approach.
A vacuum is pulled around the outside of the mold. In some cases this
either increases the casting rate or uniformity and improves the economics
or quality.

   Centrifugal casting involves spinning the mold to apply greater-than-
normal gravitational loads to make sure that the slip completely fills the
mold. This can be beneficial in the casting of some complex shapes.

Considerable sharpness and complexity of detail can be designed into a slip-casting mold. The well-known Hummel figurines are a good example. A more-engineering oriented example is illustrated in Fig. 6.8. It consists of a one-piece annular combustor of SiC fabricated by the Norton Company.

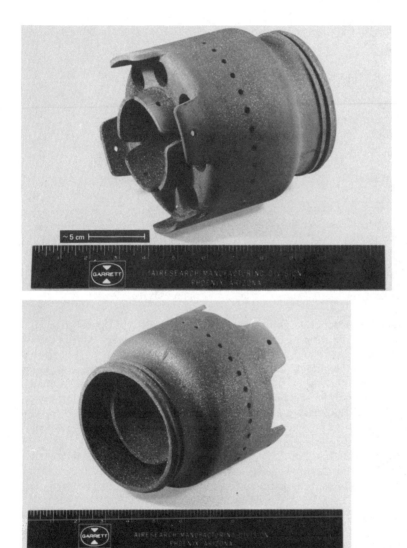

Figure 6.8 Complex annular combustor of SiC for a gas turbine engine fabricated by drain casting. (Courtesy of the Garrett Turbine Engine Company, Phoenix, Ariz., Division of The Garrett Corporation.)

A multipiece mold was required with a segmented plaster outer wall, an inner plaster mandrel, a contoured plaster base block, and nonabsorbing pins (to form the holes). The combustor shape to near-final tolerances was then achieved by drain casting.

Casting Process Control   As was illustrated in Fig. 6.6, careful process is necessary in the slip-casting process. St. Pierre [10] lists the following as critical:

Constancy of properties
Viscosity
Settling rate
Freedom from air bubbles
Casting rate
Drain properties
Shrinkage
Release properties
Strength

A detailed working knowledge with suitable specifications and certification procedures of the relative needs of each of the criteria listed above is a requirement for the manufacturer for each component. An awareness of the defects and/or effects on properties are desirable for the engineer in charge of the use of the component or its integration into a system. It usually takes interaction between the user and manufacturer to resolve a quality problem.

Constancy of properties refers to the reproducibility of the casting slip and its stability as a function of time. The slip must be easily reproduced and preferably should not be overly sensitive to slight variations in solids content and chemical composition or to storage time. The viscosity must be low enough to allow complete fill of the mold, yet the solids content must be high enough to achieve a reasonable casting rate. Too-slow casting can result in thickness and density variations due to settling. Too-rapid casting can result in tapered walls (for a drain casting), lack of thickness control, or blockage of narrow passages in the mold.

The slip must be free of entrapped air or chemical reactions that would produce air bubbles during casting. Air bubbles present in the slip will be incorporated in the casting and may be critical defects in the final densified part.

Once casting has been completed, the part begins to dry and shrink away from the mold. This shrinkage is necessary to achieve release of the part from the mold. If the casting sticks to the mold, it will usually be damaged during removal and rejected. Mold release can be aided by coating the walls of the mold with a release agent such as a silicone or olive oil.

However, it should be recognized that the coating may alter the casting rate.

The strength of the casting must be adequate to permit removal from the mold, drying, and handling prior to the firing operation. Usually, a small amount (<1%) of binder is included in the slip. Organic binders such as polyvinyl alcohol work well. With the binder present, strength comparable to or greater than blackboard chalk is achieved. Such strength is adequate for handling and also for green machining if required.

Soluble Mold Casting

Soluble mold casting is a new approach based on the much older technology of investment casting. It is also referred to as fugitive wax slip casting [12-14] and is accomplished in the following steps:

1. A wax pattern of the desired configuration is produced by injection-molding a water-soluble wax.
2. The water-soluble wax pattern is dipped in a non-water-soluble wax to form a thin layer over the pattern.
3. The pattern wax is dissolved in water, leaving the non-water-soluble wax as an accurate mold of the shape.
4. The wax mold is trimmed, attached to a plaster block, and filled with the appropriate casting slip.
5. After the casting is complete, the mold is removed by dissolving in a solvent.
6. The cast shape is dried, green machined as required, and densified at high temperature.

The application of the fugitive wax approach is illustrated in Fig. 6.9 for the fabrication of a complex-shaped stator vane for a gas turbine engine. The injection molding tool on the right produces the water-soluble wax sta- tor vane pattern. The injection molding tool on the left produces the pattern for the reservoir which will hold the slip and guide it through gating chan- nels into the stator vane mold during casting. The reservoir and vane pat- terns are bonded together by simple wax welding and are shown as the white wax assembly in the center of Fig. 6.9. Below this is the mold produced by dipping and dissolving the pattern. Below the mold is the green casting after dissolving the mold and trimming off any material remaining in the reservoir or gating area.

Very complex shapes can be fabricated by the soluble mold technique. Figure 6.10 shows integral gas turbine stators made by this approach. Individual vanes were injection molded and then welded together to form a pattern such as the one in the upper right of Fig. 6.10a. This was then dipped as a unit to form a one-piece mold.

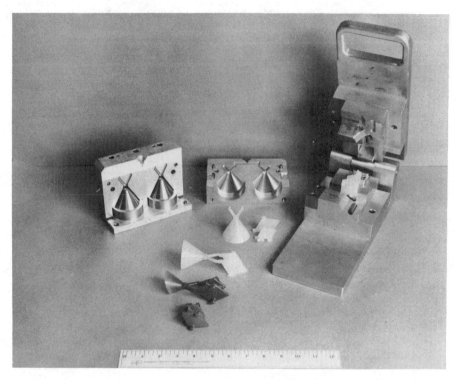

Figure 6.9 Tooling, patterns, mold, and cast part illustrating the fugitive wax slip-casting process as applied to the fabrication of a gas turbine stator vane. (Courtesy of AiResearch Casting Company, Torrance, Calif., Division of The Garrett Corporation.)

6.3 PLASTIC FORMING

Plastic forming involves producing a shape from a mixture of powder and additives that is deformable under pressure. Such a mixture can be obtained in systems containing clay minerals by addition of water and small amounts of a deflocculant, a wetting agent, and a lubricant. In systems not containing clay, such as pure oxides and carbides and nitrides, an organic material is added in place of water or mixed with water to provide the plasticity. About 25 to 50 vol % organic additive is required to achieve adequate plasticity for forming. Heat normally must be supplied simultaneously to pressure.

A major difficulty in the plastic-forming processes is removing the organic material prior to firing. In the case of a water-clay system,

(a)

(b)

Figure 6.10 Slip casting of an integral gas turbine stator using the fugitive wax process. (a) The water-soluble pattern and the resulting mold interfaced with a plaster block. (b) As-cast stators prior to trimming or machining. (Courtesy AiResearch Casting Company, Torrance, Calif., Division of The Garrett Corporation.)

substantial shrinkage occurs during drying, increasing the risk of shrinkage cracks. In the case of organic additives, the major problems are achieving high enough green density and extraction of the organic. Too rapid extraction causes cracking, bloating, or distortion. Inadequate removal results in cracking, bloating, or contamination during the later high-temperature densification process.

Plastic processes are used extensively in the fabrication of traditional ceramics such as pottery and dinnerware. The compositions contain clay and have been made workable by addition of water. Hand forming and jiggering are common fabrication techniques. These operations will not be discussed in this text. Instead, the emphasis will be on either standard techniques applied to demanding shapes or on emerging techniques applicable to modern ceramic applications. Most of the discussion will be on injection molding and extrusion.

Injection Molding

Injection molding is a low-cost, high-volume production technique for making net-shape or near-net-shape parts. A feed material consisting of or containing a thermoplastic polymer or a thermosetting polymer is preheated in the "barrel" of the injection molding machine to a temperature at which the polymer has a low-enough viscosity to allow flow if pressure is applied. A ram or plunger is pressed against the heated material in the barrel by either a hydraulic, pneumatic, or screw mechanism. The viscous material is forced through an orifice into a narrow passageway which leads to the shaped tool cavity. This helps compact the feed material and remove porosity. At the end of the passageway, the strand of viscous material passes through another orifice called the sprue and into the tool cavity. The strand then piles upon itself until the cavity is full and the material has "knit" or fused together under the pressure and temperature to produce a homogeneous part.

Injection molding is used extensively in the plastics industry to make everything from garbage cans to ice cube trays to surprisingly complex constructible toys such as model boats and airplanes. Ceramic parts are made with the same injection molding equipment, but with dies made of harder, more wear-resistant metal alloys. The ceramic powder is essentially added to the plastic as a filler. After injection molding the plastic is then removed by careful thermal treatments.

Injection Molding Parameters   Very little has been published on the parameters for injection molding of ceramics. Figure 6.11 lists the general steps in injection molding and some of the controls and inspection procedures that must be considered.

Particle size distribution is important in injection molding to achieve the densest packing and to minimize the amount of organic material. Both

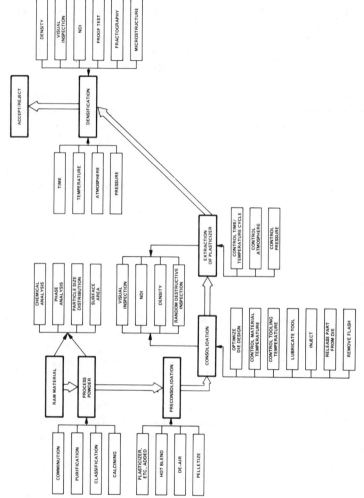

Figure 6.11 Injection molding process flow sheet.

particle packing and sizing affect viscosity. Farris [15] reports that viscosity starts increasing rapidly at about 55 vol % solids for a unimodal suspension of spheres, but that the solids loading can be increased to over 70%
before the viscosity starts increasing rapidly for a bimodal distribution
containing about 25% fine spheres. By using a graduated particle size distribution, Mangels [16] has successfully injection molded complex shapes
of silicon powder having 76.5 vol % solids. His tests were conducted on a
plunger-type injection machine at a cylinder pressure of 13.8 MPa (2000
psi) and a temperature 10°C above the melting temperature of the organic
binder.

   Whalen et al. [17] report injection molding of SiC with a solids content
of 52 vol % using a thermosetting resin. They preheated a dry ball milled
mixture of the ceramic and organic powder in the barrel of a plunger-type
injection molder at 116°C [241°F) for 1 min and then injected the mixture
into a mold preheated to 155°C (310°F).

   Injection molding can also be conducted using a thermoplastic resin.
In this case, though, the mold must be cooler than the temperature of the
preheated material.

Injection Molding Defects   A variety of defects can occur during the injection molding operation.   Table 6.2 summarizes some of these defects and
the likely causes.

   An incomplete part is usually easy to detect visually and can be rejected immediately after injection molding. Large pores intersecting the surface are also easy to detect and reject. Internal pores require nondestructive
inspection (NDI) techniques such as radiography or ultrasonics (described
in Chap. 9). However, the capabilities of these NDI techniques are currently
limited and some pores may not be detected until they cause failure of the
component. Sometimes proof testing (also described in Chap. 9) can be
successfully used to reject these defective parts.

   Knit lines are areas where the injected material does not properly fuse
together. They represent a discontinuity or a weak region in the part.
They usually have a laminar or folded appearance. Some can be severe
and are easily visible if they intersect the outer surface of the part. Others
are very subtle and difficult to detect, even with NDI techniques.

   Cracks, like knit lines, can be present in a wide range of severity and
visibility. The more severe cases are easy to see visually and can be
rejected. Others can be detected by dye penetrant inspection or by inspection at low magnification (up to about 40×) of critical areas.

   The application engineer usually does not get involved with defect detection until a part or series of parts have failed in service. It is then vital
that he determine the source of failure as soon as possible and develop a
solution. Chapter 12 describes powerful techniques for detecting the cause
of a failure.

Table 6.2 Injection Molding Defects and Causes

Description of defect	
Incomplete part	Improper feed material, poor tool design, improper material and/or tool temperature, inadequate tool lubrication
Large pores	Entrapped air, improper material flow and consolidation during injection, agglomerates, or large pockets of the organic due to incomplete mixing
Knit lines	Improper tool design or feed material, incorrect temperatures
Cracks	Sticking during removal from tool, improper tool design, improper extraction of the organic

Injection Molding of Complex Shapes   Injection molding is currently being developed for net-shape forming of complex aerodynamic rotating and stationary components for prototype gas turbine engines. Figure 6.12 shows a simple injection molding tool for fabricating individual stator vanes [14]. Two strength test bars are injected with each stator vane and accompany that vane through subsequent processing. The strength and microstructure of these bars can then be used for certification purposes to assure the manufacturer and customer that the material meets the required property specification. The other two tabs of material shown in Fig. 6.12 are overflow material. They contain any debris that may have inadvertently been in the tool passages and any excess lubricant.

Figure 6.13 shows larger, more complex prototype gas turbine parts. These were fabricated by the Ford Motor Company [17,18]. The top parts are integral stators made of reaction-bonded Si_3N_4, and the bottom parts are rotor blade rings also made of reaction-bonded Si_3N_4.

Compression Molding

Compression molding, transfer molding, and warm molding are all plastic fabrication processes similar to injection molding. In each case, the ceramic powder is in an organic carrier and is forced into a shaped tool under pressure and temperature. Compression molding also has similarities to pressing, except that a slug of material rather than a free-flowing powder is used for feed. The shape is built into the platens of the press or the faces of the tool and the mix plastically deforms under pressure to fill the cavity.

(a)

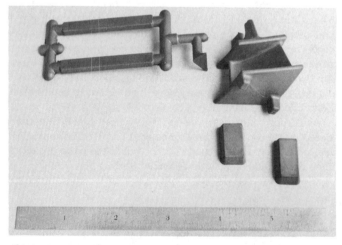

(b)

Figure 6.12 (a) Injection molding tool for fabrication of individual stator vanes for a gas turbine engine. (b) Close-up of the stator vane, test bar gating, and overflow material from a single injection. (Courtesy of AiResearch Casting Company, Torrance, Calif., Division of The Garrett Corporation.)

Figure 6.13 Complex shapes made by injection molding. (a) Integral stators. (b) Rotor blade rings. (Courtesy of Ford Motor Company, Dearborn, Mich.)

Extrusion

Extrusion is used extensively for fabrication of brick, tile, tubes, rods, and other elongated shapes that have a constant cross section. A schematic of an extruder is shown in Fig. 6.14. A plastic mix of the ceramic powder and clay or organic additives is placed in an evacuated cylinder, where it is worked or kneaded to remove air and to achieve uniform consistency. The mix is then typically carried by an auger and forced through a die. As the shaped plastic mix exits from the extruder, it is supported on a suitable flat or shaped surface to prevent distortion and is cut to the required lengths.

Extrusion of compositions containing clay and made plastic by water addition is well developed and has been in common practice for many years. Extrusion of nonclay compositions is less developed and usually requires experimental optimization of each new system. Table 6.3 summarizes examples of extrusion conditions identified by Hyde [19] for several materials.

Very complex cross sections have been extruded. One of the most remarkable is the thin-walled cellular structure shown in Fig. 6.15 fabricated by NGK in Japan. Such honeycomb structures have been fabricated of low-thermal-expansion materials such as lithium aluminum silicate and cordierite (magnesium aluminum silicate) and are being used for heat exchanger and catalyst substrate applications.

6.4 OTHER FORMING PROCESSES

Methods for forming compacted shapes of ceramic powders are limited only by the imagination of the engineer. This is especially true for modifications to current techniques to achieve improved properties, decreased rejections, and decreased cost. It is also true for the development of advanced composites. It would be interesting to speculate on some of the possibilities, but is not within the scope of this text. Discussion in this section will be limited to two important established approaches, tape forming and green machining.

Tape Forming

Some applications such as electrical substrates and plate-fin heat exchangers and exhaust emission control devices require thin strips or structures of ceramics. Tape forming has been developed as an effective means of meeting these needs.

Figure 6.16 illustrates schematically methods of forming a thin sheet or thin-walled ceramic structure by tape forming.

The doctor-blade process [20] is well established for fabrication of electronic ceramics for capacitors, for electrically insulating substrates for thick-film and thin-film circuitry, for ferrite memories, and for catalyst substrates. It consists of casting a slurry onto a moving carrier surface (usually a thin film of cellulose acetate, Teflon, Mylar, or cellophane) and

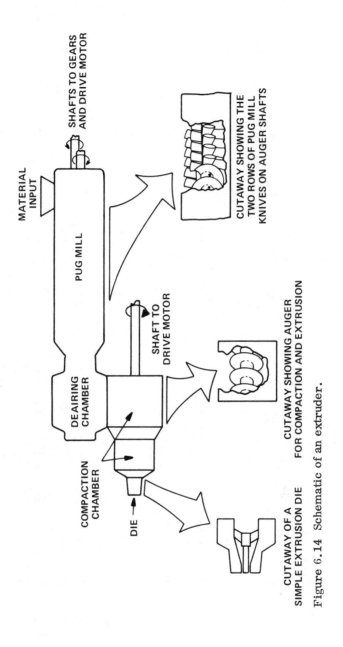

MATERIAL INPUT

SHAFTS TO GEARS AND DRIVE MOTOR

PUG MILL

CUTAWAY SHOWING THE TWO ROWS OF PUG MILL KNIVES ON AUGER SHAFTS

DEAIRING CHAMBER

SHAFT TO DRIVE MOTOR

COMPACTION CHAMBER

DIE

CUTAWAY SHOWING AUGER FOR COMPACTION AND EXTRUSION

CUTAWAY OF A SIMPLE EXTRUSION DIE

Figure 6.14 Schematic of an extruder.

Table 6.3 Examples of Extrusion of Nonclay Compositions

Composition	Additives	Conditions	Shape
Graphite	50–60 parts phenol formaldehyde emulsion	--	5/8– and 1-in. rods
Petroleum coke	Coal-tar pitch plus heavy oil	90–110°C	--
Al_2O_3–5% Cr_2O_3	Gum ghatti plus mogul starch	--	Round, square, and triangular tubes
30% Si–70% SiC	4% guar gum, 20% water, 3% silicone	--	--
BeO	10–18% mogul starch, 15–18% water–glyceryl mixture	--	Rods, tubes, thin–walled multicell tubes
MgO	30–40% flour paste	--	Vacuum-tube parts

Source: Ref. 19. Reprinted from C. Hyde, in Ceramic Fabrication Processes (W. D. Kingery, ed.), by permission of The MIT Press, Cambridge, Mass. © 1963, The MIT Press.

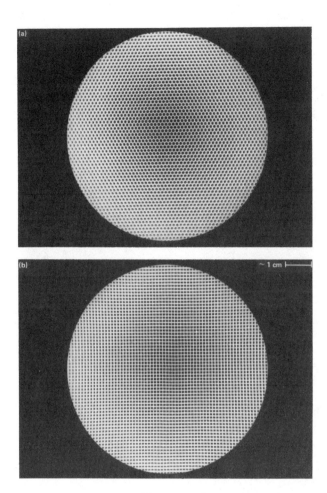

Figure 6.15 Extruded honeycomb structures for heat exchanger and emission control applications. (Courtesy of NGK Insulators, Nagoya, Japan.)

spreading the slurry to a controlled thickness with the knife edge of a blade. The slurry is then carefully dried, resulting in a thin, flexible tape that can be cut or stamped to the desired configuration prior to firing.

The doctor-blade process sounds simple, but actually requires very careful controls to avoid warpage, out-of-tolerance thickness, and other defects. This is illustrated in Table 6.4 by listing the additives required in the Western Electric Company process for making Al_2O_3 substrates [21]. Williams [20] lists a variety of other additives for both nonaqueous and aqueous doctor-blade systems.

The paper-casting process is similar to the doctor-blade process. A controlled thickness of slurry is deposited on a moving carrier. In this case the carrier is a low-ash paper that can be burned off in a later process step. The paper-casting process has been used in the manufacture of catalyst substrates and rotary regenerator heat exchangers [22].

The roll process produces a tape by mechanically reducing the thickness of a ceramic powder/organic polymer mixture. It is also used for fabrication of catalyst substrates and rotary regenerator heat exchangers.

(a) DOCTOR-BLADE PROCESS

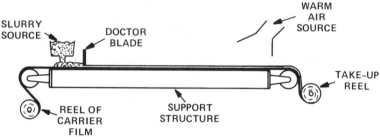

(b) PAPER-CASTING PROCESS

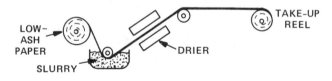

(c) ROLL PROCESS

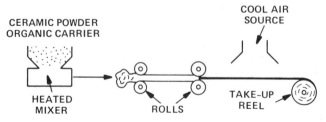

Figure 6.16 Schematics of tape-forming processes. (Adapted from Ref. 22.)

Table 6.4   Additives to Al_2O_3 for the Western Electric
Company ERC-105 Doctor-Blade Process

Material	Function
MgO	Grain-growth inhibitor
Menhaden fish oil	Deflocculant
Trichloroethylene	Solvent
Ethyl alcohol	Solvent
Polyvinyl butyral	Binder
Polyethylene glycol	Plasticizer
Octylphthalate	Plasticizer

Source: Ref. 21.

Figure 6.17 illustrates how a rotary heat exchanger is fabricated by the paper-casting and roll-forming processes [22].

Green Machining

Green machining refers to machining of a ceramic part prior to final densification while the material consists of compacted, loosely bonded powder. Such material is much softer than the ceramic in its final densified condition and can be machined much more economically since diamond tooling is not required. However, the material is relatively fragile, and great care is necessary in the design and fabrication of the tooling and fixturing so that the parts can be accurately and uniformly held during the various shaping operations. In addition, the machining parameters must be carefully controlled to avoid overstressing the fragile material and producing chips, cracks, breakage, or poor surface.

Holding of the compact for machining is typically accomplished with a combination of bee's wax and precision metal fixtures. The part must be held rigidly, but with no distortion or stress concentration. Metal fixturing should normally be located either on the outer diameter or the face of the part. This is to avoid expansion of the metal fixture into the ceramic, causing breakage when heat is applied to soften the wax during removal of the machined part from the fixture. If there is no way to locate the part except on the inside diameter, special care must be taken to heat the ceramic rapidly while keeping the metal cool.

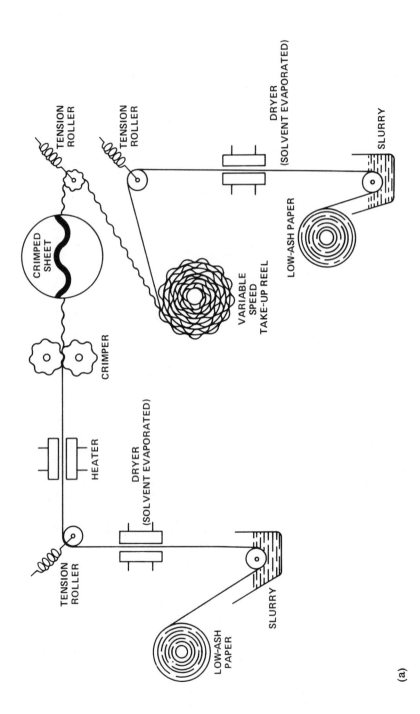

(a)

212

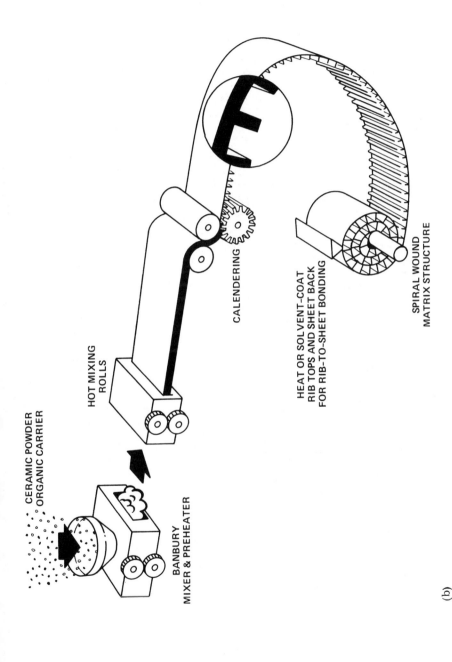

CERAMIC POWDER
ORGANIC CARRIER

BANBURY
MIXER & PREHEATER

HOT MIXING
ROLLS

CALENDERING

HEAT OR SOLVENT-COAT
RIB TOPS AND SHEET BACK
FOR RIB-TO-SHEET BONDING

SPIRAL WOUND
MATRIX STRUCTURE

(b)

Figure 6.17 Fabrication of a rotary regenerator heat exchanger structure using tape-forming processes. (a) Paper-casting process. (b) Roll-forming process. (From Ref. 22.)

Once a ceramic part has been secured rigidly in a fixture, machining can be conducted by a variety of methods—turning, milling, drilling, form wheel grinding, and profile grinding. Machining can be either dry or wet. In either case, the compact is abrasive and results in tool wear. A wear land on the cutting edge as little as 0.005 in. wide will result in a buildup of the cutting pressure and damage to the ceramic.

It is possible to machine compacts with high-speed steel or cemented carbide cutting tools, but this is not recommended for all components or all green materials. In some cases, the tool dulls so rapidly that extreme care is necessary to avoid damage to the workpiece. Figure 6.18 summarizes a study [23] of other cutting tool materials using a 5° positive rake and 10° clearance angle. The GE compact diamond cost about 10 times as much as the tungsten carbide, but resulted in a significant cost saving in terms of increased life, less time changing inserts, and reduced risk of damage to the workpiece from a dull tool. The study was conducted with single-point turning on an engine lathe. Milling with a two-flute end mill at 200 surface feet per minute (sfm) with GE compact diamond inserts showed the same life characteristics.

Green machining can also be conducted with grinding wheels containing multiple abrasive particles bonded in a resin or metal matrix. Higher

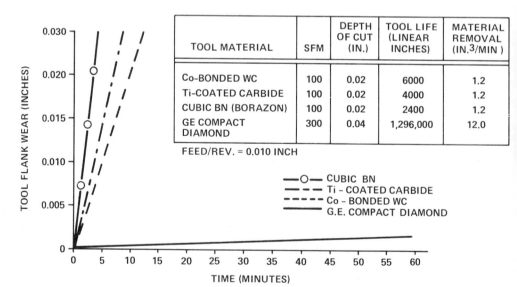

Figure 6.18 Tool wear for different tool insert materials for green machining of a presintered silicon compact in the fabrication of reaction-bonded Si_3N_4. (From Ref. 23.)

surface speed, broader contact, and decreased depth of cut are characteristic of this technique, resulting usually in a better surface and less chance
of damage. Excellent tool life can be achieved, especially if diamond abrasive is used. Furthermore, coarse abrasive can be used for roughing
passes and fine abrasive for finishing. Formed wheels can also be used to
produce a controlled and reproducible contour.

REFERENCES

1. H. Thurnauer, Controls required and problems encountered in production dry pressing, in Ceramic Fabrication Processes (W. D. Kingery,
   ed.), MIT Press, Cambridge, Mass., 1963, pp. 62-70.
2. O. J. Whittemore, Jr., Particle compaction, in Ceramic Processing
   Before Firing (G. Y. Onoda, Jr., and L. L. Hench, eds.), John Wiley
   & Sons, Inc., New York, 1978, pp. 343-355.
3. W. D. Kingery, Pressure forming of ceramics, in Ceramic Fabrication Processes (W. D. Kingery, ed.), MIT Press, Cambridge, Mass.,
   1963, pp. 55-61.
4. P. Duwez and L. Zwell, AIME Tech. Publ. 2515, Metals Trans. $\underline{1}$,
   137 (1949).
5. R. P. Seelig, in The Physics of Powder Metallurgy (W. E. Kingston,
   ed.), McGraw-Hill Book Company, New York, 1950, p. 344.
6. W. D. Kingery, Hydrostatic molding, in Ceramic Fabrication Processes
   (W. D. Kingery, ed.), MIT Press, Cambridge, Mass., 1963, pp. 70-73.
7. A. S. Michaels, Rheological properties of aqueous clay systems, in
   Ceramic Fabrication Processes (W. D. Kingery, ed.), MIT Press,
   1963, pp. 23-31.
8. F. Moore, Rheology of Ceramic Systems, MacLaren & Sons Ltd.,
   London, 1965.
9. R. E. Cowan, in Treatise on Materials Science and Technology, Vol.
   9: Ceramic Fabrication Processes (F. F. Y. Wang, ed.), Academic
   Press, Inc., New York, 1976, pp. 153-171.
10. P. D. S. St. Pierre, Slip casting nonclay ceramics, in Ceramic Fab-
    rication Processes (W. D. Kingery, ed.), MIT Press, Cambridge,
    Mass., 1963, pp. 45-51.
11. C. M. Lambe, Preparation and use of plaster molds, in Ceramic Fab-
    rication Processes (W. D. Kingery, ed.), MIT Press, Cambridge,
    Mass., 1963, pp. 31-40.
12. A. Ezis and J. M. Nicholson, "Method of manufacturing a slip cast
    article," U.S. Patent No. 4,067,943.
13. A. Ezis and J. T. Neil, Fabrication and properties of fugitive mold
    slip-cast Si_3N_4, Bull. Am. Ceram. Soc. $\underline{58}$(9), 883 (1979).
14. H. Gersch, D. Mann, and M. Rorabaugh, Slip-cast and injection-
    molding process development of reaction-bonded silicon nitride at

AiResearch Casting Company, in DARPA/Navy Ceramic Gas Turbine Demonstration Engine Program Review (J. Fairbanks and R. Rice, eds.), MCIC Report MCIC-78-36, 1978, pp. 313-340.

15. R. J. Farris, Trans. Soc. Rheol. 12(2), 281 (1968).

16. J. A. Mangels, Development of injection molded reaction bonded Si_3N_4, in Ceramics for High Performance Applications, II (J. J. Burke, E. N. Lenoe, and R. N. Katz, eds.), Brook Hill Publishing Co., Chestnut Hill, Mass., 1978, pp. 113-130. (Available from MCIC, Battelle Columbus Labs., Columbus, Ohio.)

17. T. J. Whalen, J. E. Noakes, and L. L. Terner, Progress on injection-molded reaction bonded SiC, in Ceramics for High Performance Applications, II (J. J. Burke, E. N. Lenoe, and R. N. Katz, eds.), Brook Hill Publishing Co., Chestnut Hill, Mass., 1978, pp. 179-189. (Available from MCIC, Battelle Columbus Labs., Columbus, Ohio.)

18. C. F. Johnson and T. G. Mohr, Injection molding 2.7 g/cc silicon nitride turbine rotor blade rings utilizing automatic control, in Ceramics for High Performance Applications, II (J. J. Burke, E. N. Lenoe, and R. N. Katz, eds.), Brook Hill Publishing Co., Chestnut Hill, Mass., 1978, pp. 194-206. (Available from MCIC, Battelle Columbus Labs., Columbus, Ohio.)

19. C. Hyde, Vertical extrusion of nonclay composition, in Ceramic Fabrication Processes (W. D. Kingery, ed.), MIT Press, Cambridge, Mass., 1963, pp. 107-111.

20. J. C. Williams, in Treatise on Materials Science and Technology, Vol. 9: Ceramic Fabrication Processes (F. F. Y. Wang, ed.), Academic Press, Inc., New York, 1976, pp. 173-198.

21. D. J. Shanefield and R. E. Mistler, Fine grained alumina substrates: I, the manufacturing process, Am. Ceram. Soc. Bull. Part I, 53, 416-420 (1974).

22. C. A. Fucinari, The utilization of data relating to fin geometry and manufacturing processes of ceramic matrix systems to the design of ceramic heat exchangers, in Ceramics for High Performance Applications, II (J. J. Burke, E. N. Lenoe, and R. N. Katz, eds.), Brook Hill Publishing Co., Chestnut Hill, Mass., 1978, pp. 349-365. (Available from MCIC, Battelle Columbus Labs., Columbus, Ohio.)

23. D. W. Richerson and M. W. Robare, Turbine component machining development, in The Science of Ceramic Machining and Surface Finishing, II (B. J. Hockey and R. W. Rice, eds.), NBS Special Publication 562, U. S. Government Printing Office, Washington, D.C., 1979, pp. 209-220.

Densification

In Chaps. 5 and 6 we discussed the criteria and techniques for selecting and processing ceramic powders and for forming these powders into shaped particulate compacts. In this chapter we explore the processes for densifying these particulate compacts into strong, useful ceramic components.

7.1 THEORY OF SINTERING

The densification of a particulate ceramic compact is technically referred to as sintering. Sintering is essentially a removal of the pores between the starting particles (accompanied by shrinkage of the component), combined with growth together and strong bonding between adjacent particles. The following criteria must be met before sintering can occur:

1. A mechanism for material transport must be present.
2. A source of energy to activate and sustain this material transport must be present.

The primary mechanisms for transport are diffusion and viscous flow. Heat is the primary source of energy, in conjunction with energy gradients due to particle-particle contact and surface tension.

Although ceramic materials have been used and densified for centuries, scientific understanding and control of sintering has only developed during the past 40 to 50 years. Early controlled experiments were conducted by Muller in 1935 [1]. He sintered compacts of NaCl powder for a variety of times at several temperatures and evaluated the degree of sintering by measuring the fracture strength.

Much progress in our understanding of densification has been achieved since 1935. Now sintering is studied by plotting density or shrinkage data as a function of time and by actual examination of the microstructure at various stages of sintering using scanning electron microscopy, transmission electron microscopy, and lattice imaging.

Sintering can occur by a variety of mechanisms, as summarized in Table 7.1. Each mechanism can work alone or in combination with other mechanisms to achieve densification.

Vapor-Phase Sintering

Vapor-phase sintering is important in only a few material systems and is discussed only briefly. The driving energy is the difference in vapor pressure as a function of surface curvature. As illustrated in Fig. 7.1., material is transported from the surface of the particles, which have a positive radius of curvature and a relatively high vapor pressure to the contact region between particles, which has a negative radius of curvature and a much lower vapor pressure. The smaller the particles, the greater the positive radius of curvature and the greater the driving force for vapor-phase transport. Table 7.2 shows how large an effect particle size or surface curvature can have on pressure across the curved surface and on relative vapor pressure [2].

Vapor-phase transport changes the shape of the pores and achieves bonding between adjacent particles and thus increases the material strength and decreases permeability due to open porosity. However, it does not result in shrinkage and cannot produce densification. It must be accompanied by other mechanisms that provide bulk material transport or transport of pores to external surfaces.

Table 7.1 Sintering Mechanisms

Type of sintering	Material transport mechanism	Driving energy
Vapor phase	Evaporation-condensation	Differences in vapor pressure
Solid state	Diffusion	Differences in free energy or chemical potential
Liquid phase	Viscous flow, diffusion	Capillary pressure, surface tension
Reactive liquid	Viscous flow, solution-precipitation	Capillary pressure, surface tension

Source: Adapted from Ref. 2.

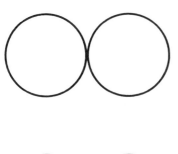

ADJACENT PARTICLES
IN CONTACT

NECK FORMATION
BY VAPOR PHASE
MATERIAL TRANSPORT

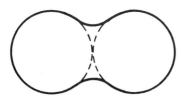

COMPLETION OF VAPOR
PHASE TRANSPORT: PARTICLES
BONDED, PORE SHAPE CHANGED,
NO SHRINKAGE

Figure 7.1 Schematic of vapor-phase material transport.

Solid-State Sintering

Solid-state sintering involves material transport by diffusion. Diffusion
can consist of movement of atoms or vacancies along a surface or grain
boundary or through the volume of the material. Surface diffusion, like
vapor-phase transport, does not result in shrinkage. Volume diffusion,
whether along grain boundaries or through lattice dislocations, does result
in shrinkage, as illustrated in Fig. 7.2.
     The driving force for solid-state sintering is the difference in free
energy or chemical potential between the free surfaces of particles and the
points of contact between adjacent particles. Mathematical models have
been derived which compare favorably with experimental data and relate
the rate of sintering. For instance, Kingery et al. [3] derived the following
equation for the mechanism of transport of material by lattice diffusion from
the line of contact between two particles to the neck region:

Table 7.2 Effect of Particle Size or Surface Curvature on the
           Pressure Difference and Relative Vapor Pressure
           across a Curved Surface

Material	Surface diameter (μm)	Pressure difference		Relative vapor pressure (P/P_0)
		MPa	psi	
Liquid water at 25°C	0.1	2.8	418	1.02
	1.0	0.28	41.8	1.002
	10.0	0.03	4.2	1.0002
Liquid cobalt at 1450°C	0.1	67.3	9750	1.02
	1.0	6.7	975	1.002
	10.0	0.67	97.5	1.0002
Silica glass at 1700°C	0.1	12.1	1750	1.02
	1.0	1.2	175	1.002
	10.0	0.12	17.5	1.0002
Solid Al_2O_3	0.1	36.2	5250	1.02
	1.0	3.6	525	1.002
	10.0	0.36	52.5	1.0002

Source: Ref. 2.

$$\frac{\Delta L}{L_0} = \left(\frac{20\gamma a^3 D^{*2/5}}{\sqrt{2}\ kT}\right) r^{-6/5} t^{2/5} \tag{7.1}$$

where $\Delta L/L_0$ is the linear shrinkage (equivalent to the sintering rate), γ the surface energy, a^3 the atomic volume of the diffusing vacancy, D^* the self-diffusion coefficient, k the Boltzmann constant, T the temperature, r the particle radius (assuming equal-size spherical starting particles), and t is time.

Equations for other volume diffusion mechanisms of sintering are similar. In each case the rate of shrinkage increases with increasing temperature and with decreasing particle radius and decreases with time.

Figure 7.3a illustrates the effects of temperature and time. Figure 7.3b shows a log-log plot of the same data. The slope of the log $\Delta L/L_0$ versus log t line is approximately two-fifths for solid-state sintering.

It is apparent from examination of equation (7.1) and Fig. 7.3 that control of temperature and particle size is extremely important, but that control of time is less important.

Finer-particle-size powder can be sintered more rapidly and at a lower temperature than coarser powder. Not apparent in the equation, but highly important to the final properties, are the uniformity of particle packing, the particle shape, and the particle size distribution. If particle packing is not uniform in the greenware, it will be very difficult to eliminate all the porosity during sintering. Agglomerates are a common source of nonuniformity, as discussed in Chap. 5. Nonuniformity can also result during

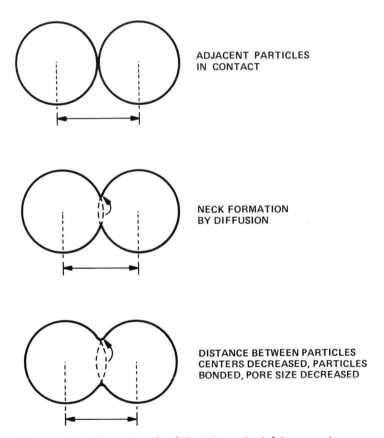

ADJACENT PARTICLES
IN CONTACT

NECK FORMATION
BY DIFFUSION

DISTANCE BETWEEN PARTICLES
CENTERS DECREASED, PARTICLES
BONDED, PORE SIZE DECREASED

Figure 7.2 Schematic of solid-state material transport.

(a)

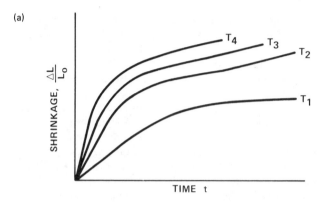

$$T_4 > T_3 > T_2 > T_1$$

(b)

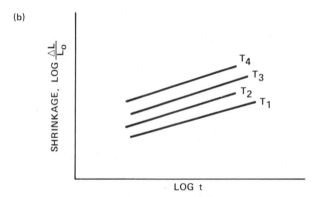

Figure 7.3 Typical sintering rate curves showing the effects of temperature and time.

shape forming due to gas entrapment, particle segregation (i.e., settling during slip casting), lamination, and fold lines (injection molding).

Particle shape can also be important. Too high a concentration of elongated or flattened particles can result in bridging during forming, producing a large or irregularly shaped pore that is difficult to remove during sintering.

Particle size distribution is also critical. Particles that are all of one size do not pack efficiently; they form compacts with large pores and a high volume percentage of porosity. Unless very uniform close packing was achieved during compacting and grain growth occurs during densification, such compacts will undergo a high percentage of shrinkage and yet will retain significant porosity. For very fine particles (500 Å) this may be acceptable and may result in very uniform properties. However, more commonly available powder has a range of particle sizes from submicron upward. Better overall packing can be achieved during compaction, but isolated pores due to bridging and agglomerates are usually quite large and result either in porosity or large grain size after sintering.

Liquid-Phase Sintering

Liquid-phase sintering involves the presence of a viscous liquid at the sintering temperature and is the primary densification mechanism for most silicate systems. Liquid-phase sintering occurs most readily when the liquid thoroughly wets the solid particles at the sintering temperature. The liquid in the narrow channels between the particles results in substantial capillary pressure, which aids densification by several mechanisms:

Rearranges the particles to achieve better packing
Increases the contact pressure between particles, which increases the rate of material transfer by solution/precipitation, creep and plastic deformation, vapor transport, and grain growth

The magnitude of capillary pressures produced by silicate liquids can be greater than 7 MPa (1000 psi). Smaller particles result in higher capillary pressure and also have higher surface energy due to the small radius of curvature and thus have more driving energy for densification than coarser particles. Materials requiring high strength and minimum porosity are generally processed from powders having an average particle size less than 5 μm and a surface area less than 5 m^2/g.

The rate of liquid-phase sintering is also strongly affected by temperature. For most compositions a small increase in temperature results in a substantial increase in the amount of liquid present. In some cases this can be beneficial by increasing the rate of densification. In other cases it can be detrimental by causing excessive grain growth (which reduces strength) or by allowing the part to slump and deform. The amount of liquid present at a selected temperature can be predicted with the use of phase equilibrium diagrams. The following discussion reviews briefly the principles of simple phase equilibrium diagrams and illustrates how they can be used to predict liquid content versus temperature, optimum compositions and temperatures for sintering, and the resulting high-temperature properties of the densified material.

Simple Binary Eutectic Diagram Figure 7.4 shows a simple binary eutectic phase equilibrium diagram. Considerable information is available from this diagram. With no calculations at all, the following information can be read directly:

The melting temperatures of the pure components A and B are T_A and T_B, respectively.

The melting temperature decreases as B is added to A or A is added to B.

The lowest melting temperature for this system occurs for a composition of 30% A and 70% B and is referred to as the binary eutectic T_e.

Regions are defined which tell whether solid, liquid, or a mixture is present for each temperature and composition when the system is in equilibrium.

With simple calculations considerable additional information is available from the phase equilibrium diagram as illustrated by Fig. 7.5 and the following examples.

EXAMPLE 7.1 What percentage liquid and solid are present in Fig. 7.5 for composition C_1 at temperature T_1? What are the compositions of the liquid and solid? We can use the simple lever rule to calculate the percentages:

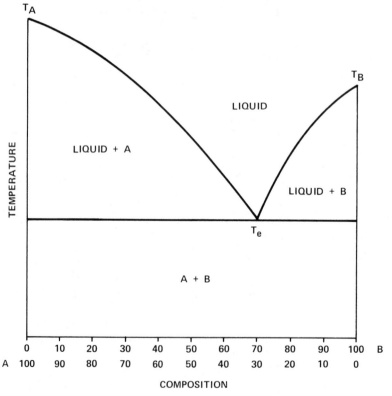

Figure 7.4 Simple binary eutectic diagram.

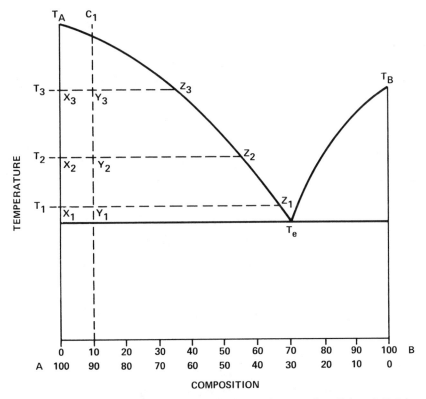

Figure 7.5 Binary eutectic diagram used with Examples 7.1 and 7.2 to describe methods of calculating percentages of solid and liquid for composition C_1 at various temperatures.

$$\% \text{ Solid} = \frac{X_1 Z_1 - X_1 Y_1}{X_1 Z_1} = \frac{67 - 10}{67} = 85^*$$

$$\% \text{ Liquid} = \frac{X_1 Z_1 - Y_1 Z_1}{X_1 Z_1} = \frac{67 - 57}{67} = 15$$

The composition of the liquid is defined by the intersection of the T_1 line and the liquidus (the curve that extends between T_e and T_A and thus separates

---

*Which is all A because the composition C_1 is in the liquid + A region of the diagram.

the region of all liquid from liquid + A) and is thus Z_1: Composition of liquid = Z_1 = 67% B and 33% A. We can now check to see if our calculations are correct by adding up all the A and B in the liquid and the solid to see if they equal our starting composition C_1:

	Percent in solid	Percent in liquid	Total
A	85	$33 \times 0.15 = 5$	90
B	0	$67 \times 0.15 = 10$	10

Composition C_1 is 90% A and 10% B, so our calculations are correct.

EXAMPLE 7.2 What percentages of liquid and solid are present for composition C_1, at temperatures T_2 and T_3? What are the corresponding liquid compositions?

$$\% \text{ Solid at } T_2 = \frac{X_2 Z_2 - X_2 Y_2}{X_2 Z_2} = \frac{55 - 10}{55} = 82\%$$

$$\% \text{ Liquid at } T_2 = \frac{X_2 Z_2 - Y_2 Z_2}{X_2 Z_2} = \frac{55 - 45}{55} = 18\%$$

Composition liquid at T_2 = Z_2 = 55% B and 45% A

$$\% \text{ Solid at } T_3 = \frac{X_3 Z_3 - X_3 Y_3}{X_3 Z_3} = \frac{35 - 10}{35} = 71\%$$

$$\% \text{ Liquid at } T_3 = \frac{X_3 Z_3 - Y_3 Z_3}{X_3 Z_3} = \frac{35 - 25}{35} = 29\%$$

Composition liquid at T_3 = Z_3 = 35% B and 65% A

Besides showing how useful calculations can be made with simple binary eutectic diagrams, the examples above illustrate some important considerations in liquid-phase sintering. First, small additions of a second component (B in this case) can result in a substantial quantity of liquid at a temperature well below the melting temperature of the pure component A. This aids in sintering, but if too much liquid is present can result in distorting, bloating, reaction with setter plates in the furnace, and insufficient high-temperature properties for the intended application. Therefore, the composition must be carefully selected and controlled to achieve a suitable amount of liquid at the selected sintering temperature.

Second, by comparing Examples 7.1 and 7.2, one can see that temperature has a strong effect on the amount of liquid present, and therefore must also be carefully selected and controlled. Another related consideration is that the liquid present at high temperature normally solidifies to a glass during cooling. Softening of this glass during subsequent usage of the part can result in creep and slow crack growth which will limit the operating temperature and life of the part. However, since one can estimate from the phase equilibrium diagrams the composition and percentage of this glass, one can design the starting composition and sintering process to satisfy the objectives of the application. Another option is to select a glass composition that can be crystallized by an appropriate heat treatment before the part is put in service. Even though the crystalline phase may have the same composition as the glass, it will not be subject to intermediate temperature softening typical of the glass.

One can also use Fig. 7.5 to understand better what happens during solidification or crystallization of a ceramic composition. Let us look again at composition C_1. At temperatures above the liquidus, only a liquid is present having a composition of 90% A and 10% B. As soon as the temperature cools down to the liquidus, crystals of component A begin to form in the liquid. The liquid correspondingly becomes increasingly richer in component B as the temperature is decreased and more A crystallizes. By the time the eutectic temperature T_e is reached, only about 14% liquid remains and is a mixture of 70% B and 30% A. This crystallizes at T_3 as a mixture of A and B crystals. Below T_e only a solid mixture of A and B is present.

The binary eutectic represents one of the simplest interactions of two components. The two components affect each other by reducing melting temperatures, but otherwise do not interact either chemically or structurally. Other systems are more complex and have solid solution between components or chemical interactions that result in intermediate compositions. These features are also included in the phase equilibrium diagram and can provide the engineer with very useful information on the nature of the materials he or she is dealing with.

Other Binary Phase Equilibrium Diagrams   Figure 7.6a shows a binary system with an intermediate congruently melting compound (AB). It is basically just two binary eutectic diagrams attached, each having its own eutectic (e_1 and e_2). The term congruent melting means that the compound is solid until it reaches its melting temperature. A, B, and AB all melt congruently. Figure 7.6b shows a binary system with an intermediate incongruently melting compound. In this case, the intermediate compound AB does not melt directly, but decomposes to a liquid plus B. In this case the eutectic is replaced by what is referred to as a peritectic (p_1 in Fig.

7.6b). Both congruently melting and incongruently melting system are common for ceramics.

Some compounds are stable only over a limited temperature range. This information is also available in the phase equilibrium diagram, as shown in Fig. 7.7. Such dissociating compounds are not common, but may be encountered and can affect sintering.

Much more important and widespread and critical to sintering is the effect of solid solution. Solid solution involves the ability of one atom or group of atoms to substitute into the crystal structure of another atom or group of atoms without resulting in a change in structure. Solid solution must be distinguished from mixtures. In a mixture two or more components are present, but they retain their own identity and crystal structure. In a solid solution one component is "dissolved" in the other component such that only one continuous crystallographic structure is detectable. This concept was discussed in Chap. 1 under bonding and crystal structure. Crystallographic substitutions take place most easily if two atoms are

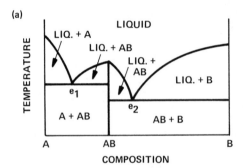

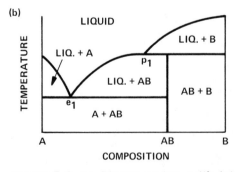

Figure 7.6 (a) Binary system with intermediate congruently melting compound. (b) Binary system with intermediate incongruently melting compound.

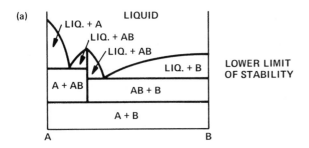

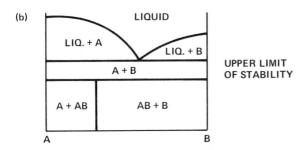

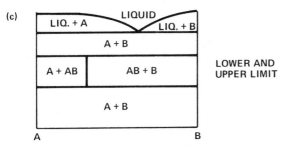

Figure 7.7 Representation of dissociating compounds on a phase equilibrium diagram.

similar in size and valence. For instance, Mg^{2+}, Co^{2+}, and Ni^{2+} are all similar in size and can readily replace each other in the cubic rock salt structure. In fact, each can replace the other up to 100% in the oxide, resulting in continuous solid solution. Figure 7.8a is the phase equilibrium diagram for the system MgO-NiO, showing complete solid solution between the MgO and NiO [4]. This is the most common type of continuous solid solution. Figure 7.8b and c show less common types in which either a

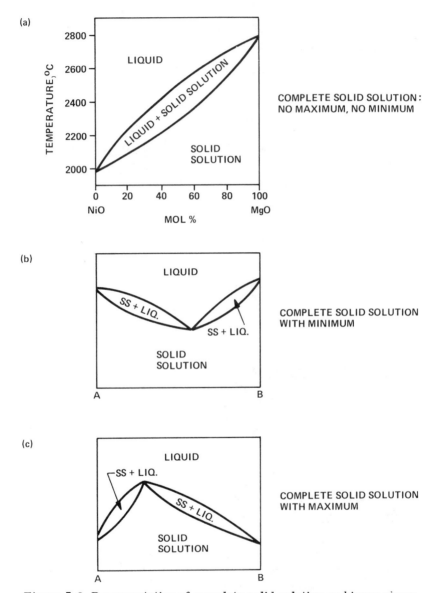

Figure 7.8 Representation of complete solid solution on binary phase equilibrium diagrams. (From Ref. 4.)

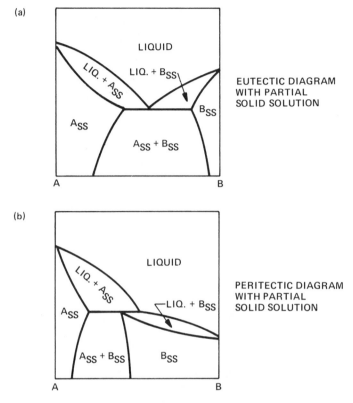

Figure 7.9 Binary phase equilibrium diagrams showing partial solid solution.

maximum or minimum is present. These maxima and minima are neither compounds nor eutectics, just limits in melting temperature for the solid solution.

Solid solution does not have to be complete between two different components and generally is not. Usually, one chemical component will have limited solid solubility in the other. The limits are determined by the similarity in the crystal structures and the size of ions or atoms. Figure 7.9 illustrates partial solid solution for a binary eutectic system and a binary peritectic system. For the eutectic system in Fig. 7.9a, component A can contain up to about 40% B in solid solution (at the eutectic temperature) and B can contain up to about 17% A. For the peritectic system shown, A can contain up to 20% B and B up to 60% A.

The presence of solid solution can have a dramatic effect on the sintering behavior of a ceramic system by significantly changing the percent liquid available at a given temperature. This is illustrated by comparing Fig. 7.5 with Fig. 7.10 and Examples 7.1 and 7.2 with Examples 7.3 and 7.4.

EXAMPLE 7.3   Figure 7.10 represents a binary eutectic system identical to the one shown in Fig. 7.5 (same melting temperatures of A and B and same eutectic temperature and composition), except that the system in Fig. 7.10 has partial solid solubility of both A in B and B in A. What are the percentages of liquid and solid for composition C_1, at the same temperatures evaluated in Examples 7.1 and 7.2 (T_1, T_2, and T_3)?

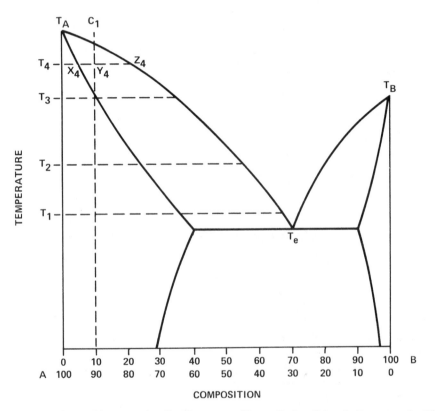

Figure 7.10 Binary eutectic diagram with partial solid solution, used with Examples 7.3 and 7.4 to describe methods of calculating percentages of solid and liquid versus composition and temperature and for comparison with similar calculations with the binary eutectic diagram without solid solution shown in Fig. 7.5.

T_1 = 100% solid, 0% liquid

T_2 = 100% solid, 0% liquid

T_3 = 100% solid, 0% liquid

It is obvious from this example that different sintering conditions would be required for the system with solid solution than for the system with no solid solution. The system without solid solution had 15% liquid at composition C_1 and temperature T_1 and could have probably been easily sintered by liquid-phase sintering at any temperature above the eutectic temperature. The system with solid solution would not have any liquid present under equilibrium conditions until slightly above temperature T_3 and thus would require a very high temperature for liquid-phase sintering. However, once sintered, the solid solution system would have much better high-temperature properties.

EXAMPLE 7.4 What are the percentages of solid and liquid for composition C_1 at temperature T_4 in Fig. 7.10? What are the compositions of the liquid and solid?

$$\% \text{ Solid} = \frac{X_4 Z_4 - X_4 Y_4}{X_4 Z_4} = \frac{(21 - 6) - (10 - 6)}{(21 - 6)} = 73\%$$

$$\% \text{ Liquid} = \frac{X_4 Z_4 - Y_4 Z_4}{X_4 Z_4} = \frac{15 - 11}{15} = 27\%$$

Composition of solid = X_4 = 94% A and 6% B

Composition of liquid = Z_4 = 79% A and 21% B

To check our calculations:

	Percent in solid	Percent in liquid	Total
A	$0.73 \times 94 = 69$	$0.27 \times 79 = 21$	90
B	$0.73 \times 6 = 4$	$0.27 \times 21 = 6$	10

The total adds up to 90% A and 10% B, which is the correct composition for C_1.

Another important material property which can affect sintering is polymorphic transformation. This information also shows up on phase equilibrium diagrams, as illustrated in Fig. 7.11 for a simple eutectic binary and a eutectic binary with solid solution of component B in each of the component A polymorphs.

Ternary Systems In many applications the ceramic is composed of or fabricated from three chemical components which can be represented by a

(a)

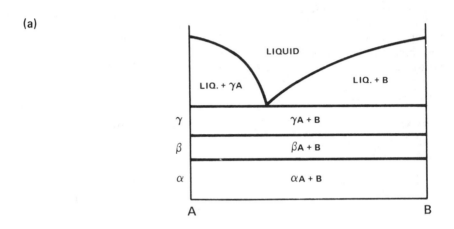

(b)

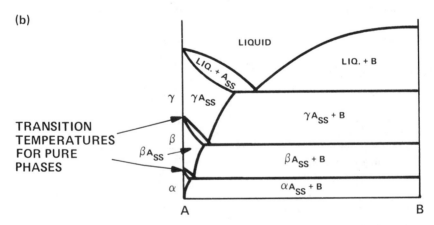

Figure 7.11 (a) Simple binary eutectic showing polymorphic transformations. (b) Binary eutectic with solid solution and polymorphic transformation.

ternary phase equilibrium systems. For instance, lithium aluminum silicate can be represented by the ternary system $LiO-Al_2O_3-SiO_2$ and cordierite by the system $MgO-Al_2O_3-SiO_2$.

The phase equilibrium diagrams for ternary systems can be useful in
predicting or understanding sintering behavior in the same fashion that the
binary system diagrams are. The same information regarding melting
temperatures, compositions, percent liquid or solid, solid solution, and
polymorphism are present.

A simple eutectic ternary diagram is illustrated in Fig. 7.12a. It is
the top view of the liquidus surface formed by placing the three binary
diagrams from the system end to end as shown in Fig. 7.12b. The corners
of the triangular diagram represent the melting temperatures of the three
pure components. The dashed lines show contours of constant melting
temperature and thus allow one to estimate the melting temperature for any
composition in the ternary system. The eutectics of each binary system
show up along the lines connecting the corners of the triangle and are
labeled e_1, e_2, and e_3 in Fig. 7.12. Melting temperatures typically decrease with further progression toward the interior of the ternary diagram,
reaching the minimum value at the ternary eutectic labeled E. If intermediate compounds are present in any of the binary systems, more than
one ternary eutectic will occur in the ternary system.

It is beyond the scope of this text to go into further detail on ternary
systems. For further study, the reader is referred to Refs. 5 through 7.
However, the key points being made in this brief discussion are that much
information relevant to sintering behavior is available in ternary diagrams,
and particularly that a third chemical component further reduces the temperature capability of the system. This latter point is especially important
in selection of starting powders and their processing. Impurities can act
as modifying chemical components which substantially reduce melting temperatures and result in more liquid phase at the sintering temperature than
the manufacturer predicted and not as good high-temperature properties as
the user requires.

Reactive Liquid Sintering

Reactive liquid sintering is also referred to as transient liquid sintering.
A liquid is present during sintering to provide the same types of densification driving forces as discussed for liquid-phase sintering, but the liquid
either changes composition or disappears as the sintering process progresses
or after it is completed. Since the liquid phase is consumed in the reaction.
the resulting material can have extremely good high-temperature properties
and in some cases can even be used at temperatures above the sintering
temperature.

One means of achieving reactive liquid sintering is to select starting
powders or additives that go through a series of chemical combinations or

(a)

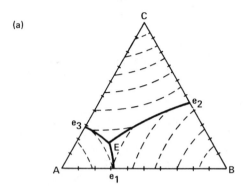

(b)

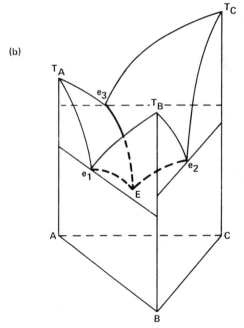

Figure 7.12 (a) Simple ternary triangular composition diagram showing
the three components, the binary eutectics, the ternary eutectic, and constant melting-temperature contours. (b) Three-dimensional view showing
how the ternary is related to its three constituent binaries.

reactions before the final stable compound is formed, with one or more of
the intermediate compounds being liquid and the final compound being solid.
Another is to use starting powders that will form a solid solution at equilibrium but will pass through a liquid stage before equilibrium is reached.
A third approach (which is not really reactive liquid sintering, but gives the
same results) is to liquid-phase sinter, cool to yield a glass at the grain
boundaries, and then heat treat to crystallize the glass.

Sintering Problems

A variety of conditions can result in improper sintering and have a deleterious effect on the material properties. Normally, the manufacturer will
detect these problems either during processing or during quality control
inspection. However, sometimes defective or inferior material is not detected by the manufacturer and is shipped to the user, where the defect
does not show up until the component fails prematurely in service. Under
these circumstances the source of the problem and a feasible solution must
be found quickly. The responsible engineer will have a distinct advantage
if he or she knows generally how the ceramic was processed and knows what
possible problems to look for. The following paragraphs describe some of
the problems that can occur during sintering and some of the artifacts in the
ceramic component that will help the engineer to identify the cause.

Warpage Warpage is a common problem and usually is detected before
the part is put into service. It increases reject rate and hence the cost per
part. It also can cause delays if it arises intermittently. Warpage usually
results from inadequate support during sintering or from density variations
in the greenware. The former can be corrected by shifting the orientation
of the part in the furnace or by supporting the part with saggers (refractory,
nonreactive ceramic pieces which restrict the component from deforming
during sintering). The latter can be corrected only by solving the problem
in an earlier processing step that caused the inhomogeneity. The two
sources of warpage can usually be distinguished from each other by dimensional inspection or by examination of a polished section of the microstructure. Warpage due to sagging will not show variations in thickness or
microstructure across the cross section, but warpage due to density variation will.

Overfiring Overfiring is another of the more common sintering problems
with ceramics. It can cause warpage, reaction with surrounding furnace
structures, bloating, or excessive grain growth. The first three are usually easy to detect visually. Excessive grain growth is more difficult to
detect during routine inspection and may require preparation of a polished
surface, etching to accentuate the grain structure, and examination by
reflected light microscopy.

However, the presence of large grains is readily visible on a fracture surface at low magnification and can provide the engineer with valuable insight into the cause of a component failure. As discussed in Chap. 3, an increase in grain size usually results in a decrease in strength. This is true even if only a portion of the grains have increased size. Sometimes overfiring results in exaggerated grain growth, where a few grains preferentially grow very large compared to other grains in the microstructure and compared to the optimum grain size required for the intended application.

Burn-Off of Binders   As discussed in Chap. 5, binders are often added to the ceramic powder prior to compaction. These are usually organic and can leave a carbon residue in the ceramic during sintering if the time/temperature/atmosphere parameters are not properly controlled. If large percentages of binders are present, such as in injection molded ceramics, the binder may have to be removed very slowly as a gas or liquid by thermal decomposition or capillary extraction. Too-rapid removal results in formation of cracks in the component.

The author once conducted experiments on sintering compacts of glass powders. A variety of binders were evaluated in glass compositions having a wide range in melting temperature. If the glass started to soften before the binder was completely burned off, discoloration would result. In one case the binder subsequently decomposed to produce a gas after the glass had partially sintered and expanded the glass into a porous foam having many times the volume of the original compact.

Proper binder removal is normally accomplished by slowly raising the temperature to a level at which the binder can volatilize, and holding at this temperature until the binder is gone. The temperature can then be safetly increased to the sintering temperature. However, if the temperature is increased before the binder has completely volatilized, the portion remaining will char and leave a residue of carbon.

In some materials, the carbon will be relatively inert, but in others it can cause severe chemical reactions during sintering. In one case the carbon resulted in localized reducing conditions in the core of a part causing bloating and severe dark discoloration. The surface of the same part was white and sintered properly.

Decomposition Reactions   Ceramics are frequently prepared using a different starting composition than the final composition. For instance, carbonates, sulfates, nitrates, or other salts are often used rather than the oxides, even though the final product is an oxide. There are a variety of reasons for doing this. The salts are often purer or more reactive or can be mixed more uniformly. However, during sintering the salt must decompose to the oxide and react with other constituents to form the desired final composition. If the salt does not decompose early enough, the component can be damaged by gas evolution. If the salt does not decompose completely,

an off-composition or inhomogeneous condition can result. The degree of
sensitivity of a component to this is dependent on the sintering temperature,
the time-temperature schedule, and the decomposition temperature and
kinetics of the salt. Problems are usually not encountered with hydrates
and nitrates because they have low decomposition temperatures. Carbonates tend to have higher decomposition temperatures, but usually do not
pose a problem if the sintering temperature is above 1000°C (1832°F).
Some sulfates do not completely decompose until 1200 to 1300°C (2200 to
2372°F).

Polymorphic Transformations Polymorphic transformations do not usually
cause problems during sintering, but can cause problems during cooldown
after sintering if a sudden volume change is involved. A good example is
ZrO_2. ZrO_2 is monoclinic from room temperature up to about 1000°C
(1832°F), at which point it transforms abruptly to a tetragonal form. No
problem occurs during heat-up because the individual ZrO_2 particles are
not constrained. However, after sintering the original particles are now
grains that are solidly bonded to adjacent grains and are thus restrained.
They are also randomly oriented. Now, when the component goes through
the transformation temperature, the grains are not free to move. Very
high internal stresses result at the grain boundaries and many cracks are
initiated, significantly weakening the material.

The problem with ZrO_2 has been resolved by controlled additions of
CaO, MgO, or Y_2O_3 which produce a cubic form of ZrO_2 that does not undergo a transformation.

Many ceramic materials undergo polymorphic transformations which
can decrease the strength of sintered material either during cooldown or
by further thermal cycling. Quartz and cristobalite forms of SiO_2 both
have displacive transformations accompanied by substantial volume change.

An engineer has several ways of determining if a material is susceptible
to damage by polymorphic transformation. First, he or she can determine
what crystallized compositions are present in the material by x-ray diffraction analysis. Then the engineer can look up phase equilibrium diagrams
for the material and its constituents. These diagrams will show if polymorphic phases are present. Figure 7.11 showed simplified examples
having three different polymorphs. Unfortunately, the equilibrium diagram
provides no information about the volume change during transformation.
This can be obtained by thermal expansion measurement.

7.2 MODIFIED DENSIFICATION PROCESSES

So far only conventional sintering where a powder compact is densified
under the influence of temperature has been discussed. Other processes
are also available which achieve densification, deposition of a solid phase,

or strong bonding. In this section attention is given to hot pressing, reaction sintering, chemical vapor deposition, liquid particle deposition, and cementicious bonding.

Hot Pressing

Hot pressing is analogous to sintering except that pressure and temperature are applied simultaneously [8]. Hot pressing is often referred to as pressure sintering. The application of pressure at the sintering temperature accelerates the kinetics of densification by increasing the contact stress between particles and by rearranging particle positions to improve packing. It has been established that the energy available for densification is increased by greater than a factor of 20 by the application of pressure during sintering, providing several processing and property advantages:

1. Reduces densification time
2. Can reduce densification temperature, often resulting in less grain growth than would occur with pressureless sintering
3. Minimizes residual porosity
4. Results in higher strength than can be achieved through pressureless sintering, due to the minimization of porosity and grain growth

The following sections describe how hot pressing is conducted, some of the unique properties, and some of the interesting applications.

Hot-Pressing Equipment   Figure 7.13 shows a simple schematic of a uniaxial hot-pressing setup. It consists of a furnace surrounding a high-temperature die with a press in line to apply a controlled load through the die piston. The type of furnace is dependent on the maximum temperature and uniformity of the hot zone required. Induction heating, with water-cooled copper coils and a graphite susceptor, is most commonly used and has a temperature capability greater than 2000°C (3632°F). The furnace must either be evacuated or back-filled with N_2, He, or Ar during operating to minimize oxidation of the graphite. Furnaces with graphite or other resistance heating elements can also be used for hot pressing. The author is not aware of gas-fired furnaces in use for hot pressing.

The source of pressure is usually a hydraulic press with a water-cooled platen attached to the ram. However, this does not provide adequate cooling to extend the ram into the furnace, so blocks of graphite or other refractory material are used. Obviously, the size of the press is dependent on the size of the part being hot pressed and the pressure required. Most hot pressing is done in the range 6.9 to 34.5 MPa (1000 to 5000 psi).

The die material is perhaps the most important element of the hot press. It must withstand the temperature, transient thermal stresses, high hot-pressing loads, and be chemically inert to the material being hot pressed.

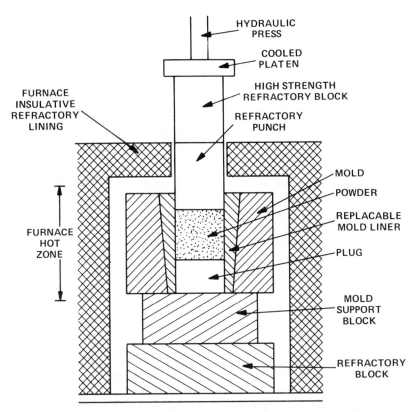

Figure 7.13 Schematic showing the essential elements of a hot press.

Graphite is the most widely used die and piston material. It has high-temperature capability, its strength increases with temperature, and it has low friction. It does not react with most materials and can be coated with a boundary layer such as boron nitride (BN) to prevent direct contact with material it might interact with. As with the graphite susceptor, though, graphite does oxidize and must be used under a protective environment.

Refractory metal dies such as molybdenum, tantalum, and the molybdenum alloy TZM have been used in limited cases. However, they are expensive, have high reactivity, and deform easily at high temperatures. Leipold [9] recommends TZM coated with $MoSi_2$ or a composite die consisting of a molybdenum jacket surrounding an Al_2O_3 liner. This latter approach takes advantage of the strength of the molybdenum and the abrasion resistance, creep resistance, and lower thermal expansion coefficient of the Al_2O_3.

Superalloys have also been used for hot-pressing dies for ceramics, but only at temperatures below 900°C (1652°F) and loads below 104 MPa (15,000 psi). A major problem with these materials is high thermal expansion. If the expansion of the die is higher than that of the material being hot pressed, the die will essentially shrink fit around the material during cooling and make ejection extremely difficult.

Ceramic dies, especially Al_2O_3 and SiC, are used frequently for hot pressing. They have reasonably low thermal expansion, are nonreactive, and have excellent resistance to galling and abrasion. Al_2O_3 can be used to approximately 1200°C (2192°F), dense SiC to about 1400°C (2552°F).

Reactivity is a special concern of die assemblies. Many of the carbides, ferrites, and other materials are very susceptible to property alteration through variations in stoichiometry and must be hot pressed under very controlled conditions. Graphite dies are often lined with a "wash" or spray coating of BN or Al_2O_3. Dies used for ferrites and some other electronic ceramics are often lined with ZrO_2 or Al_2O_3 powder.

The nature of the powder to be hot pressed is equally important to correct selection of the die material. The same type of fine-grained powders suitable for pressureless sintering are usually acceptable for hot pressing. In most cases a densification aid or a grain-growth inhibitor is added to achieve maximum density and minimum grain size. Table 7.3 summarizes sintering aids and grain-growth modifiers for a variety of oxides, carbides, nitrides, and borides. Specific references on the hot pressing of each of these ceramic materials are listed in Refs. 9 and 10.

Hot pressing is typically conducted at approximately half the absolute melting temperature of the material [9], which is usually a lower temperature than the material can be densified by pressureless sintering. Time at temperature is also reduced. The reduced temperature and time at temperature combine to minimize grain growth, thus providing better potential for improved strength.

Powder to be hot pressed can be loaded directly into the die or can be precompacted separately into a powder preform or compact that is then loaded into the die. Loading powder directly into the die is the most common procedure. However, the problems with this method are the difficulty in achieving uniformity and the pickup of contamination. Another disadvantage involves the low packing density of the loose powder and the resulting increase in the stack height to achieve a given part thickness. This reduces the number of parts that can be produced in a hot-pressing run and also increases the die wall friction. Increased die wall friction increases the variation of pressure within the compact and increases the chances for nonuniformity in the final part. The author encountered another problem with loose powder die loading while scaling up the hot pressing of Si_3N_4 from 7.6-cm (3-in.)-diameter development samples to 15.2-cm (6-in.)-diameter pilot production billets. The 7.6-cm (3-in.) samples had a strength of 897 MPa (130,000 psi) and were uniform across the diameter. The early

Table 7.3 Densification Aids and Grain-growth Modifiers

Material	Densification aids	Grain-growth inhibitors	Modifiers, enhancers
Al_2O_3	LiF	Mg, Zn, Ni, W, BN, ZrB_2	H_2, Ti, Mn
MgO	LiF, NaF	MgFe, Fe, Cr, Mo, Ni, BN	Mn, B
BeO	Li_2O	Graphite	--
Si_3N_4	MgO, Y_2O_3, $BeSiN_2$	--	--
SiC	B, Al_2O_3, Al	--	--
TaC, TiC, WC	Fe, Ni, Co, Mn	--	--
ZrB_2, TiB_2	Ni, Cr	--	--
ThO_2	F	Ca	--
ZrO_2	--	H_2, Cr, Ti, Ni, Mn	--
$BaTiO_3$	--	Ti, Ta, Al/Si/Ti	--
Y_2O_3	--	Th	--
$Pb(ZrTi)O_3$	--	Al, Fe, Ta, La	--

Source: Refs. 9 and 10.

243

15.2-cm (6-in.) billets were near theoretical density around the edges, but of decreased density in the interior. The overall density was within specification. The strength also appeared within specification since it was being measured on material sliced from the edge of the part. From all appearances the billets were of equivalent quality to the smaller development samples and were acceptable for delivery to a customer. However, when further testing was conducted, which included an evaluation of the billet interior, it was found that this region had density below specification and a strength of less than 690 MPa (100,000 psi). The source of the problem turned out to be a combination of loose powder loading and nonuniform temperature distribution. The loose powder had a very low thermal conductivity such that the edges in close proximity to the graphite die heated up faster than the interior and began to sinter. This physically shifted material from the center toward the edge and ultimately resulted in the density and strength gradient. The lesson is that flaws may result in a part that are not readily detectable, but if the engineer is aware of the mechanisms of processing and of some of the things that can go wrong, he or she will have a better chance of solving a problem that occurs or producing a quality control specification that will minimize such occurrences. The problem was resolved by precompacting the powder better and by modifying the time-temperature profile during hot pressing. Recurrence was prevented by initiating a more rigid density specification and strength certification procedure.

Unique Hot-Pressed Properties   Hot pressing permits achieving near-theoretical density and very fine grain structure, which result in optimization of strength. It also permits reduction of the amount of sintering aid required to obtain full density. This can result in orders-of-magnitude improvement in high-temperature properties such as creep and stress rupture life.

Table 7.4 compares the properties of several sintered and hot-pressed Si_3N_4 compositions developed during the past 10 years or currently under development. Similar differences exist between sintered and hot-pressed varieties of other materials such as Al_2O_3, SiC, spinel, and mullite.

Hot pressing can cause preferred orientation of the grain structure of some materials and result in different properties in different directions. This occurs predominantly when powders with a large aspect ratio such as rods or needles are used. It can also occur due to flattening of agglomerates or laminar distribution of porosity perpendicular to the direction of hot pressing. Figure 7.14 illustrates the strength variations measured for specimens cut from various orientations from a hot-pressed Si_3N_4 billet. The strength was greatest in the plane perpendicular to the direction of hot pressing. This was thought to be due to a combination of preferred orientation of Si_3N_4 grains and laminar density contours.

Table 7.4 Comparison of Densities and Strengths Achieved by Hot Pressing Versus Sintering

Material	Sintering aid	Density (% theoretical)	RT MOR[a]		1350°C MOR	
			MPa	kpsi	MPa	kpsi
Hot-pressed Si_3N_4[b]	5% MgO	>98	587	85	173	25
Sintered Si_3N_4[b]	5% MgO	~90	483	70	138	20
Hot-pressed Si_3N_4[c]	1% MgO	>99	952	138	414	60
Sintered Si_3N_4[d]	$BeSiN_2$ + SiO_2	>99	560	81	--	--
Sintered Si_3N_4[e]	6% Y_2O_3	~98	587	85	414	60
Hot-pressed Si_3N_4[e]	13% Y_2O_3	>99	897	130	669	97

[a]Room-temperature modulus of rupture.
[b]G. R. Terwilliger, J. Am. Ceram. Soc. 57(1), 48–49 (1974).
[c]D. W. Richerson, Am. Ceram. Soc. Bull. 52, 560–562, 569 (1973).
[d]C. D. Greskovich and J. A. Palm, U.S. DOE Conf.–791082, 1979, pp. 254–262.
[e]Data from C. L. Quackenbush, GTE Laboratories, Waltham, Mass.

Strength test specimens are normally cut from the plane perpendicular to the hot-pressing direction. This is usually the strongest direction (if anisotropy is present) and may give the engineer false confidence in the material. The engineer should be aware that the strength and other properties in the other directions may be inferior and adjust the material qualification testing accordingly.

Hot-Pressing Limitations   The major limitation of hot pressing is shape capability. Flat plates, blocks, or cylinders are relatively easy to hot press. Long cylinders, nonuniform cross sections, and intricate or contoured shapes are difficult and often impossible by conventional uniaxial techniques. Figure 7.15 and the following paragraphs describe the nature of the problem.

The starting powder goes into the die as a relatively uniform stack of powder or as a uniform preform. During densification the powder or preform will compact in the axial direction of pressure application until the porosity has been eliminated and near-theoretical density achieved. The amount of compaction required to go from the loose powder or preform to the pore-free part is referred to as the compaction ratio. The compaction ratio for a well-compacted preform usually ranges from 2:1 to 3:1 and can be even higher for loose powder having a very fine particle size. For instance, one batch of Si_3N_4 powder had a compaction ratio of 8:1.

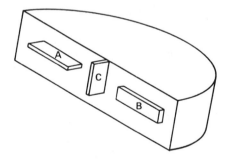

	A	B	C
AVERAGE 4-PT. BEND STRENGTH	876 MPa (127 Kpsi)	762 MPa (110 Kpsi)	713 MPa (103 Kpsi)
STANDARD DEVIATION	105 MPa (15.3 Kpsi)	142 MPa (20.6 Kpsi)	92 MPa (13.3 Kpsi)

Figure 7.14 Variations in the strength of hot-pressed Si_3N_4 as a function of direction.

(a)

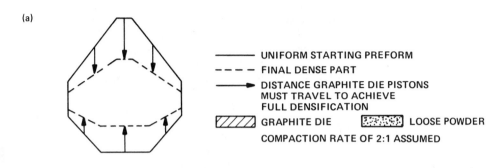

UNIFORM STARTING PREFORM
FINAL DENSE PART
DISTANCE GRAPHITE DIE PISTONS
MUST TRAVEL TO ACHIEVE
FULL DENSIFICATION
GRAPHITE DIE   LOOSE POWDER
COMPACTION RATE OF 2:1 ASSUMED

(b)   (c)

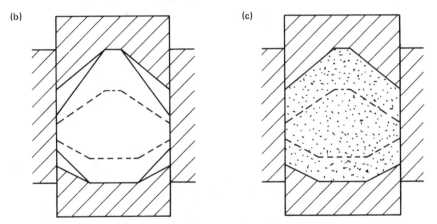

Figure 7.15 Problems associated with complex shape hot pressing.
(a) Assuming a uniform starting preform or powder stack, the preform will
have to shrink different amounts to achieve uniform final density and the
required shape. (b) A preform with the correct powder distribution has a
different shape than the die. (c) Loose powder fill requires a powder redistribution during hot pressing.

Figure 7.15a illustrates the shape of a preform having a compaction
ratio of 2:1 that would be required to make a fully dense part of an arbitrary
nonuniform cross section. The shape of the preform is different than the
final shape and the required movement of the graphite die punches is greater
for thick sections than for thin sections. For instance, in the example in
Fig. 7.15, the total shrinkage in the thick section of the part is four times
greater than the shrinkage in the thin section, even though the percentage
is the same in each case. And this is only for a minimal compaction ratio
of 2:1. The shrinkage difference is greater for higher compaction ratios.

How can one design rigid graphite tooling to accommodate the differences in distance and still achieve the required shape? Usually, it cannot be done. One either has to make the preform a different shape than the graphite tooling (as shown in Fig. 7.15b) and hope the preform does not break up prematurely and alter the powder distribution, or one has to load loose powder to fill the die cavity (as shown in Fig. 7.15c) and hope that the powder will redistribute to the required distribution during hot pressing.

The latter approach has been used with success for some materials and shapes and is worth trying because the tooling is usually not prohibitively expensive. The former approach requires two sets of tooling and has not yet been developed, but may also be worth considering.

Another approach to uniaxial hot pressing of shapes having nonuniform cross section is the use of nonrigid tooling. Two concepts are illustrated in Fig. 7.16. The first is referred to as pseudoisostatic hot pressing. A preform is prepared by cold pressing, slip casting, or other approach. The dimensions are selected (with knowledge of the compaction ratio for the specific material) such that the required shape will result after densification. The preform is embedded in loose powder in the hot-press die cavity. The loose powder is selected so that it will not densify and will not chemically react with the preform being hot pressed. Hexagonal boron nitride and graphite powders have both been used successfully and work especially well because of their self-lubricating character and excellent chemical stability. During hot pressing the loose powder transmits pressure from the die punches to the preform. A true isostatic pressure distribution is not achieved, but enough pressure is apparently transmitted to the preform to allow densification. Most shapes hot pressed by this approach have achieved near-theoretical density but have undergone some distortion during pressing. However, once the distortions are accounted for in the preform, near net shape can probably be reproducibly achieved.

Figure 7.16b illustrates the second nonrigid tooling approach. In this case three preforms having the same compaction ratio are required. The center preform will densify to become the required shaped part. The other two preforms are simply conforming layers between the flat die punch surfaces and the contoured part surface. A nonreactive boundary layer such as boron nitride is placed between the preforms so that they can be separated after hot pressing. This approach simulates hot pressing of a flat plate of uniform cross section and results in uniform density and properties in the finished part.

Hot Isostatic Pressing Recently, substantial progress has been made in hot isostatic pressing (HIP) of ceramics. Apparatus for HIP consists of a high-temperature furnace enclosed in a water-cooled autoclave capable of withstanding internal gas pressures up to about 32,000 kg/mm^2 (45,000 psi) and providing a uniform hot-zone temperature up to about 2000°C (3632°F) [11]. Pressurization gas is either argon or helium. Heating is usually by molybdenum or graphite resistance-heated elements.

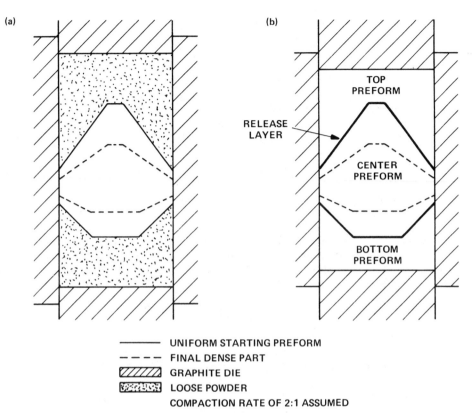

(a)

(b)

TOP
PREFORM

RELEASE
LAYER

CENTER
PREFORM

BOTTOM
PREFORM

——————— UNIFORM STARTING PREFORM
– – – – FINAL DENSE PART
///// GRAPHITE DIE
LOOSE POWDER
COMPACTION RATE OF 2:1 ASSUMED

Figure 7.16 Nonrigid tooling approaches for hot pressing. (a) Pseudoisostatic approach. (b) Multiple preforms simulating a flat plate approach.

To achieve densification of a ceramic preform, the preform must first
be evacuated and then sealed in a gas-impermeable envelope. If any high-
pressure gas leaks into the preform, pressure is equalized and the preform
cannot hot press. The earliest HIP studies encapsulated the ceramic in
tantalum or other refractory metal, depending on the temperature required
for densification. However, this severely limited the shape capability.
Current studies are directed toward the use of glass encapsulation. The
glass can be applied as a preformed envelope, which is sealed around the
part under vacuum and then collapsed at high temperature to conform to
the ceramic preform shape. This has worked for relatively simple shapes.
ASEA in Sweden and Battelle in the United States have developed techniques
to apply the glass as a particulate coating. The preform with this coating
is then placed in the HIP autoclave and evacuated. The temperature is
raised until the glass softens and forms a continuous layer on the surface.
The pressure and temperature are then increased to the levels required to

accomplish densification of the ceramic preform. Battelle has demonstrated that individual gas turbine rotor blades of Si_3N_4 can be fabricated by
this approach. ASEA has fabricated small [~12.7 cm (5 in.)] integral axial
rotors of Si_3N_4 by this approach.

HIP has the potential of resolving some of the major limitations of uniaxial hot pressing. It makes possible net shape forming because the pressure is equally applied from all directions. This also results in greater
material uniformity by eliminating die wall friction effects and preferred
orientation, resulting in higher strength and Weibull modulus. Also, much
higher pressures and temperatures can be used, making possible more
complete densification and greater flexibility in selection of composition.
For instance, the higher pressure and temperature may permit densification of compositions containing less sintering aid and having dramatically
improved stress rupture life and oxidation resistance.

HIP has been used to improve the strength and wear resistance of some
MgZn and NiZn ferrites, yttrium iron garnet, and $BaTiO_3$, especially for
applications such as magnetic recording heads. The ceramic is first sintered to closed porosity and is then further densified by HIP without a requirement for encapsulation [12].

Reaction Sintering

Unique processes have been developed in England for achieving moderately
strong Si_3N_4 and SiC shapes without undergoing the large shrinkage of conventional sintering [13,14]. These processes are referred to as reaction
sintering or reaction bonding.

Reaction-Sintered Silicon Nitride   Reaction-sintered Si_3N_4 is fabricated
from silicon powder. The silicon powder is processed to the desired particle size distribution and formed into the required shape by pressing, slip
casting, injection molding, or other suitable process. The compacted Si
shape is then placed in a furnace under a nitrogen or mixed nitrogen/hydro-
gen or nitrogen/helium atmosphere and heated initially to about 1200 to
1250°C (2200 to 2282°F). The nitrogen permeates the porous Si compact
and begins to react with the Si to form Si_3N_4. Initially, α-Si_3N_4 fibers
grow from the Si particles into the pores. As the reaction progresses, the
temperature is slowly raised to approximately 1400°C (2552°F), near the
melting temperature of Si. As the temperature increases, the reaction
rate increases and primarily β-Si_3N_4 is formed. Great care is necessary
in controlling the rate of temperature increase and nitrogen flow. The
reaction of N_2 and Si is exothermic and, if allowed to proceed too fast, will
cause the silicon to melt and exude out of the surface of the part. A typical
nitriding cycle in which the exotherm is controlled and no exuding occurs
is on the order of 7 to 12 days, depending on the volume of material in the
furnace and the green density of the starting Si compacts.

Approximately 60% weight gain occurs during nitriding, but less than 0.1% dimensional change. This makes possible excellent dimensional control. Bulk densities up to 2.8 g/cm^3 (0.101 lb/in.^{3}) have been achieved (compared to a theoretical density for Si$_3$N$_4$ of about 3.2 g/cm^3).

The earliest reaction bonded Si$_3$N$_4$ had a density of about 2.2 g/cm^3 and a strength under 138 MPa (20,000 psi). Current 2.8-g/cm^3 material has 4-point flexure strength in the range 345 MPa (50,000 psi).

Another advantage of the reaction-sintered Si$_3$N$_4$ is its creep resistance. No sintering aids are added to achieve densification, so no glassy grain boundary phases are present. Strength is retained to temperatures greater than 1400°C (2552°F) and the creep rate is very low. The strength versus temperature and creep rate were compared with other materials in Figs. 3.11 and 4.6.

The primary disadvantage of reaction-bonded Si$_3$N$_4$ is its porosity. The porosity is interconnected and can result in internal oxidation and accelerated surface oxidation at high temperature. The internal oxidation appears to affect the thermal stability of a component. For instance, Ford Motor Company observed that experimental turbine engine parts fractured when the weight gain due to oxidation reached about 2%. The mechanism was not determined, but could have been associated with internal stresses induced by the thermal expansion mismatch between the Si$_3$N$_4$ and the cristobalite formed during oxidation.

Reaction-sintered Si$_3$N$_4$ has a relatively low elastic modulus and coefficient of thermal expansion and a relatively high thermal conductivity (considering its porosity). These properties, combined with a moderately high strength give reaction-sintered Si$_3$N$_4$ good thermal shock resistance and make it a feasible candidate for such applications as welding nozzle tips, gas turbine static structure components, and aluminum metal processing.

Reaction-Sintered Silicon Carbide   Reaction-sintered SiC is processed from an intimate mixture of SiC powder and carbon. This mixture is formed into the desired shape and exposed at high temperature to molten or vapor-phase Si. The silicon reacts with the carbon to form in situ SiC, which bonds the original SiC particles together. The remaining pores are filled with metallic Si. The resulting material is basically a nonporous Si-SiC composite and can have a broad range of strength and elastic modulus, depending on the particle size distribution and the percent Si. Table 7.5 summarizes data for several reaction-sintered SiC materials [15,16].

Reaction-sintered SiC materials have a relatively flat strength versus temperature curve nearly up to the melting temperature of silicon, at which point the strength drops off rapidly. A typical strength versus temperature curve is shown in Fig. 7.17 compared to reaction-bonded Si$_3$N$_4$, hot-pressed Si$_3$N$_4$, and sintered SiC [15].

Reaction-sintered SiC has similar advantages to reaction-bonded Si$_3$N$_4$ for complex shape fabrication; it undergoes dimensional change of less than

1% during densification. The initial shape can be formed by casting, plastic molding, pressing, extrusion, and any of the other processes applicable to ceramics. It is especially suitable for the plastic processes such as extrusion and compression molding. The plasticizer can be a thermosetting resin such as a phenolic. Instead of having to remove the plasticizer after molding, as is required with other ceramics, the plasticizer is simply charred to provide the carbon source for reaction with the silicon.

Another interesting method of fabricating reaction-sintered SiC is to start with woven carbon fibers or felt, as reported by Hillig et al. [16]. Laminae of carbon fiber weave are laid up much the same as fiberglass to form the desired shape. This can then be reacted with molten silicon to form a composite of SiC-Si or C-SiC-Si. By controlling the type of fibers and the weave, a complete range of composites with varying SiC-to-Si ratios can be engineered. Ones with high Si have low elastic modulus ($\sim$30 $\times$ 10^6) and relatively low strength (<30,000 psi); ones with high SiC have high elastic modulus (>50 $\times$ 10^6 psi) and high strength ($\sim$70,000 psi). The range of composites can be further expanded by pitch impregnation or deposition of pyrolytic carbon or glassy carbon prior to siliconizing.

Vapor Deposition

Vapor deposition is a generic term that can refer to a variety of techniques that deposit essentially nonporous ceramic coatings on a substrate. Chemical

Table 7.5 Properties of Reaction Sintered SiC Materials Illustrating Variations Achievable by Varying Microstructure and Silicon Content

Material	Volume % silicon	Young's modulus, E (psi)	Flexure strength, σ (psi)
NC-435[a]	$\sim$20	50.7 $\times$ 10^6	57,200
Refel[b]	--	57.5 $\times$ 10^6	44,900
Type TH[c]	15-20	57 $\times$ 10^6	70,000
Type THL[c]	55-60	44 $\times$ 10^6	48,000
Type F[c]	75-80	29 $\times$ 10^6	30,000

[a] Norton Company, Worcester, Mass.
[b] British Nuclear Fuels, Ltd., U.K.
[c] General Electric Company, Schenectady, N.Y.
Source: Refs. 15 and 16.

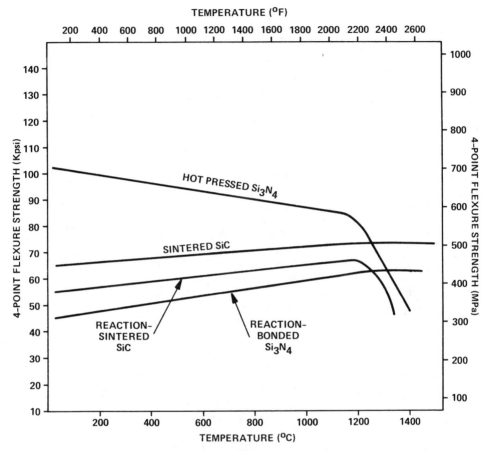

Figure 7.17 Strength of reaction-sintered SiC compared to other Si_3N_4 and SiC ceramics. (From Ref. 15.)

vapor deposition (CVD) and sputtering are perhaps the most widely used and will be discussed briefly.

CVD is typically accomplished by heating the part to be coated in a vacuum chamber and passing a controlled gas or gas mixture over the part. The gas mixture is selected such that it will react or decompose when it comes in contact with the part preheated to a specific temperature. Deposition rates are usually less than 250 μm (0.010 in.) per hour and careful control is required to obtain a uniform coating. However, the coatings are very fine grained and impervious and are usually of higher purity and hardness than can be achieved by other ceramic fabrication processes.

Figure 7.18b shows a CVD Si_3N_4 coating on reaction-bonded Si_3N_4. Note the difference in porosity between the reaction-bonded Si_3N_4 and the CVD coating. Also note the columnar structure of the coating, with the grains oriented perpendicular to the surface. This is typical of CVD coatings and is usually not desirable. Figure 7.17a shows a CNTD SiC coating on hot-pressed SiC. Note the extremely fine grain size and the lack of columnar growth. CNTD stands for Controlled Nucleation Thermochemical Deposition and was developed by Chemetal, Inc., specifically to eliminate the deficiencies of columnar growth. Tungsten carbide with strength approaching 3450 MPa (500,000 psi) has been deposited by the CNTD approach.

A second important technique for depositing ceramic coatings is sputtering. The part to be coated is placed in an evacuated chamber in close proximity to a flat plate of the coating material. This flat plate is called the target and is bombarded by a beam of electrons. The electrons essentially knock atoms off the target onto the surface of the part facing the target. Only the portion of the part directly exposed to the target gets coated. Coating rates are very slow and only thin coatings can be produced effectively. The major advantages are that the purity of the coating can be controlled and that the part is not heated during coating. An application where this might be important is in a miniaturized electrical device where a ceramic coating would provide protection and electrical insulation.

Molten Particle Deposition [17]

Molten particle deposition is most commonly referred to as flame spray or plasma spray. It consists of impacting small molten particles of ceramic against the surface to be coated. Almost any oxide, carbide, boride, nitride, or silicide that does not sublime or decompose can be applied by molten particle techniques. Coatings are most often applied, but free-standing parts can also be made by using a removable mandrel or form.

The first widely used molten particle approach was the oxyacetylene powder gun, more frequently referred to as the flame-spray gun. Ceramic powder is aspirated into the oxyacetylene flame and melts. The molten particles exit the gun through a nozzle and strike the substrate to be coated at a velocity of about 45 m/sec (150 ft/sec). By moving either the substrate or the gun, a uniform coating can be built up having approximately 10 to 15 vol % porosity and a surface finish of 150 to 300 μin. rms.

A similar approach is the oxyacetylene rod gun. Instead of using ceramic powder, a sintered rod of the coating material is fed into the oxyacetylene flame. Molten ceramic at the tip of the rod is carried to the substrate by bursts of air traveling at about 180 m/sec (600 ft/sec). Surface finishes are similar to those produced by the powder gun, but porosity is usually lower (6 to 10%) because the particles are completely molten and the impact velocity is higher.

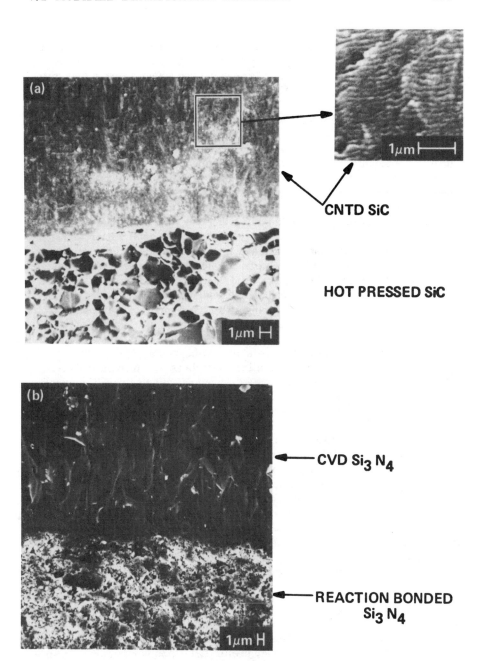

Figure 7.18 (a) CNTD SiC coating on hot-pressed SiC. (b) CVD Si_3N_4 coating on reaction-bonded Si_3N_4. (Courtesy of the Garrett Turbine Engine Company, Division of Garrett Corporation, Phoenix, Ariz.)

The oxyacetylene guns are widely used. Another widely used approach
is the arc-plasma gun. A high-intensity direct-current arc is maintained
in a chamber. Helium or argon is passed through the chamber, heated by
the arc, and expelled through a water-cooled copper nozzle as a high-temperature high-velocity plasma. Ceramic particles are injected into the
plasma, where they are melted and directed against the substrate. Velocities as high as 450 m/sec (1500 ft/sec) have been obtained, yielding coatings with porosity as low as 3% and surfaces with a finish in the range 75 to
125 μin. rms. The major difficulty of the arc-plasma gun is the temperature of the plasma. Ceramic substrates may have to be preheated to avoid
thermal shock damage and metal substrates may have to be cooled to avoid
melting.

Deposition rates for molten particle spray are much higher than for
chemical vapor deposition, in the kilograms per hour range rather than in
grams or milligrams per hour. However, for some applications this is not
enough. One technique has been developed which has demonstrated 2 to 4
kg/min and is projected to scale up to 200 kg/min. The ceramic powder is
mixed with fuel oil to form a slurry and is burned in oxygen in a water-
cooled gun. Al_2O_3, mullite, and SiO_2 have been successfully sprayed by
this method to form refractory linings for high-temperature furnaces such
as the oxygen converters in steel mills.

Molten particle spray techniques have been used extensively to deposit
wear-resistant and chemically-resistant coatings on a wide variety of metal
and ceramic products. One interesting example is the spraying of chromium
oxide (Cr_2O_3) on the propeller shafts of large seagoing ships. The chromium
oxide greatly reduces erosive wear, provides a good surface to seal against
(after surface grinding to achieve a suitable surface finish), and inhibits
seawater corrosion.

Coatings are also applied to provide thermal protection. Stabilized
ZrO_2 has a very low thermal conductivity and emissivity and is applied to
stainless steel and superalloy parts as a thermal barrier coating. Although
other oxide ceramics have similar thermal properties, ZrO_2 was selected
because it has a coefficient of expansion similar to the metals. This is one
case for a ceramic where a high coefficient of thermal expansion is beneficial.

An important advantage of molten particle spray techniques is that a
wide range of size and shape of substrates can be coated. On-site repairs
are even feasible. Flame spray has great versatility and an engineer should
be aware of the various techniques, sources, and capabilities.

Cementicious Bonding

All of the densification processes discussed so far involve high-temperature operations to achieve a strong, useful part. Another important approach
is cementicious bonding, in which an inorganic ceramic adhesive bonds
together an aggregate of ceramic particles. The adhesion results primarily

from hydrogen bonding. The resulting materials are not of high strength, but are adequate for many wear-resistance, building, and refractory applications. Major advantages are that the cement can be poured, troweled, or gunned into place, has little dimensional change during setting, and can be repaired on-site.

Many different cements have been developed, ranging from common concrete to very high temperature furnace linings. The cements can be classified according to the mechanism of bind formation: hydraulic bonds, reaction bonds, and precipitation bonds [18].

Hydraulic Cements Hydraulic cements set by interaction with water. The most common hydraulic cement is portland cement, which is primarily an anhydrous calcium silicate. It is slightly soluble in water and sets by a combination of solution-precipitation and reaction with water to form a hydrated composition. The reaction is exothermic and care must be taken to ensure that adequate water is initially present and that the heat of reaction does not dry out the cement prematurely. This explains why a competent cement contractor keeps the surface of freshly laid concrete damp.

The ratio of water to cement in the initial mix has a primary effect on the final strength of the cement. As long as adequate water is added for hydration and for workability, the lower the water-to-cement ratio, the higher the resulting strength.

Calcium aluminate cements are also hydraulic setting, but have much higher temperature capability than portland cement and are thus used for refractory applications such as furnace linings. Other hydraulic cements include natural lime-silica cements, barium silicate and barium aluminate cements, slag cements, and some ferrites.

Similar to hydraulic cements are gypsum cements such as plaster of paris and Keene's cement. They set by a hydration reaction but are much more soluble than the hydraulic cements and recrystallize to a highly crystalline structure that has little adhesion. Rather than being used for bonding aggregates, plaster of paris is used alone to make wallboard, plaster molds for slip casting and metal casting, and decorative knick-knacks and statuettes.

Reaction Cements Reaction cements are formed by a chemical reaction between two constituents other than water. One of the most common is monoaluminum phosphate, formed by the reaction of aluminum oxide powder with phosphoric acid. This cement sets in air at room temperature, but is usable over a very broad temperature range. Above 310°C (500°F) the cement is dehydrated to form the metaphosphate, which is then stable up to very high temperature. In fact, the strength increases as the temperature is increased.

Most metal oxides form phosphate cements when reacted with phosphoric acid. In addition to Al_2O_3, the following metal oxides have been shown by

Kingery [19] to form phosphate cements: BeO, CdO, Fe_2O_3, Y_2O_3, ZnO, ZrO_2, CrO_3, CuO, CO, ThO_2, V_2O_5, and SnO. Compositions based on ZnO have been widely used for dental cements.

Most of the ceramic cements are porous and brittle and require an aggregate to provide durability. An interesting reaction cement that has some resiliency is magnesium oxychloride cement. It has been used for floors, building facings, signs, and a variety of other applications.

Precipitation Cements Precipitation cements are primarily gels formed by precipitating colloidal suspensions by adjusting the pH or ion concentration. Sodium silicate is perhaps the best known and most widely used precipitation cement. It is inexpensive and its composition can be controlled to achieve setting by drying, heating, or chemical means. Chemical setting is achieved by addition of acid salts, especially sodium silicofluoride. The setting rate can be controlled by the amount and grain size of the silicofluoride and the amount of water in the cement. Organic materials such as esters (ethyl acetate) and alcohols also precipitate alkali silicate cements.

Another important precipitation cement is prepared from ethyl orthosilicate. It is precipitated by dehydration. Reaction can be accelerated by addition of magnesia or by heating.

Precipitation cements are used extensively in applications where acid resistance is critical. They are also used for some abrasion resistance applications, for bonding low-temperature refractories, and for forming of foundry molds for metal casting.

Development of ceramic cements is very competitive and most compositions and specific processing approaches are proprietary. A variety of cements are available commercially and can be purchased ready to mix or ready to use. Sources can be obtained by referring to the Ceramic Products Issue included each year in the Bulletin of the American Ceramic Society. More detailed technical information can be found in Ref. 18.

REFERENCES

1. H. G. Muller, Zur Natur der Rekristallisationsvorgange, Z. Phys. 96, 279 (1935).
2. W. D. Kingery, H. K. Bowen, and D. R. Uhlmann, Introduction to Ceramics, 2nd ed., John Wiley & Sons, Inc., New York, 1976, p. 187.
3. W. D. Kingery, H. K. Bowen, and D. R. Uhlmann, Introduction to Ceramics, 2nd ed., John Wiley & Sons, Inc., New York, 1976, p. 476.
4. H. V. Wartenberg and E. Prophet, Schmelzdiagramme Hochstfeuerfester Oxyde, Z. Anorg. Allg. Chem. 208, 379 (1932).
5. E. M. Levin, C. R. Robbins, and H. F. McMurdie, Phase Diagrams for Ceramists, American Ceramic Society, Columbus, Ohio, 1964.

6. A. Findlay, A. N. Campbell, and N. O. Smith, The Phase Rule and Its Applications, 9th ed., Dover Publications, Inc., New York, 1951.

7. F. E. W. Wetmore and D. J. LeRoy, Principles of Phase Equilibrium, McGraw-Hill Book Company, New York, 1951.

8. T. Vasilos and R. M. Spriggs, Proc. Br. Ceram. Soc. 3, 195-221 (1967).

9. M. H. Leipold, in Treatise on Materials Science and Technology, Vol. 9: Ceramic Fabrication Processes (F. F. Y. Wang, ed.), Academic Press, Inc., New York, 1976, pp. 95-134.

10. R. J. Brook, in Treatise on Materials Science and Technology, Vol. 9: Ceramic Fabrication Processes (F. F. Y. Wang, ed.), Academic Press, Inc., New York, 1976, p. 361.

11. E. S. Hodge, Mater. Des. Eng. 61, 92-97 (1965).

12. K. H. Hardtl, Gas isostatic hot pressing without molds, Am. Ceram. Soc. Bull. 54, 201 (1975).

13. N. L. Parr, Silicon nitride, A new ceramic for high temperature engineering and other applications, Research (Lon.) 13, 261-269 (1960).

14. P. Popper, The preparation of dense self-bonded silicon carbide, in Special Ceramics (P. Popper, ed.), Academic Press, Inc., New York, 1960, pp. 209-219.

15. D. C. Larsen, Property Screening and Evaluation of Ceramic Turbine Engine Materials, Final Report for period July 1, 1975-Aug. 1, 1979, AFML-TR-79-4188, Oct. 1979.

16. W. B. Hillig, R. L. Mehan, C. R. Morelock, V. J. DeCarlo, and W. Laskow, Silicon/silicon carbide composites, Bull. Amer. Ceram. Soc. 54[12], 1054-1056 (1975).

17. Ceramic Processing, Natl. Acad. Sci. Publ. 1576, Washington, D.C., 1968, pp. 105-111.

18. J. F. Wygant, Cementicious bonding in ceramic fabrication, in Ceramic Fabrication Processes (W. D. Kingery, ed.), MIT Press, Cambridge, Mass., 1963, pp. 171-188.

19. W. D. Kingery, Fundamental study of phosphate bonding in refractories, J. Amer. Ceram. Soc. 33(8), 239-250 (1950).

8
Final Machining

Some ceramic parts can be fabricated to net shape by the methods described in Chap. 7. However, more frequently, machining of some of the surfaces is required to meet dimensional tolerances, achieve improved surface finish, or remove surface flaws. This machining can represent a significant portion of the cost of fabrication and thus should be minimized and conducted as efficiently as possible.

8.1 MECHANISMS OF MATERIAL REMOVAL

Ceramic materials are difficult and expensive to machine due to their high hardness and brittle nature. Machining must be done carefully to avoid brittle fracture of the component. Most ceramics cannot be successfully machined with the type of cutting tools used for metal because these tools are either not hard enough to cut the ceramic or because they apply too great a local tensile load and cause fracture. The tool must have a higher hardness than the ceramic being machined and must be of a configuration that removes surface stock without overstressing the component.

Ceramic material can be removed by mechanical, thermal, or chemical action. Mechanical approaches are used most commonly and are discussed first. They can be divided into three categories: mounted abrasive, free abrasive, and impact.

Mounted Abrasive Machining

Mounted abrasive tools consist of small, hard, abrasive particles bonded to or immersed in a softer matrix. The abrasive particles can be SiC, Al_2O_3, $Al_2O_3-ZrO_2$, or other hard ceramic material, and the matrix can be rubber, organic resin, glass, or a crystalline ceramic composition softer than the abrasive particles. Good examples are the wide variety of grinding wheels used extensively in home workshops and industry. For machining very hard ceramics such as Al_2O_3, Si_3N_4, and SiC, diamond is usually the most efficient abrasive, mounted in a matrix of soft metal or organic resin.

Mounted abrasive tools can be made in a wide variety of configurations and compositions. Coarse abrasives are used for rough machining, where rapid stock removal is desired. Finer abrasives are used for final machining, where close tolerances and smooth surface finishes are required.

Stock removal is achieved by moving the tool in relation to the ceramic workpiece while simultaneously applying pressure. The abrasive particles are small and irregular in shape such that a sharp corner of the particle is usually in contact with the ceramic. This small contact area produces high localized stress concentration and the particle plows a microscopic groove across the surface of the ceramic. The larger the abrasive, the larger the groove and the greater the depth of damage in the ceramic. This will be discussed in more detail later in a section on effects of machining on material strength, including suggested procedures to minimize strength reduction.

Free Abrasive Machining

Free abrasive machining consists of the use of loose abrasive and is usually used for achieving the final surface finish with very fine particle size abrasive. Lapping is the most commonly used free abrasive approach. The fine abrasive is placed on a soft material such as cloth or wood, which is then moved in relation to the ceramic being lapped. Because of the fine abrasive size, material removal rate is very slow. However, very smooth surfaces with flatness measured in wavelengths of light can be achieved. These surfaces result in very low friction and have been used in bearing and seal applications.

Free abrasive is also used in trepanning. This is a technique used for drilling circular holes in a ceramic. The tool consists of a thin-walled hollow cylinder of a soft metal such as brass. The loose abrasive is placed between the tool and the workpiece together with a coolant such as water or oil, and pressure is applied simultaneously with rotation to achieve grinding action.

Impact Abrasive Machining

The two previous machining methods remove stock essentially by sliding motion. Material can also be removed by impact. Sandblasting is the primary impact approach. Abrasive particles are carried by compressed air through a nozzle which directs them at high velocity against the workpiece. Rate of material removal for a ceramic increases with particle size, hardness of the abrasive and velocity, and for angles of impingement approaching 90°. Al_2O_3 and SiO_2 are the most commonly used abrasives.

Sandblasting has been used in sculpture and for fabrication of tomb-stones, but is not used often for fabrication of ceramic components for engineering applications. The primary drawbacks are difficulty in achieving close tolerances or a uniform surface.

Sandblasting is used for cleaning the surface of ceramics and also as a test method for evaluating the wear resistance of materials.

A second type of impact machining makes use of ultrasonics. However, the nature of the impact is significantly different than that which occurs during sandblasting. In ultrasonic machining the abrasive is suspended in a water slurry which flows over the surface of the tool. The tool is vibrated at high frequency, which accelerates each abrasive particle over a very short distance to strike the workpiece. Impact occurs only where the tool is in close proximity to the workpiece, so that close tolerances can be achieved by control of the tool and abrasive dimensions. No pressure is applied between the tool and abrasive so that very little strength-limiting damage to the ceramic results. Boron carbide is frequently used as the abrasive because it is harder than most other ceramics (except dia-mond), is less expensive than diamond, and is available in narrow size ranges.

Chemical Machining

Chemical machining is used primarily to achieve improved surface finish and thus increased strength or decreased friction. It is achieved by im-mersing the surface to be machined into a liquid in which the ceramic is soluble. Most silicate glass compositions can be etched or chemically machined with hydrofluoric acid (HF). Al_2O_3 can be etched with molten $Na_2B_4O_7$. These treatments are often referred to as chemical polishing because they produce such a smooth surface.

Photoetching

Some glass compositions can be chemically machined into very complex geometries using photoetching [1]. One such glass contains Ce_2O_3 and Cu_2O. A mask or photo negative is placed on the glass and irradiated with ultraviolet light. In the unmasked areas the Cu_2O is reduced by the Ce_2O_3 by the reaction $Ce^{3+} + Cu^+ \longrightarrow Ce^{4+} + Cu$. The glass is then exposed to a controlled heat treatment in which the Cu particles act as nucleation sites for localized crystallization. The crystallized material can be etched in hydrofluoric acid at a rate 15 times the rate of the original glass.

Some very intricate configurations have been produced by this photo-etching technique. An example is a 600-mesh sieve.

Electrical Discharge Machining

Electrical discharge machining (EDM) can be performed only with electrically conductive materials. A shaped tool is held in close proximity to the part being machined, retaining a constant predetermined gap with the use of a servomechanism that responds to change in the gap voltage. A dielectric liquid is flowed continuously between the tool and workpiece. Sparks produced by electrical discharge across this dielectric erode the ceramic by a combination of vaporization, cavitation, and thermal shock produced by the intense local heating [2].

EDM has been used successfully with conductive carbides, silicides, borides, and nitrides. The advantages are that no mechanical load is applied during EDM and that holes, recesses, and outer dimensions can be produced in the same types of shapes that could be formed in metal by stamping or in dough by a cookie cutter. The disadvantages are the slow rate of cut, the limitation to conductive materials, and the relatively poor surface finish achieved. The surface is typically pitted and microcracked and results in substantial strength reduction.

Laser Machining

Only a few studies have been reported on laser machining of ceramics. Lumley [3] reports the use of laser machining to score Al_2O_3 electronic substrates to allow them to be fractured to the desired size. The mechanism of material removal appeared to be localized thermal shock spalling.

Copley et al. [4] have reported machining of SiC, Si_3N_4, and SiAlON using a CO_2 laser. In this case material removal was apparently by evaporation, since all three of these materials decompose rather than melt. A hot-pressed Si_3N_4 cylinder containing 1/4 in. × 20 threads was laser machined from rectangular stock. A SiAlON cylinder was also machined. Surface finish measurement determined that the maximum nonsmoothness from peak to valley was 7.5 μm. This suggests that laser machining of ceramics may be a feasible approach and should be evaluated further. In particular, the effects on material strength should be assessed.

8.2 EFFECTS ON STRENGTH

To understand the effects of machining on the strength of a ceramic material, we must examine the interactions that occur at the tool-workpiece interface and define the flaws that are initiated in the ceramic. First, let us consider a single mounted abrasive particle plowing a furrow in a ceramic workpiece. Material directly in the path of the abrasive particle sees very high stress and temperatures and is broken and deformed. Material adjacent

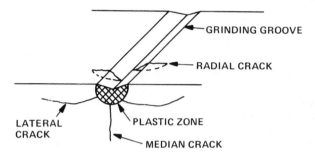

Figure 8.1 Schematic showing the cracks and material deformation that occurs during grinding with a single abrasive particle.

to the abrasive particle is placed in compression and may also deform plastically. After the abrasive particle passes, this material rebounds and either cracks or spalls off, due to the resulting tensile stresses [5,6]. Thus the size of the machining groove for most ceramics is larger than the size of the abrasive particle.

Figure 8.1 shows schematically the types of cracks that can form adjacent to the grinding groove [7-10]. The median crack is parallel to the direction of grinding and perpendicular to the surface, and results from the high stresses at the bottom of the grinding groove. Because it is parallel to the direction of grinding, it has also been called a longitudinal crack. It is usually the deepest crack and produces the greatest strength reduction.

Lateral cracks are parallel to the surface and extend away from the plastic zone. They result from the high tensile stress that exists at the edge of the plastic zone and extend as the material relaxes immediately after the abrasive particle passes. Lateral cracks tend to curve toward the surface and often result in a chip spalling off. Because lateral cracks are parallel to the surface, they do not result in stress concentration during subsequent mechanical loading and thus do not significantly reduce the strength of the material. However, they do account for a substantial portion of stock removal during grinding.

Radial cracks normally result from single particle impact or indentation and extend radially from the point of impact. They are perpendicular to the surface, but are usually shallow and do not degrade the strength as much as a median crack. The cracks shown in Fig. 8.1 which are perpendicular to the grinding groove are analogous to radial cracks. There can be many of these along the length of the grinding groove. They have been referred to in the literature as transverse, chatter, or crescent cracks, but their mechanism of formation has generally not been discussed. It is most likely that they are initiated by the high tensile stress that arises at the trailing edge of the contract of the abrasive particle and the workpiece. This biaxial stress mechanism for high-friction situations was discussed in Chap. 3.

Effect of Grinding Direction

Most ceramics are machined with tools containing many abrasive particles rather than just one single point. However, it is likely that the resulting surface flaws are similar for both types of tools and that the median and radial cracks control the strength. Which flaw controls the strength depends on the orientation of the grinding grooves to the direction of stress application. This is shown schematically in Fig. 8.2 for specimens loaded in bending.

As the load is applied and the specimens begin to bend, stress concentration will occur at the tips of cracks perpendicular to the stress axis but not at cracks parallel to the stress axis. Thus for specimens ground in the longitudinal direction, stress concentration will occur at the radial (transverse) cracks, and for specimens ground in the transverse direction, stress concentration will occur at the median (longitudinal) cracks. Since the median cracks are usually the most severe, one would expect the strength to be lowest for transverse grinding, where the grooves and median cracks are perpendicular to the tensile stress axis. This can be seen in Table 8.1.

From Table 8.1 it is obvious that substantial differences in load-bearing capability for a ceramic component can result, depending on the orientation of grinding with respect to the stress distribution in the component. This anisotropy of strength is an important consideration for an engineer designing a component that must withstand high stress.

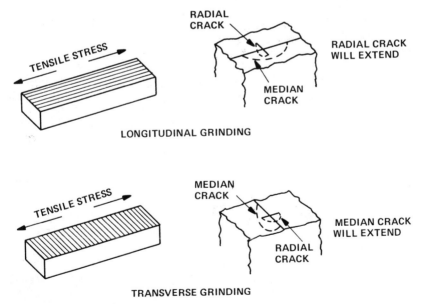

Figure 8.2 Grinding direction and crack distribution versus tensile stress axis.

Table 8.1   Strength Versus Grinding Direction Orientation with Respect to
Tensile Stress Axis

Material	Longitudinal grinding		Transverse grinding	
	MPa	psi	MPa	psi
Hot-pressed Si_3N_4	669	97,000	428	62,000
Soda-lime glass	97	14,100	68	9,900
Mullite	319	46,300	259	37,600
MgF_2	87	12,600	53	7,700
B_4C	374	54,200	154	22,300

Source: Refs. 11 (Si_3N_4 data) and 12 (data for other materials).

Effects of Microstructure

The microstructure of the ceramic material has a pronounced effect on the
rate of machining and on the residual strength after machining. Rice [5]
reports that fine-grained ceramics require higher grinding force and longer
time to slice or machine. This is shown for several ceramic materials in
Fig. 8.3. Uniformly distributed porosity increases the rate of machining,
but also decreases the smoothness of surface finish that can be achieved.
    The degree of strength reduction resulting from machining is dependent
on a comparison of the size of flaws initially present in the ceramic to those
produced by machining. Machining has very little effect on the strength of
ceramics containing high porosity or large grain size because the flaws
introduced during machining are no larger than the microstructure flaws
initially present. On the other hand, the strength of fine-grained ceramics
such as most Si_3N_4 and Al_2O_3 can be reduced significantly.

Effects of Grinding Parameters

The parameters selected for machining a ceramic have a large effect on
the rate of machining and tool wear and on the resulting properties of the
ceramic. Table 8.2 summarizes general trends associated with variations
in individual machining parameters. It should be emphasized that these
are just trends and that they may not hold true for all ceramic materials
and all levels of variation. Also, most of these parameters are interactive
and may have a different effect when combined than when considered individually.

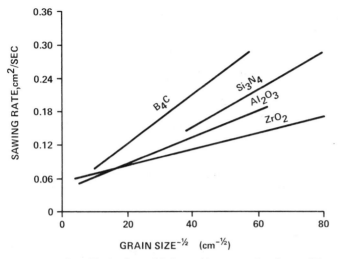

Figure 8.3 Effect of machining rate on grain size. (From Ref. 5.)

Large abrasive particles result in a greater depth of grinding damage and are used for roughing operations. Finer abrasives are used to remove the subsurface damage produced by the large abrasive during roughing and to achieve the final tolerances and surface finish.

To achieve maximum strength, grinding is usually done in several steps, decreasing the abrasive size in each step and removing enough surface stock to remove subsurface damage resulting from the prior step. Sometimes, intermediate machining steps are skipped to save time and decrease cost. The desired tolerances and surface finish can be achieved, but the strength requirements may not, and the component may fail in service.

Harder ceramic materials are more difficult to cut and grind than softer materials and require higher force. This results in increased wheel wear, higher interface temperatures, and greater danger of damage to the ceramic. Increased wheel speed decreases the required force and usually results in lower tool wear, lower temperatures, and less surface damage. The use of water or other suitable lubricant or coolant provides similar benefits .

Minimizing Machining Effect on Properties

The deleterious effects of machining on ceramic properties can be decreased by experimental optimization of the machining parameters. This includes selecting the appropriate abrasive and wheel bond, wheel speed, downfeed, coolant, and abrasive size sequence. Postmachining procedures

Table 8.2 Effects of Machining Parameters

Parameter variation	Effect on machining rate[a]	Effect on tool wear[a]	Effect on material strength[a]
Increasing abrasive size	Increases rate	Depends more on other factors	Decreases strength
Increasing downfeed or table speed	Increases rate	Increases wear	Decreases strength
Increasing wheel speed	Increases rate	Usually decreases wear	Increases strength
Use of lubricants and coolants	May increase rate	Usually decreases wear	Usually increases strength
Increasing abrasive hardness	Increases rate	Decreases wear	Depends more on other factors
Changing abrasive concentration and wheel bond	Different optimum for different materials and configurations	Can be optimized	Can be optimized

[a]Relative to an arbitrary baseline.

have been developed to obtain further improvement in the properties. These include the following:

Lapping
Annealing
Oxidation
Chemical etching
Surface compression
Flame polishing

Each of these approaches either removes surface and subsurface flaws resulting from machining or reduces the stress concentration due to the flaw.

Lapping Lapping involves the use of very fine free abrasive particles suspended in a slurry and applied to the workpiece surface with a soft tool surface such as cloth or wood. It is equivalent to mechanical polishing used in preparation of metallographic samples for microstructure examination. The size of the abrasive determines the final surface finish that can be achieved. However, as mentioned before, the strength does not necessarily increase as the surface smoothness increases. To achieve strength increase, machining and lapping must be done in a diminishing abrasive size sequence such that each step successfully removes the worst surface damage produced by the prior step. A suitable sequence might be to rough machine with 200-grit diamond, finish machine with 320- and 600-grit diamond, rough lap with 30- and 9-μm Al_2O_3 or CeO_2, and finish lap with 3-, 0.3-, and 0.06-μm Al_2O_3 or CeO_2. Final lapping can also be done with diamond.

The degree of lapping or surface finishing is obviously dependent on the cost restraints and the criticality of the application. Because of the fine grain structure and hardness of advanced ceramic materials, excellent surface finish and tolerances can be achieved. For example, the following capabilities have been reported for dense Al_2O_3 by Western Gold and Platinum Co. (WESGO) in their Technical Circular No. L-779:

"Flat lapping to half a light band (0.000006 in. or 0.000152 mm flatness)
Parallelism to 0.000010 in. (0.00025 mm)
Dimensional tolerances of 0.000010 in. (0.00025 mm)
Cylindrical outside diameters between 0.060 in. (1.524 mm) and 3 in. (76.2 mm) lapped to 0.000005 in. (0.000127 mm) dimensional tolerance and roundness to 0.000005 in. (0.000127 mm)
Stepped diameters lapped to 0.000050 in. (0.00127 mm) concentricity
Outside-inside mating diameters lapped to 0.000050 in. (0.00127 mm) tolerance of mating clearance
Cylindrical inside diameters between 0.060 in. (1.524 mm) and 2 in. (50.8 mm) lapped to 0.000005 in. (0.000127 mm) dimensional tolerance and roundness to 0.000005 in. (0.000127 mm)

Blind holes and intersecting holes lapped to 0.000005 in. (0.000127 mm)
dimensional tolerance and roundness to 0.000005 in. (0.000127 mm)
Radial and spherical concave and convex surfaces lapped to within 0.000010
in. (0.00025 mm) of true radius and 0.000010 in. (0.00025 mm) roundness."

Most applications do not require such precise lapping. Some of the
more critical applications in terms of surface finish or tolerances include
optical glass, laser ceramics, bearings, seals, some papermaking components and thread guides.

Annealing Since ceramic materials are normally processed at very high
temperature, internal stresses often result during cooldown. Sometimes
these residual stresses can improve strength, but often they reduce strength.
Annealing at high temperature followed by slow cooldown can often relieve
these stresses. Annealing can also relieve surface stresses resulting from
machining and actually heal subsurface flaws such as median cracks. Annealing can also crystallize glass phases and achieve improved strength or
stability.

Oxidation Oxidation has been demonstrated to increase the strength of
machined hot pressed Si_3N_4 [11] and other nitrides and carbides. In the
case of Si_3N_4, the reaction $Si_3N_4 \longrightarrow 3SiO_2 + 2N_2$ can increase the strength
by completely removing the depth of surface containing the residual machin-
ing cracks or by rounding the crack tip and reducing stress concentration.
This approach was used effectively for improving the load-bearing capability
of hot-pressed Si_3N_4 rotor blades [11]. The blade machining process resulted in transverse grinding perpendicular to the tensile stress axis such
that the median crack was strength controlling. The as-machined strength
was about 428 MPa (62,000 psi). Oxidation at 960°C (1800°F) for 50 hr
increased the strength to 635 MPa (92,000 psi).

Chemical Etching Chemical etching has been used for many years for removing machining or other surface damage, especially the use of hydrofluoric acid with glass. Strength increases greater than tenfold have been
achieved both for glass compositions and for oxide ceramics.

Surface Compression Placing the surface in compression obviates the effects of surface flaws by preventing concentration of tensile stresses at the
crack tip. The surface compressive stress must first be exceeded by an
applied tensile stress before the stress concentration will begin to build up
and lead to crack propagation.
    Perhaps the most common example of surface compression is in safety
glass. Surface compression is achieved in glass most commonly by either
ion exchange or quenching. In the former case, glass is exposed at elevated

temperature to positive ions that are larger than those initially in the glass. Since the glass structure is expanded at the elevated temperature, these larger ions are able to trade places with smaller ions near the surface. When the glass is cooled, these ions no longer fit, but are trapped in the structure and result in surface compression. An example is that of exchanging calcium ions for sodium ions.

Ion exchange can also be applied to crystalline materials. The exchange ion can cause compression by size difference or by producing a surface composition with a lower coefficient of thermal expansion [13].

Quenching has also been known for a long time as a method of improving strength. Kirchner [13] discusses in detail the use of quenching to strengthen Al_2O_3, TiO_2, spinel, steatite ($MgSiO_3$), forsterite (Mg_2SiO_4), SiC, and Si_3N_4. Quenching Al_2O_3 rods from 1600°C into a silicone oil resulted in an increase in the average strength from 331 MPa (47,900 psi) to 694 MPa (100,600 psi).

For quenching to be effective, the material must be heated to a temperature such that some plasticity is present. A temperature of 1500 to 1600°C is appropriate for Al_2O_3, but 1900 to 2000°C is required for SiC. During quenching the surface cools very rapidly and is placed in compression as the interior cools more slowly.

Surface compression can also be achieved with glazes (glass surface coatings). The glaze composition is selected to have a lower coefficient of expansion than the matrix material. For instance, Kirchner used a glaze with a thermal expansion of 5.3×10^{-6} per °C for coating Al_2O_3 with a thermal expansion of 6.5×10^{-6} per °C. Glazes can be used in conjunction with quenching or ion exchange to achieve additional benefits. Kirchner achieved a strength of 767 MPa (111,200 psi) for Al_2O_3 that was glazed and quenched, compared to 331 MPa (47,900 psi) for as-received material.

Flame Polishing Flame polishing is used primarily for reducing the size and quantity of surface flaws in small-diameter rods or filaments, especially of sapphire or ruby (single-crystal Al_2O_3). Flame polishing is conducted by rotating the rod or filament and passing it through a H_2-O_2 flame such that the thin surface layer melts. Noone and Heuer [14] report bend strengths for flame-polished ruby and sapphire in the range 4000 to 5000 MPa (580,000 to 724,000 psi) compared to approximately 300 MPa (43,500 psi) for as-ground specimens. Stokes [15] compares the tensile strength of flame polished single-crystal Al_2O_3 with other surface preparations. His results are summarized in Table 8.3.

Stokes [15] also discussed the effects of machining on properties other than strength. The shape of the hysteresis loop in magnetic ferrites is significantly changed by near-surface stresses resulting from machining. Polishing, annealing, and etching procedures are routinely used to attain reproducibly the desired loop shape. Electrical and optical properties are also affected strongly by machining and surface condition.

Table 8.3 Surface Preparation Versus Tensile Strength for
           Single-Crystal Al_2O_3

| | Tensile strength | |
	MPa	psi
Surface preparation		
As-machined	440	60,000
Polished by centerless grinding	590	90,000
Annealed in oxygen after polishing by centerless grinding	1,040	150,000
Chemical polished with molten borax	6,860	1,000,000
Flame polished	7,350	1,100,000
Pristine whiskers	15,900	2,300,000

Source: Ref. 15.

REFERENCES

1. S. D. Stookey, Ind. Eng. Chem. 45, 115-118 (1953).
2. D. W. Lee and G. Feick, The techniques and mechanisms of chemical, electrochemical and electrical discharge machining of ceramic materials, in The Science of Ceramic Machining and Surface Finishing (S. J. Schneider and R. W. Rice, eds.), NBS Special Publication 348, U.S. Government Printing Office, Washington, D.C., 1972, pp. 197-211.
3. R. M. Lumley, Controlled separation of brittle materials using a laser, Am. Ceram. Soc. Bull. 48(9), 850-854 (1969).
4. S. M. Copley, M. Bass, and R. G. Wallace, Shaping silicon compound ceramics with a continuous wave carbon dioxide laser, in The Science of Ceramic Machining and Surface Finishing, II (B. J. Hockey and R. W. Rice, eds.), NBS Special Publication 562, U.S. Government Printing Office, Washington, D.C., 1979, pp. 283-292.
5. R. W. Rice, Machining of ceramics, in Ceramics for High Performance Applications (J. J. Burke, A. E. Gorum, and R. N. Katz, eds.), Brook Hill Publishing Co., Chestnut Hill, Mass., 1974, pp. 287-343. (Available from MCIC, Battelle Columbus Labs., Columbus, Ohio.)
6. P. J. Gielisse and J. Stanislao, Dynamic and Thermal Aspects of Ceramic Processing, University of Rhode Island, Kingston, Naval Air Systems Command Contract Report, Nov. 15, 1969-Nov. 15, 1970 (AD 728 011), Dec. 1970.
7. A. G. Evans, Abrasive wear in ceramics: an assessment, in The Science of Ceramic Machining and Surface Finishing, II (B. J. Hockey

and R. W. Rice, eds.), NBS Special Publication 562, U.S. Government Printing Office, Washington, D.C., 1979, pp. 1-14.

8. J. T. Hagan, M. V. Swain, and J. E. Field, Nucleation of median and lateral cracks around Vickers indentations in soda-lime glass, in The Science of Ceramic Machining and Surface Finishing, II (B. J. Hockey and R. W. Rice, eds.), NBS Special Publication 562, U.S. Government Printing Office, Washington, D.C., 1979, pp. 15-21.

9. H. P. Kirchner, R. M. Gruver, and D. M. Richard, Fragmentation and damage penetration during abrasive machining of ceramics, in The Science of Ceramic Machining and Surface Finishing, II (B. J. Hockey and R. W. Rice, eds.), NBS Special Publication 562, U.S. Government Printing Office, Washington, D.C., 1979, pp. 23-42.

10. A. B. Van Groenou, N. Maan, and J. B. D. Veldkamp, Single-point scratches as a basis for understanding grinding and lapping, in The Science of Ceramic Machining and Surface Finishing, II (B. J. Hockey and R. W. Rice, eds.), NBS Special Publication 562, U.S. Government Printing Office, Washington, D.C., 1979, pp. 43-60.

11. D. W. Richerson, J. J. Schuldies, T. M. Yonushonis, and K. M. Johansen, ARPA/Navy ceramic engine materials and process development summary, in Ceramics for High Performance Applications, II (J. J. Burke, E. N. Lenoe, and R. N. Katz, eds.), Brook Hill Publishing Co., Chestnut Hill, Mass., 1978, pp. 625-650. (Available from MCIC, Battelle Columbus Labs., Columbus, Ohio.)

12. R. W. Rice and J. J. Mecholsky, The nature of strength-controlling machining flaws in ceramics, in The Science of Ceramic Machining and Surface Finishing, II (B. J. Hockey and R. W. Rice, eds.), NBS Special Publication 562, U.S. Government Printing Office, Washington, D.C., 1979, pp. 351-378.

13. H. P. Kirchner, Strengthening of Ceramics, Marcel Dekker, Inc., New York, 1979.

14. M. J. Noone and A. H. Heuer, Improvements in the surface finish of ceramics by flame polishing and annealing techniques, in The Science of Ceramic Machining and Surface Finishing (S. J. Schneider and R. W. Rice, eds.), NBS Special Publication 348, U.S. Government Printing Office, Washington, D.C., 1972, pp. 213-232.

15. R. J. Stokes, Effect of surface finishing on mechanical and other physical properties of ceramics, in The Science of Ceramic Machining and Surface Finishing (S. J. Schneider and R. W. Rice, eds.), NBS Special Publication 348, U.S. Government Printing Office, Washington, D.C., 1972, pp. 348-352.

9

Quality Control

Quality control (QC) is required throughout processing of any material or product, and ceramics are no exception. The degree of QC is determined by the criticality of the application. Most applications require a specification or a written manufacturing procedure and one or more certification tests to assure that the manufacturing procedure has been followed and the specification met. More critical or demanding applications may require destructive sampling, proof testing, or nondestructive inspection (NDI). In this chapter these various aspects of QC are explored from the perspective of both the manufacturing engineer and the applications engineer.

9.1 IN-PROCESS QC

The fundamental step in in-process QC is the preparation of a formal written manufacturing procedure. This document describes each operation required in the process from procurement of raw materials through final shipping. It also normally defines the paperwork that must accompany the part through the process, lists the signatures that are required on this paperwork to certify that each operation has been followed as specified, and describes the procedures that must be followed if a process change is considered. The signed paperwork that follows the part through the process is kept on file as traceability for that part and as a record that the part was processed by the designated fixed process.

In-process QC starts with procurement of raw materials. It may consist simply of checking the chemical analysis and particle size distribution analysis submitted for the material by the supplier to ensure that they are within the specification of the manufacturing operation document. Or it may involve additional analyses, depending on how critical the raw material characteristics are to achieving the desired characteristics of the final product. For electrical, magnetic, optical, and many structural applications, the purity and particle size aspects of the raw materials are extremely important and careful QC is justified.

In-process QC continues into the next step of the fabrication process, powder processing. Again, purity and particle sizing are usually most important, so that QC consists of chemical analysis (such as emission spectroscopy, x-ray fluorescence, infrared spectroscopy, x-ray diffraction, or atomic absorption) and particle size distribution analysis (such as x-ray sedimentation, Coulter counter, BET surface area, and screening).

The nature of in-process QC changes with the shape-forming and densification steps of the manufacturing process. There is less interest in the bulk chemistry and more interest in isolated forming defects, dimensions, and properties. Forming defects such as cracks, pores, inclusions, laminations, and knit lines can be detected by visual inspection and NDI. Dimensions can be determined by gauges, shadow graphs, and other standard inspection instruments.

Electrical, magnetic, optical, and physical properties can be determined by direct measurement, either on each part or by random sampling. Mechanical properties are usually not as easy to qualify. The final product shape is usually not of a configuration readily tested for mechanical properties. The options are to process appropriate mechanical property test shapes with the product, to cut test bars out of random samples of the product, or to conduct a proof test on the product.

Obviously, the objective of in-process QC is to reject unacceptable components and to accept good components. The accept/reject criteria result from experience, primarily joint experience between the manufacturer and the user. This usually leads to a specification prepared by the user and accepted by the manufacturer, plus a QC or certification procedure prepared by the manufacturer and accepted by the user.

9.2 SPECIFICATION AND CERTIFICATION

At some point in their career, most engineers will have to prepare or use a specification. Specifications are not required only for ceramic components, but also for metal and organic components and for systems. In fact, most products manufactured have a specification and certification procedure to assure that the specification is met.

Tables 9.1 and 9.2 illustrate how a specification and certification procedure are integrated into a manufacturing operation procedure to produce the hypothetical ceramic component shown in Fig. 9.1. The objective is to conduct certification tests at each step in the process that will allow rejection of faulty parts as early as possible. Of course, there will be a tradeoff based on the economics and service conditions of the part being manufactured. More in-process QC and certification testing can be justified for an expensive or especially critical component than for a very low cost component or one that has a loose specification.

Table 9.1 Specification for the Part Shown in Fig. 9.1.

1.0 Material composition

2.0 Properties

    2.1 Density >98.5% of theoretical

    2.2 Hardness >2200 kg/mm^2 Knoop$_{500}$

    2.3 Average room temperature MOR[a] for eight specimens must be >280 MPa (40,000 psi)

    2.4 Average 1000°C (1832°F) MOR[a] for four specimens must be >207 MPa (30,000 psi)

    2.5 Stress rupture life for same specimen configuration under 140 MPa (20,000 psi) load at 1000°C is >100 hr

3.0 Dimensions

    3.1 ID 1.000 + 0.002-0.000 in. with a 2-μin. surface finish

    3.2 OD$_1$ 1.500 ± 0.005 in.

    3.3 OD$_2$ 3.000 ± 0.005 in.

    3.4 h$_1$ 0.750 ± 0.010 in.

    3.5 h$_2$ 0.500 ± 0.002 in. with surface flat and parallel to 0.001 in.

---

[a] Specimen cross section 0.32 × 0.64 cm (0.125 × 0.25 in.) to be tested in 3-point bending over a span of 3.8 cm (1.5 in.).

The specification in Table 9.1 is quite tight both in terms of properties and dimensions and thus will require substantial certification. The component is apparently for a high-temperature application where strength and creep resistance are critical. Since creep and high-temperature strength are often controlled by composition and impurity levels, the earliest opportunity in the process for QC and certification is chemical analysis of the starting powder, or at least review of the supplier's analysis. The component manufacturer typically has background experience in the levels of variation in the chemical composition that can be tolerated and has an in-house specification (separate from the component specification of the user). Therefore, chemical analysis becomes an effective certification test for the manufacturer, even though the chemical composition is not listed in the user's specification.

    A similar situation exists for the processed powder. The manufacturer has an in-house specification relating parameters such as particle size distribution to properties such as strength and hardness and can then use a certification test to accept or reject a batch of processed powder.

The next step in the process is shape forming. Emphasis now shifts to dimensions. Generally, one or two dimensions are most critical for a component and can be measured quickly in the green state. This will be an automatic procedure if a portion of the shaping is achieved by green machining.

After densification the certification procedure usually shifts back to properties as shown in Table 9.2. Based on the specification in Table 9.1, bulk density would be measured first (for parts not showing any visual defects). If the density were acceptable, the other critical properties listed in the specification would be measured as part of the certification testing.

The final process step is often finish machining. The tolerances in the hypothetical component are tight and would require accurate dimensional measurements as part of the certification procedure.

Components passing all the certification tests would then be delivered to the customer, usually along with written documentation of the certification test results.

Table 9.2 Integration of QC and Certification into the Manufacturing Process

Process steps	In-process and final QC	Certification tests
Starting powder	Chemical analysis X-ray diffraction Particle size distribution	Chemical analysis
Powder processing	Chemical analysis Particle size distribution	Particle size distribution
Shape forming	Visual and dimensional Radiography Green density	Visual and dimensional
Densification	Visual NDI Final bulk density Critical property measurement	Visual Final bulk density Critical property measurement
Final machining	Dimensional	Dimensional

TOP VIEW

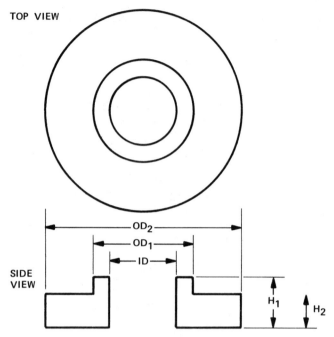

Figure 9.1 Hypothetical ceramic component used to illustrate integration of specification and certification into the manufacturing operation. Dimensions and tolerances are listed in Table 9.1.

9.3 PROOF TESTING

Many components are proof-tested, i.e., exposed to conditions comparable to or exceeding those of the service environment. Proof tests can be conducted for critical electrical, mechanical, or other properties using parameters known from prior application experience to control the acceptability or life of the component in the application. Proof tests are applied to metals, organics, and ceramics. For example, superalloy metal rotors for small gas turbine engines are typically spin-tested to a predetermined overspeed prior to being installed into engines. Similarly, ceramic grinding wheels are proof-tested by spin testing. Ceramic electrical insulators for high-voltage application are usually electrically proof tested to make sure that they will not undergo dielectric breakdown.

To be a true proof test, the test condition must exactly simulate the service conditions, including temperature and stress distributions. This is extremely difficult to achieve without actually putting the component in real service. Hence many tests referred to as proof tests really are not.

Instead, most of these tests only expose the material or component to the most critical aspect of the service environment and are more properly referred to as qualification or screening tests.

A major concern with proof or qualification testing is that no damage is done to the material that might reduce its service life. This is especially important with ceramics in mechanical loading. The ceramic can withstand the proof load, but be damaged subcritically either by the load or during unloading. This is especially likely to occur where a room-temperature overload condition is applied to qualify a part for a high-temperature application. Damage that will not cause failure at room temperature may cause failure at elevated temperature or grow subcritically and cause failure after only a short time.

Proof or qualification testing can dramatically improve the reliability of a ceramic component, especially for applications requiring high mechanical strength [1,2]. As shown in Chap. 3, the strength of a ceramic material is dependent on the type of loading and the volume and area under stress. This, plus the wide flaw size distribution in a single ceramic component, results in a significant probability of failure over a wide range of applied load. A properly applied proof test can truncate this distribution and substantially reduce the probability of failure under service conditions. This is illustrated in Fig. 9.2 using a Weibull plot of probability of failure versus applied stress. Without proof testing, 8 components out of 1000 would fail at the service load of 50,000 psi and 3 of 100 at an accidental overload condition of 60,000 psi. By proof testing at 55,000 psi, the strength distribution is truncated and no failures would occur at the service load of 50,000 psi and less than 1 in 100 at an overload condition of 60,000 psi.

9.4 NONDESTRUCTIVE INSPECTION

Nondestructive inspection (NDI) involves examination of a component or test specimen by instrumentation that detects surface or subsurface flaws in the material but does not result in material damage. Such techniques are also referred to as nondestructive evaluation (NDE) and nondestructive testing (NDT) and are applicable to ceramics, metals, and organics. In this section the administration of various NDI techniques and the nature of their strengths and limitations are reviewed. Emphasis is on techniques that show the most promise for ceramics.

X-Ray Radiography

Conventional X-ray Radiography   Conventional x-ray radiography consists of passing a beam of x-rays through the part being examined to expose a

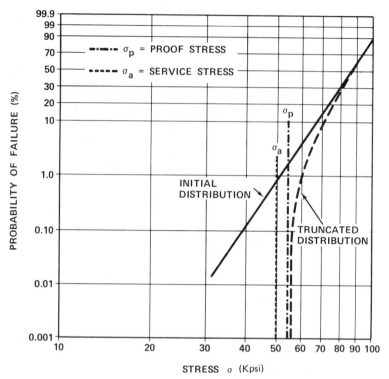

Figure 9.2 Truncation of the material strength distribution to achieve improved reliability and increased operating margin. (Adapted from Ref. 1.)

sheet of film, as illustrated schematically in Fig. 9.3a. The material absorbs a portion of the x-rays. If the material is of constant thickness and contains no flaws, the film adjacent to the part will be uniformly exposed and will be uniformly gray in color after developing, surrounded by fully exposed black where the film was not shielded by the part. If the part is of a nonuniform thickness, the thicker portions will absorb more of the x-rays and the film in this region will be less exposed (lighter gray in color). Density variations in the material will show up in a similar manner, higher density showing up lighter and lower density showing up darker. Similarly, as shown in Fig. 9.3b, a void or hole will show up darker and a high-density inclusion that absorbs x-rays more than the matrix will show up lighter. Obviously, if a positive photographic print is made from the film negative, the opposite will be true; voids will appear lighter than the matrix and a high-density inclusion will appear darker.

The size of defect that can be detected by x-ray radiography depends on a combination of factors:

(a) SIDE VIEW

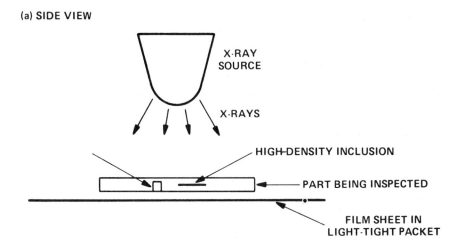

X-RAY
SOURCE

X-RAYS

HIGH-DENSITY INCLUSION

PART BEING INSPECTED

FILM SHEET IN
LIGHT-TIGHT PACKET

(b) TOP VIEW OF
    FILM AFTER
    DEVELOPING

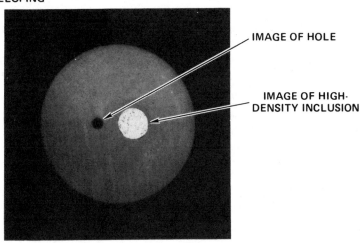

IMAGE OF HOLE

IMAGE OF HIGH-
DENSITY INCLUSION

Figure 9.3 (a) Schematic of conventional x-ray radiography setup.
(b) Resulting image on the developed film.

The thickness of the part and its x-ray absorption
The size of the flaw compared to the thickness of the part
The difference in x-ray absorption between the flaw and the part
The orientation of the flaw

Elements of low atomic weight and structures of low packing density have relatively low absorption of x-rays and can thus be inspected by x-ray radiography in greater thicknesses than materials having high absorption owing to their content of high-atomic-weight elements. Al_2O_3, graphite, Si_3N_4, SiC, and other similarly low-atomic-weight ceramics have low absorption for x-rays and can be inspected over a broad thickness range. WC has high atomic weight, and most metals have close packing and therefore have higher absorption for x-rays, resulting in restriction of the thickness that can be inspected.

Other factors can also affect x-ray absorption. Different elements preferentially absorb specific wavelengths of x-rays.

The difference in x-ray absorption between the part and the inclusions or voids it contains determines the sensitivity or resolution capability of the material. Small voids are difficult to detect in materials like Si_3N_4 and SiC because the difference in absorption between the void and the material is relatively small. On the other hand, inclusions of WC or iron in Si_3N_4 or SiC can be resolved even in very small sizes because the x-ray absorptions are so different.

The resolution capability or specified requirement is often stated in terms of the thickness of the defect compared to the thickness of the part. Inspection to a 2T level means that the technique or apparatus used must be able to detect flaws larger than 2% of the thickness of the part. Kossowsky [3] reported the following limits of resolution for defects in hot-pressed Si_3N_4 using conventional x-ray radiography: voids (holes or cracks), 3% of thickness; BN-filled cavity, 4% of thickness; steel particle, 0.7% of thickness; and WC particle, 0.5% of thickness. Similar sensitivities were reported by Richerson et al. [4] using microfocus x-ray.

The final factor affecting detection sensitivity is the orientation of the flaw. A tight crack perpendicular to the x-ray source will not be detected, whereas one that is parallel or at a low angle to the source will have a better chance of detection. It is for this reason that radiographs are normally taken with the part in several different orientations. Also, two radiographs are usually taken for each orientation, so that artifacts due to film blemishes, developing, or surface contamination can be distinguished from defects in the material.

Conventional x-ray radiography is widely used for all types of materials and is a quick, convenient, and cost-effective way of detecting internal flaws in components. A major advantage of radiography is that it can be used effectively with complex shapes, which is a limiting factor for most other NDI techniques. However, as with other NDI techniques, radiography

requires standards in order to quantify the size and type of defect being evaluated.

Microfocus X-ray Radiography   Microfocus x-ray radiography is a relatively new technique made possible by development at Magnaflux of their
MXK-100M x-ray tube, which has a small focal spot of 0.05 mm (0.002 in.).
This tube provides improved resolution and geometric sharpness. Because
of the fine focal spot, closer working distances are possible and direct
radiographic enlargements up to 36× can be achieved without reduction in
sensitivity due to parallax and secondary radiation effects [5,6].

Microfocus x-ray is especially useful for complex shapes where a small
region is especially critical, such as the leading and trailing edges of rotor
blades and stator vanes for gas turbine engines. The MXK-100M tube is
small and portable (weighs under 4 kg) and can be easily maneuvered to
inspect the desired region at the appropriate orientation.

The microfocus x-ray tube has also been modified to permit panoramic
radiography [5] of a hollow cylindrical object.

Image Enhancement   Once a photographic negative has been produced by
radiography, it must be examined with back lighting and judgments made as
to which indications on the film represent defects in the material. Such
film interpretation is very subjective and requires an experienced individual.
For instance, Fig. 9.4 shows radiographs of graphite, iron, and tungsten
carbide in hot-pressed Si_3N_4. Which coloration differences actually represent flaws? Are there four graphite inclusions in Fig. 9.4a?

Image enhancement technology, developed initially for evaluation of
satellite photos, has been adapted to radiography. In general, the radiograph film is back-lighted and the image picked up by a television camera.
The image is divided into discrete elements (pixels) and entered into a digital computer. In the study conducted by Schuldies and Spaulding [6], the
image was divided into an array of 480 × 512 pixels and each pixel assigned
a gray-level value ranging from 0 (black) to 255 (white). This is far more
gray levels than can be distinguished by the human eye.

Once the image data are digitized and stored in the computer, a variety
of computer programs can be run to achieve greater contrast and thus enhance the image. The enhanced image is displayed on a black-and-white
TV screen which can be photographed to provide a permanent record.

Figure 9.5 shows a block schematic of the image enhancement system.
Besides the computer enhancement, visual enhancement can be achieved
directly with the scanner to obtain improved contrast. Color enhancement
can also be used where gray levels can be replaced by colors. In addition,
a video cursor is tied into the computer and display system to permit physical distance measurements and digital readout.

Image enhancement is a significant aid to interpretation of radiographs.
It helps reduce the subjectivity of the operator and provides more objective

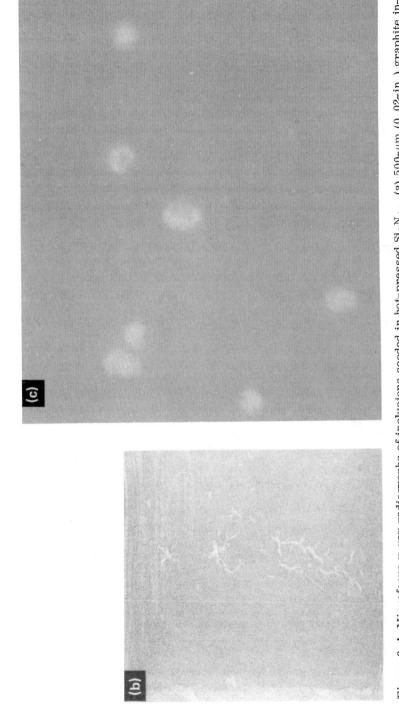

Figure 9.4 Microfocus x-ray radiographs of inclusions seeded in hot-pressed Si_3N_4. (a) 500–μm (0.02-in.) graphite inclusions. (b) 250–μm (0.01-in.) iron inclusions. (c) 500–μm (0.02-in.) WC inclusions. (Courtesy of Garrett Turbine Engine Company, Phoenix Ariz., Division of the Garrett Corporation.)

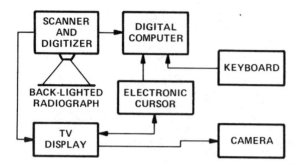

Figure 9.5 Schematic of the image enhancement system.

evaluation of the radiograph. However, it should be recognized that film anomalies such as graininess, scratches, water marks, and processor defects will also be enhanced to the same degree as images of defects in the part, so it still ends up that the individual has to make judgments.

Figures 9.6 through 9.8 are examples of the benefits of image enhancement. Figure 9.6 shows enhancement of the radiograph shown previously in Fig. 9.4 for graphite inclusions in Si_3N_4. The edge-enhancement computer algorithms have dramatically isolated the inclusions. Now we can answer the question we raised before. There are more than four graphite inclusions. If only the fifth inclusion had been present, it would have been missed by the film examiner without the aid of enhancement.

Figure 9.7 shows enhancement of the radiograph shown previously in Fig. 9.4b for iron inclusions in Si_3N_4. The iron appears to be present as dendrites. Since it was initially seeded in the material as equiaxed particles [7], the iron evidently recrystallized or reacted with the Si_3N_4 matrix during hot pressing. This provides us with information about the ceramic processing that is not normally available from NDI data. The circular inclusion in the right-hand photograph of Fig. 9.7 turned out to be WC. It had been seeded as a near-spherical particle and did not change its shape during processing, suggesting that it was more stable or inert during processing than the iron was.

Figure 9.8 shows enhancement of the radiograph shown in Fig. 9.4c for WC inclusions in Si_3N_4. As in the other examples, enhancement has substantially increased the detectability of the inclusions.

Ultrasonic NDI

Ultrasonic NDI is another important technique for detecting subsurface flaws in materials. It has been used extensively with metals and is now being developed for use with ceramics. A simple schematic illustrating the basics of the technique is shown in Fig. 9.9. The part to be inspected is immersed in water. A piezoelectric transducer in close proximity to the surface of the part is stimulated by an electric current to emit acoustic waves of a known amplitude and wavelength. These waves pass into and through the part. Each discontinuity (the surfaces plus any internal defects) perturbs the acoustic waves, resulting either in scattering or in reflected secondary waves. A receiver picks up the secondary waves and the electronics of the equipment converts the signal into a graphical representation. The receiver can either be opposite the emitting transducer and pick up the transmitted waves, or be on the same side and pick up reflected waves. The latter approach is commonly used and is referred to as the pulse-echo technique. As was the case with radiography, flaw orientation affects detectability and inspection should be conducted in multiple directions if possible.

Ultrasonic inspection is most easily conducted on material having a flat surface and a constant cross section. The transducers are scanned across the part and results are plotted with a pen-type XY-recorder. The electronics are adjusted so that an electronic window eliminates the wave reflections for the two surfaces. This approach is known as C-scan. Since the surface reflections must be filtered out, C-scan does not detect near-surface flaws. The closeness to the surface that can be evaluated depends partially on how accurately the electronic window can be set. It is also affected by transducer noise and electronic signal damping limitations. Surface irregularities and variations in cross-sectional thickness further restrict the ability of the operator to set the window close to the surface and makes C-scan of complex shapes very difficult. In the past, this has been the primary disadvantage of ultrasonic inspection. However, new techniques are currently being developed for complex shapes. These include the use of microprocessors to accurately control movement of the transducers, use of computers to analyze the data, and use of arrays of transducers instead of scanning with a single transducer.

Figure 9.10 shows the C-scan printout for a 0.64-cm (0.25-in.)-thick flat plate of hot-pressed Si_3N_4 containing various sizes of inclusions and voids [8]. The resolution of both inclusions and voids is quite good. However, such success was not achieved on the first attempt. A variety of transducers and electronic gating procedures were tried before optimum conditions were defined. This reemphasizes the importance of standards. The Si_3N_4 plate had originally been prepared as a standard with seeded defects specifically to evaluate and optimize the resolution capabilities of different NDI techniques [7]. Without standards, the operator has difficulty optimizing equipment parameters and interpreting printout results.

The ultrasonic C-scan inspection detected more of the defects in the hot-pressed Si_3N_4 plate than could be detected by other techniques. X-ray radiography had shown up only the high-density WC and Fe inclusions and the largest graphite inclusion. Neutron radiography showed only the BN.

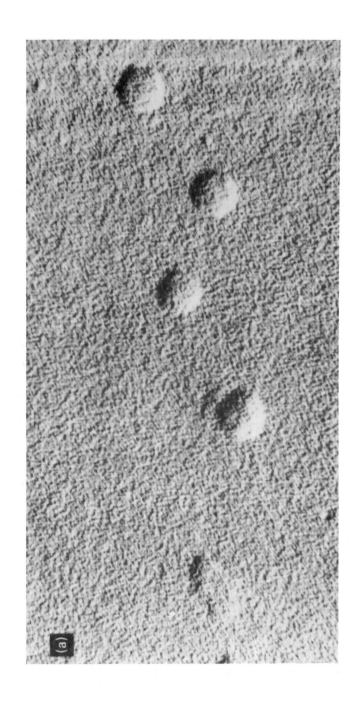

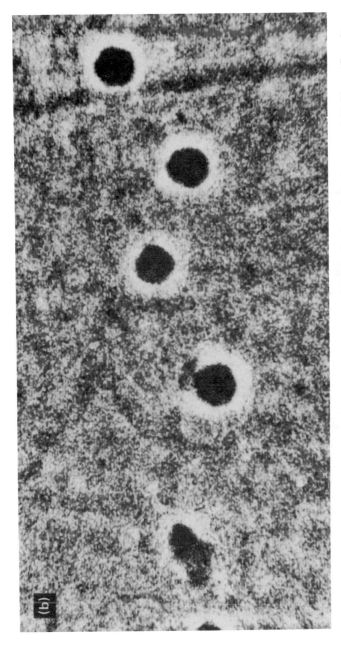

Figure 9.6 Image enhancement of 500-μm (0.02-in.) graphite inclusions in hot-pressed Si_3N_4. (Courtesy of Garrett Turbine Engine Company, Phoenix, Ariz., Division of the Garrett Corporation.)

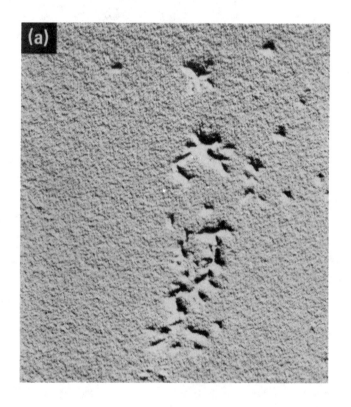

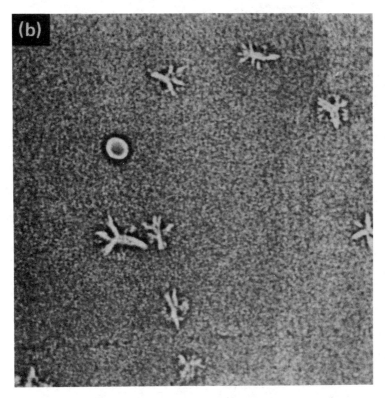

Figure 9.7 Image enhancement of 250-μm (0.01-in.) iron inclusions in hot-pressed Si_3N_4. (Courtesy of Garrett Turbine Engine Company, Phoenix, Ariz., Division of the Garrett Corporation.)

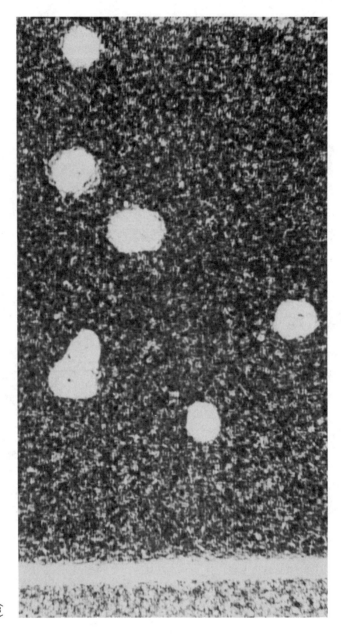

(a)

(b)

Figure 9.8 Image enhancement of 500-μm (0.02-in.) WC inclusions in hot-pressed Si$_3$N$_4$. (Courtesy of Garrett Turbine Engine Company, Phoenix, Ariz., Division of the Garrett Corporation.)

Ultrasonic NDI appears to have excellent potential for the inspection of ceramic materials. However, it must be emphasized that development is only beginning and that success to date has been only on flat plates having machined parallel sides. Complex shapes and parts with rough as-processed surfaces represent a much more difficult problem.

Another limitation of ultrasonics is the loss in intensity in the scattering of the waves as they pass through the material. This is called <u>attenuation</u> and limits the thickness of the part that cab be inspected. Attenuation is accentuated by porosity or other microstructural features that cause scattering (second-phase distributions, microcracking, etc.). Attenuation is also affected by the frequency of the transducer. Increasing the frequency increases the sensitivity of detecting smaller flaws, but also increases scattering and decreases the thickness that can be effectively inspected.

The best potential for near-term improvement in the resolution and shape inspection capability for internal defect detection of ceramics appears

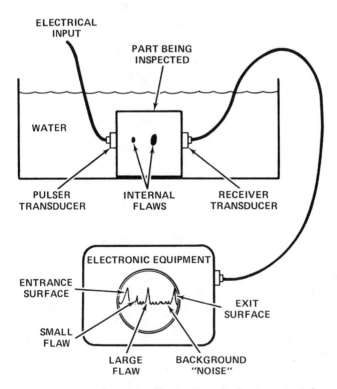

Figure 9.9 Schematic illustrating the basic principles of ultrasonic NDI.

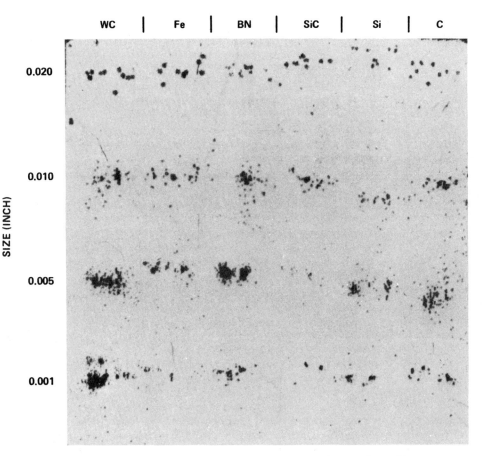

Figure 9.10 Ultrasonic C-scan with a 25-MHz transducer of a 0.64-cm
(0.25-in.)-thick hot-pressed Si_3N_4 plate. (Courtesy of Garrett Turbine
Engine Company, Phoenix, Ariz., Division of the Garrett Corporation.)

to be computer-aided ultrasonics. In conventional systems, resolution
sensitivity is reduced by system noise (from the transducer and electronics)
and material noise (wave scattering by microstructure and surfaces). Seydel [9] has shown that both sources of noise can be reduced significantly by
digitizing the ultrasonic pulses and using a minicomputer for signal averaging. A simple schematic of a computerized system is shown in Fig. 9.11.
    Another technique being developed for inspection of shapes is the use
of an array of very small transducers which essentially "coat" the surface
of the part rather than using a single transducer to scan the part. The signals from this array of transducers can then be analyzed by computer to
locate internal defects.

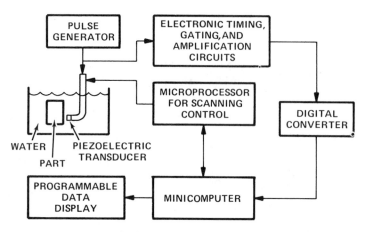

Figure 9.11   Schematic of computer-aided ultrasonic NDI system.

Another ultrasonic method has potential for detecting surface or near-surface flaws. The transducer is placed at a low angle to the surface. The acoustic waves travel along the surface rather than penetrating the interior of the part and interact with surface or near-surface discontinuities. Since the strength of ceramic materials is so sensitive to surface flaws, this may be a useful method to consider.

Penetrants

Penetrants are used extensively for the detection of surface flaws. Usually, a three-step procedure is used: (1) the part is soaked in a fluorescent dye, (2) the part is dried or cleaned in a controlled fashion to remove the dye from smooth surfaces but not from surface defects, (3) the part is examined under ultraviolet light. Surface defects such as cracks and porosity which retain dye show up brilliantly under the ultraviolet light.

The use of penetrants for inspection is widespread and is frequently included as part of a specification. Penetrants are categorized into classes according to their sensitivity and are usually identified in a specification only according to their sensitivity category.

Penetrants are effective for most metals and for nonporous ceramics. If a ceramic has open porosity, the penetrant will usually enter all the pores and result in fluorescence of the whole part, preventing detection of other surface flaws.

Not all penetrants are fluorescent dyes. The KET process exposes the part to radioactive krypton gas and subsequently detects flaws by wrapping the part in film. Radioactive krypton retained in cracks or other defects locally exposes the film. It is important in this technique to have

the film as close as possible to the surface of the part, making complex
shape inspection difficult.

Laser Holographic Interferometry

Laser holographic interferometry is another technique that has potential
for NDI of ceramics and other materials, but has not been extensively evaluated. To understand how interferometry works, we must first understand
what a hologram is. A simple photograph records only the amplitude of
light reflected from an object and is only a two-dimensional image. A
hologram records both the amplitude and phase of the reflected light and
thus produces a three-dimensional image. To achieve a hologram the
object must be illuminated by coherent light (all the light waves in phase).
   Holographic interferometry involves a comparison of the part in the
stressed and unstressed condition. Generally, a hologram is first taken
of the unstressed part. The part is then stressed and its holographic image
superimposed on the original hologram. The slight distortions due to the
stressing result in the reflected light being slightly out of phase. This
shows up as an interference pattern of dark lines or fringes which can be
quantitatively analyzed.
   A uniform material will have a uniform interference pattern when
stressed. A material containing a localized defect or discontinuity will
have a locally distorted interference pattern. The limitations of size and
type of defects detectable by this approach for ceramic materials have not
been determined.

Acoustic Holography

Acoustic holography is similar to laser holography except that coherent
sound waves are used instead of coherent light waves [10]. The schematic
of a typical setup is shown in Fig. 9.12. Ultrasonic waves of a selected
frequency are produced by a transducer and allowed to pass through the
object being inspected. A series of waves of the same frequency are produced by another transducer and aimed such that they meet the first set of
waves at the surface of the liquid in which both transducers and the object
are immersed. The intersecting waves form an interference pattern and
produce a ripple pattern at the liquid surface. Waves that were not affected
by discontinuities in the object produce a strong ripple pattern and waves
affected by a discontinuity produce a weak ripple pattern. The liquid surface is illuminated with coherent light from a laser. Strong ripple patterns
result in much diffraction of the light and weak patterns in little diffraction,
resulting in a range of reflected light intensity from the various portions
of the liquid surface which are proportional to the range in intensity of the
ultrasonic waves transmitted through the object. The reflected light is

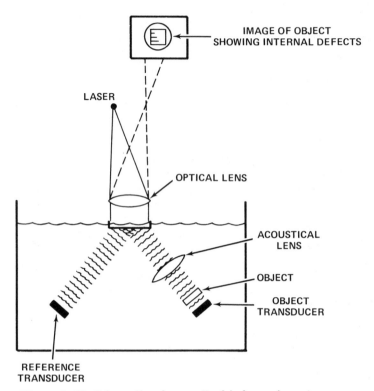

Figure 9.12 Schematic of acoustical holography setup.

captured optically and the image viewed on a screen. What results is a
real-time image showing the internal defects or structure of the object,
similar to the image seen by x-ray.

   Acoustic holography is used for medical research and diagnostics, for
inspection of welds, for locating delaminations or nonbonds in laminates,
and for many other NDI applications. It has been used for metals, organics,
and ceramics and is especially useful for inspection of sections that are
too thick for penetration by x-rays. Small flaws 10 in. below the surface
have been detected in metals such as aluminum, steel, and titanium.

   The flaw size detection capability of acoustic holography for ceramics
has not been extensively studied. Like other NDI techniques, this would
have to be conducted for the specific material and application by preparation
and inspection of standards containing flaws of known size.

Emerging NDI Approaches

The NDI procedures discussed so far have all been demonstrated successfully and are being used in industry for material and component inspection.

However, in many cases these procedures do not meet the inspection requirements of components for advanced applications. This is especially
true for complex advanced ceramic components for heat engine applications,
where stresses in service may result in fracture from internal defects
smaller than 100 μm (0.004 in.) and from surface flaws under 50 μm (0.002
in.). The following paragraphs describe briefly some of the NDI approaches
being developed that have potential for these applications.

High-Frequency Ultrasonics Metals can tolerate much larger defects than
ceramics can tolerate. In many applications a metal can have visible cracks
and still have many hours of life without danger of catastrophic failure,
even under high mechanical or thermal loads. Ultrasonic inspection at low
frequencies (5 to 10 MHz) is usually adequate. However, the small critical
flaw size in ceramics requires higher frequencies. The work reported
earlier was conducted at 25 MHz. Other studies are being conducted at frequencies in the range 150 to 300 MHz in an effort to detect flaws as small
as 10 μm (0.0004 in.) [11].

Initial 300-MHz transducers were made by sputtering an 8-μm-thick
film of zinc oxide (ZnO) on a sapphire (Al_2O_3) rod. The sapphire rod would
then be placed in contact with the object being inspected and act as a waveguide in addition to supporting the piezoelectric ZnO transducer. This system was reported to have detected 25-μm (0.001-in.) inclusions of BN, WC,
Fe, graphite, SiC, and Si in a 0.64-cm (0.25-in.)-thick flat plate of hot-
pressed Si_3N_4. However, considerable development will be necessary before the system is capable of inspecting complex shapes.

A more efficient high-frequency transducer has been produced by bonding single-crystal $LiNbO_3$ to a Si_3N_4 waveguide [11].

Microwave Many ceramics are transparent to microwaves and can therefore be inspected nondestructively in a fashion similar to ultrasonic C-scan.
The object being examined by microwaves does not have to be immersed in
water.

Bahr [12] examined the hot-pressed Si_3N_4 seeded-defect billet described
in Ref. 7 using a cross-polarized transmission technique with 91- to 98-
GHz microwaves. All of the 125-μm (0.005-in.) inclusions were detected.
So also were the 25-μm (0.001-in.) Si inclusions. It appears that microwaves may be one of the better methods of detecting Si inclusions in Si_3N_4.

X-ray Tomography Tomography is better known in the medical field,
where it is commonly referred to as brain scan or body scan. A sequence
of radiographs are taken at many different angles and then reconstructed
by a computer to provide views of the internal structure. This approach
is also being evaluated for inspection of complex-shaped ceramic components [13].

<u>Acoustic Emissions</u> Acoustic waves are emitted when a material is
stressed to the extent that internal or surface flaws are perturbed, formed,
or begin to propagate. For ceramic materials, an increase in acoustic
emission activity occurs prior to fracture and provides a potential means
of either prechecking a component for unacceptable large flaws or monitoring it to detect when failure is imminent.

Acoustic emissions are detected by either attaching a transducer to the
component or by attaching the transducer to a waveguide that is in contact
with the component. The piezoelectric transducer converts wave pulses
that strike it into electrical impulses that are amplified and displayed.

<u>Other Techniques</u> A variety of other advanced NDI techniques are being
developed. Many of these are variations on laser or acoustic holograph
and ultrasonics. If the reader is interested in information on these techniques, a good starting point is the <u>Proceedings of the ARPA/AFML Review</u>
<u>of Progress in Quantitative NDE</u>, from which Refs. 11 and 12 were taken.

REFERENCES

1. L. P. Wynn, D. J. Tree, T. M. Yonushonis, and R. A. Solomon,
    Proof testing of ceramic components, in <u>Proceedings of the 1977</u>
    <u>DARPA/NAVSEA Ceramic Gas Turbine Demonstration Engine Program</u>
    <u>Review</u> (J. W. Fairbanks and R. W. Rice, eds.), MCIC Report MCIC-
    78-36, 1978, pp. 493-516.
2. S. Wiederhorn, Reliability, life prediction, and proof testing of
    ceramics, in <u>Ceramics for High Performance Applications</u> (J. J.
    Burke, A. E. Gorum, and R. N. Katz, eds.), Brook Hill Publishing
    Co., Chestnut Hill, Mass., 1974, pp. 635-663. (Available from MCIC,
    Battelle Columbus Labs., Columbus, Ohio.)
3. Defect detection in hot-pressed Si_3N_4, in <u>Ceramics for High Per-</u>
    <u>formance Applications</u> (J. J. Burke, A. E. Gorum, and R. N. Katz,
    eds.), Brook Hill Publishing Co., Chestnut Hill, Mass., 1974, pp.
    665-685. (Available from MCIC, Battelle Columbus Labs., Columbus,
    Ohio.)
4. D. W. Richerson, J. J. Schuldies, T. M. Yonushonis, and K. M. Jo-
    hansen, ARPA/Navy ceramic engine materials and process development
    summary, in <u>Ceramics for High Performance Applications</u> (J. J.
    Burke, A. E. Gorum, and R. N. Katz, eds.), Brook Hill Publishing
    Co., Chestnut Hill, Mass., 1974, pp. 625-650. (Available from MCIC,
    Battelle Columbus Labs., Columbus, Ohio.)
5. D. J. Cassidy, NDE techniques used for ceramic turbine rotors, in
    <u>Ceramics for High Performance Applications, II</u> (J. J. Burke, E. N.
    Lenoe, and R. N. Katz, eds.), Brook Hill Publishing Co., Chestnut
    Hill, Mass., 1978, pp. 231-242. (Available from MCIC, Battelle
    Columbus Labs., Columbus, Ohio.)

6. J. J. Schuldies and W. H. Spaulding, Radiography and image enhancement of ceramics, in Proceedings of the 1977 DARPA/NAVSEA Ceramic Gas Turbine Demonstration Engine Program Review (J. W. Fairbanks and R. W. Rice, eds.), MCIC Report MCIC-78-36, 1978, pp. 403-428.
7. J. J. Schuldies and D. W. Richerson, NDE approach, philosophy and standards for the ARPA/NAVSEA ceramic turbine program, in Proceedings of the 1977 DARPA/NAVSEA Ceramic Gas Turbine Demonstration Engine Program Review (J. W. Fairbanks and R. W. Rice, eds.), MCIC-78-36, 1978, pp. 381-402.
8. J. J. Schuldies and T. Derkacs, Ultrasonic NDE of ceramic components, in Proceedings of the 1977 DARPA/NAVSEA Ceramic Gas Turbine Demonstration Engine Program Review (J. W. Fairbanks and R. W. Rice, eds.), MCIC Report MCIC-78-36, 1978, pp. 429-448.
9. J. A. Seydel, Improved discontinuity detection in ceramic materials using computer-aided ultrasonic nondestructive techniques, in Ceramics for High Performance Applications (J. J. Burke, A. E. Gorum, and R. N. Katz, eds.), Brook Hill Publishing Co., Chestnut Hill, Mass., 1974, pp. 697-709. (Available from MCIC, Battelle Columbus Labs., Columbus, Ohio.)
10. B. B. Brenden, Recent developments in acoustical imaging, Materials Research and Standards, MTRSA II(9), 16 (1971).
11. G. S. Kino, B. T. Khuri-Yakub, Y. Murakami, and K. H. Yu, Defect characterization in ceramics using high frequency ultrasonics, in Proceedings of the ARPA/AFML Review of Progress in Quantitative NDE, AFML-TR-78-25, 1979, pp. 242-245.
12. A. J. Bahr Proceedings of the ARPA/AFML Review of Progress in Quantitative NDE, AFML-TR-78-25, 1979, pp. 236-241.
13. D. J. Cassidy and M. F. Elgart, X-ray Radiography of Gas Turbine Ceramics, Annual Report No. 1 under ONR Contract N00014-78-C-0714, 1979.

III

DESIGN WITH CERAMICS

Part I discussed the relationships among atomic bonding, crystal structure, and properties of ceramics as compared to other materials. It was shown that the intrinsic properties are controlled largely by the nature of the bonding and structure, but that the extrinsic or actual properties are controlled by such factors as structural defects, impurities, and fabrication flaws.

Part II reviewed the fabrication processes for ceramic materials and components, defined potential sources in these fabrication processes of property-limiting flaws, and described techniques for detecting and limiting the occurrence of these flaws.

The objective of Part III is to apply the property, fabrication, and inspection principles learned in Parts I and II to the selection and design of ceramic components for advanced engineering applications. Chapter 10 discusses design considerations, such as requirements of the application, property limitations, fabrication limitations, cost limitations, and reliability requirements. Chapter 11 considers design approaches. The approach is normally based on the design considerations and can range from empirical to deterministic to probabilistic. Chapter 12 explores the importance and techniques of failure analysis. If a ceramic component fails, often the only means of determining whether the failure was design oriented or material oriented is by examination of the fractured pieces.

The final chapter, Chap. 13, reviews a range of ceramic applications. Emphasis is on the criteria for selecting the best material for each application based on the design considerations and design approaches described in Chaps. 10 and 11.

10

Design Considerations

The selection of a material and a fabrication process for a component for an engineering application is governed by a variety of factors, not just the material properties. The shape and cost limitations of the fabrication process must be considered. The requirements of the application, including such factors as load distribution, environment, and tolerances, must be considered, as also must be the reliability requirements, such as life expectancy, the risk of premature failure, and the effects of premature failure on the rest of the system and personnel.

10.1 REQUIREMENTS OF THE APPLICATION

The first step in design of a ceramic component or any other component is to define clearly and prioritize the requirements of the application. Usually, one or two characteristics will be most critical and allow an initial selection of candidate materials. For instance, a primary characteristic of a wear-resistant material is hardness. However, if wear resistance is required in a severe chemical environment or at high temperature, other characteristics become critical and must be considered on an equal or nearly equal basis to hardness. Table 10.1 lists some of the design characteristics that an engineer must consider for an application.

To get a better feeling for the thought process that an engineer goes through in defining and prioritizing the critical design requirements for an application, consider two examples: a grinding wheel and a gas turbine rotor. These both rotate at high speeds and must have similar design requirements. Right? Not necessarily.

The grinding wheel and rotor do have some important requirements in common:

1. They must have suitable strength to remain intact at their respective design speeds.

Table 10.1 Examples of Design Characteristics That Must Be Considered

Load	Tolerances
Stress distribution	Surface finish
Attachment	Stability to radiation
Interfaces	Life requirement
Friction	Safety requirements
Chemical environment	Toxicity
Temperature	Pollution
Thermal shock	Electrical property requirements
Creep	Magnetic property requirements
Strain tolerance	Optical property requirements
Impact	Cost
Erosion	Quantity

2. They must have an acceptable margin of safety as defined by industry and government standards.
3. They must be fabricated such that they are in balance when rotating.

Other critical design requirements of the grinding wheel include controlled surface breakdown to expose fresh abrasive grains, impact resistance, low cost, and adaptability to mass production. Other critical design requirements for the rotor include high strength and oxidation/corrosion resistance at high temperature, resistance to extreme thermal shock, and complex shape fabrication to close tolerances. The differences in design requirements result in very different material design selections with corresponding design, manufacturing, and quality control choices. The grinding wheel is best made from a composite material with hard abrasive particles bonded by a softer matrix. The turbine rotor requirements have not yet been met reliably by a ceramic material and apparently will require further material and design development, although there is evidence that the problem may soon be solved.

Design requirements can be determined in many ways. For existing applications, where an alternative material is being sought to achieve benefits such as lower cost, longer life, or improved performance, a specification usually exists defining quantitatively the critical design requirements. This can be a good starting point. However, one must remember that ceramics have different properties than other materials and that modification or redesign may be necessary. The engineer should especially consider thermal expansion mismatch (if the component is to be used over a temperature range) and the implication of point loading or flexural loading.

For new applications, design requirements will either have to be assumed based on the best estimates of service conditions, estimated by analogy with similar applications, determined experimentally or predicted analytically. This can result in a multiphase program in which the first phase would be design analysis and material property screening; the second phase would then be fabrication of prototypes; the third phase, component testing, will overlap with the second phase and allow iteration back and forth between prototype fabrication, component testing, and redesign.

10.2 PROPERTY LIMITATIONS

The second step in design of a ceramic component is to compare the properties of candidate ceramic materials with the requirements of the application. This is usually hampered by lack of property data at the design conditions, especially if an adverse service environment is involved. However, an initial set of candidates that have the closest fit with the design requirements can usually be defined. These candidates can then be included in screening tests to isolate the best candidate.

The method and extent of property evaluation varies according to the nature of the application. Some materials may clearly satisfy the property requirements so that no measurements are necessary. Such is the case in many room-temperature wear-resistance applications, where technical ceramics such as polycrystalline sintered Al_2O_3 or hot-pressed B_4C have strengths a factor of 10 higher than design loads and more than adequate hardness. In this case, factors such as cost and large quantity availability are usually more important and determine the final selection.

In other applications extensive property and QC measurements are required. This is especially true of electrical and magnetic ceramics, where properties must be precisely controlled. It is also true of optical applications, where index of refraction, absorption, and color more often must be controlled to a tight tolerance.

Various approaches can be pursued in evaluating the suitability of a material's properties. If the shape is simple and the part can be fabricated quickly and inexpensively, it may be best to make the part to print and test it directly in the system being developed. This has the potential of leading to commercialization with a minimum of time and development cost. However, the engineer must carefully assess the consequences of a failure during this testing. Will a failure damage much more costly components in the system? Will it endanger personnel or facilities? Will initial test parts be of high-enough quality to provide a meaningful component test, or should material development and property verification be conducted first?

For many advanced applications, no existing material is clearly suitable. In fact, at the current time, we are design limited in most advanced materials applications. This means that engineers have already identified approaches to improving overall systems, but do not have materials with acceptable properties. Therefore, these applications are dependent on material development; often the project engineer has the responsibility to get this development done. Two examples where material development is required follow.

Fiber-optic communication has great potential. Small fiber bundles transmitting coherent laser light can carry many times the information that can currently be carried by conventional wire cables of much larger diameter. The current problem is loss of intensity of the light signal due to absorption and scattering within a fiber, limiting the distance that can be transmitted without an amplifier station. Glass fiber improvements are required to eliminate the imperfections that cause the losses. Success in this continuing development endeavor will revolutionize the communications industry.

Ceramics are currently being evaluated for gas turbine components to allow increased operating temperatures. By increasing operating temperatures from current metal-limited levels of 1800 to 2100°F to 2500°F or greater, fuel savings from 10 to 25% could be achieved. The feasibility has been demonstrated [1-3], but present ceramic materials do not yet have the predictably reproducible strength to provide long-term reliability [4].

As discussed in previous chapters, property limitations frequently result from fabrication limitations. The property-controlling material defects occur during the various steps of processing. Often, design needs can be met simply by increased care during processing. Sometimes this can be achieved by a minor modification in the processing specification. Other times, iterative development will be necessary.

Another factor that affects properties is the quantity of parts being manufactured. Industry experience has shown that part-to-part variation is usually high in prototype or small production quantities, but decreases substantially when high-volume production is reached.

10.3 FABRICATION LIMITATIONS

Comparison of the design requirements with the property limitations dictates the fabrication requirements. At this point, two primary questions will be asked: (1) Will existing fabrication experience and technology achieve the required properties? (2) Can existing fabrication experience and technology achieve the required configuration in the necessary quantity at an acceptable cost?

If the answer to the first question is "yes," the engineer can concentrate on the second question. If the answer is "no," then the following options need to be considered.

1. Achieve the required properties by improvement of an existing commercial material or fabrication process.
2. Continue development of an emerging or developmental material or fabrication process.
3. Develop a new material fabrication process or material system (such as a composite).

Obviously, the difficulty, time, and cost will increase substantially if item 2 or 3 is the only feasible option rather than item 1. It is the engineer's responsibility to assess which level of development is required and whether the program resources are adequate to implement the development. Many programs have failed or experienced substantial cost overruns because an engineer did not make an adequate assessment of the material property and fabrication limitations.

Shape capability is the next critical fabrication concern. Once shape and tolerances have been defined for the application by the design analysis, the engineer must evaluate the fabrication approaches and manufacturing sources. This is usually best done by direct discussion with the material suppliers; however, finding the appropriate supplier to talk to is the first step. The following are potential sources of information:

1. Thomas's Register.
2. Ceramic publications such as the Bulletin and the Journal of the American Ceramic Society and Ceramic Industry; both the Bulletin and Ceramic Industry publish a yearly directory of suppliers of materials, services, and finished components.
3. The library, especially reference books such as Ceramic Abstracts and Chemical Abstracts.
4. Special abstracting services such as Chemical Abstracts at Columbus, Ohio; Materials and Ceramics Information Center at Battelle Columbus Laboratories; and National Technical Information Service, Springfield, Va.
5. Ceramic consultants, usually listed in the classified ads of monthly ceramic publications.
6. Professors at universities that offer degrees in ceramic engineering or materials science, such as Massachusetts Institute of Technology, Pennsylvania State University, University of Washington, University of Utah, Ohio State University, Iowa State University, University of Illinois, University of California at Los Angeles, University of California at Berkeley, State University of New York at Alfred, Virginia Polytechnic Institute, and University of Michigan.
7. Research Institutes such as Battelle Columbus Laboratories and IIT Research Institute.
8. Annual meeting and exposition of the American Ceramic Society.

The first contact with a supplier involves a description of the required component together with critical considerations such as service environment, quantity required, and key properties. If this first discussion is encouraging, a set of prints or drawings is sent to the potential supplier for further evaluation. This is usually followed by meetings during which the final procurement decision is made and program details are negotiated.

An engineer with a knowledge of the various ceramic fabrication processes has a pronounced advantage in evaluating the fabrication limitations associated with a new design. Processes such as uniaxial pressing and extrusion are very good for reproducibly fabricating large quantities of simple parts. Injection molding can produce more complex parts in large quantity, but greater care is necessary in tool design and quality control because of the increased likelihood of fabrication flaws. Slip casting can also produce complex parts, but in lesser quantity than pressing or injection molding. For high-strength, high-reliability requirements, hot pressing might be considered, but one must remember the difficulties and cost of achieving complex shape by this process.

A development program is usually required to fabricate a new ceramic component. A typical flowchart is shown in Fig. 10.1. The steps usually consist of tool design, tool fabrication, fabrication of initial parts, evaluation of the dimensions and integrity of these parts, tool redesign and rework as required, fabrication of parts, inspection, testing of the parts in the application or a simulation rig, and iteration as required. Frequently, prototype parts will be made by a different fabrication process than the one intended for production. This is done to minimize program cost, especially tool cost. Once feasibility has been demonstrated and a workable design configuration verified, it is much easier to justify large capital outlay for expensive production tooling. For instance, a complex injection molding tool can cost more than $100,000. The program would be quite expensive

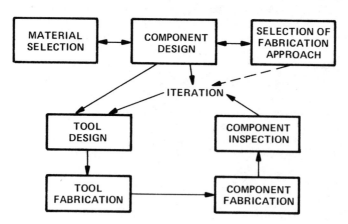

Figure 10.1 Schematic of program approach to develop fabrication capability for a new ceramic part.

if several retoolings were required to determine the optimum component design. However, making the initial parts by cold isostatic pressing and green machining or by slip casting could limit initial tooling cost to perhaps $10,000.

10.4 COST CONSIDERATIONS

Cost is an important design consideration and must be evaluated concurrently with other factors such as property, fabrication, and reliability requirements. Although it must be realized that initial prototype parts will be expensive and that adequate information may not be available to project production costs, an initial cost analysis should be conducted and a strategy defined for obtaining the necessary information and achieving the ultimate production-cost objectives. There have been many programs in which an engineer has ignored cost considerations and ultimately made a component work, only to find that there is no way of reducing the system cost to a marketable level. Conversely, there have been other programs not started or terminated prematurely because an engineer took high prototype costs too seriously and did not adequately evaluate production-cost projections.

Cost projection has many pitfalls. The individual engineer should not assume the whole responsibility, but should seek other individuals with as much experience as possible. The optimum consultant would be a person who has solved and commercialized a different ceramic component of the same material for a similar application. In cost projection there is no equivalent for experience and technical understanding of the specific material and process.

10.5 RELIABILITY REQUIREMENTS

The reliability requirements are also part of the initial requirements of the application and may be ultimately written into a specification or a warranty. The term "reliability" is really rather ambiguous and varies dramatically depending on the application [3]. For instance, the heat shield tiles on the space shuttle must be 100% reliable for the required time. If only one tile fails, burnthrough could result and lead to destruction of the vehicle. On the other hand, breakage of a household floor or wall tile causes some inconvenience, but does not jeopardize life or equipment. A similar comparison could be made between the glass windows in a deep-sea submergence vehicle and those in an automobile or in a house. Each has its own definition of reliability.

The following are some of the factors that must be considered when evaluating reliability requirements.

1. The acceptable failure rate for the application
2. The type of warranty for the system and its subcomponents
3. Expectations of the potential customer
4. Safety requirements defined by industry or government regulations

10.6 SUMMARY

The probability of success of a new ceramic component can be effectively increased by using a systematic design approach which first quantitatively defines the requirements of the application and then evaluates candidate materials in terms of property and fabrication limitations as well as cost and reliability requirements. The probability of success can be further improved by an iterative, overlapping program in which close liaison is maintained between designers and manufacturers throughout the development and demonstration program.

REFERENCES

1. J. E. Harper, ARPA/NAVAIR ceramic gas turbine engine demonstration program, in Ceramics for High Performance Applications, III (J. J. Burke, E. N. Lenoe, and R. N. Katz, eds.), Plenum Publishing Co., New York, 1982.
2. A. G. Metcalfe, in Ceramics for High Performance Applications, III (J. J. Burke, E. N. Lenoe, and R. N. Katz, eds.), Plenum Publishing Co., New York, 1982.
3. R. N. Katz, Science 208, 841-847 (May 23, 1980).
4. Reliability of Ceramics for Heat Engine Applications, prepared by the Committee on the Reliability of Ceramics for Heat Engine Applications, Natl. Acad. Sci. Publ. NMAB-357, Washington, D.C., 1980.

11
Design Approaches

In Chap. 10 we discussed briefly some of the important considerations of component design in general and ceramic design in particular. In this chapter we discuss in more detail the design approaches for ceramics. For the purposes of this discussion, design approaches can be divided roughly into five categories:

1. Empirical
2. Deterministic
3. Probabilistic
4. Linear elastic fracture mechanics
5. Combined

11.1 EMPIRICAL DESIGN

Empirical design is a trial-and-error approach that emphasizes iterative fabrication and testing and deemphasizes mathematical modeling and analysis. It can be the most effective approach in cases where a ceramic is already in use and is only being modified and in cases where mechanical loads are minimal. It can also be the optimum approach when the available property data for the candidate ceramic material are too limited for the more analytical approaches. Finally, empirical design may be the only approach, or may be required in addition to analytical approaches, where the survival of a component is strongly affected by environmental factors such as chemical attack or erosion.

Historically, most ceramic design has been empirical, especially with traditional ceramics. Only recently, with the advent of ceramics in demanding structural applications, has it become necessary to use analytical approaches.

11.2 DETERMINISTIC DESIGN

Deterministic design is a standard "safety-factor" approach. The maximum stress in a component is calculated by finite element analysis or closed-form mathematical equations [1]. A material is then selected which has a strength with a reasonable margin of safety over the calculated peak component stress. The margin of safety is usually determined from prior experience, so that this approach is really a combination of analytical and empirical.

The deterministic approach is routinely used with design of metals. It works well, largely because metals have relatively low property scatter. Often, metals can be designed within a small margin of their ultimate strength and used with confidence that they will not fail prematurely. This is not true with current structural ceramic materials. Ceramics have wide strength scatter and the measured strength is affected by the volume and area of material under stress (as discussed in detail in Chap. 3). Deterministic design does not account for the effects of flaw distribution and variations in strength with volume and area stressed. Thus when deterministic design is used with highly stressed ceramic components, large safety factors are required to assure a reasonably low risk of failure.

Figure 11.1 compares the typical strength data for a high-strength metal and a high-strength ceramic. The solid lines represent the average strength and the dashed lines represent plus and minus three times the standard deviation. Normal statistics [2] defines the average strength $\bar{\sigma}$ as

$$\bar{\sigma} = \frac{\sum_{i=1}^{N} \sigma_i}{N} \tag{11.1}$$

and the standard deviation S as

$$S = \left[\frac{\sum_{i=1}^{N} (\sigma_i - \bar{\sigma})^2}{N} \right]^{1/2} \tag{11.2}$$

where σ_i is the bend strength of individual test bars and N the number of test bars.

The data for the Inconel 713LC are unpublished data from the Garrett Turbine Engine Company based on specimens machined from as-cast axial gas turbine rotors. The curve represents 96 uniaxial tensile tests. The Si_3N_4 data are from Ref. 3. The 4-point bend tests were conducted on

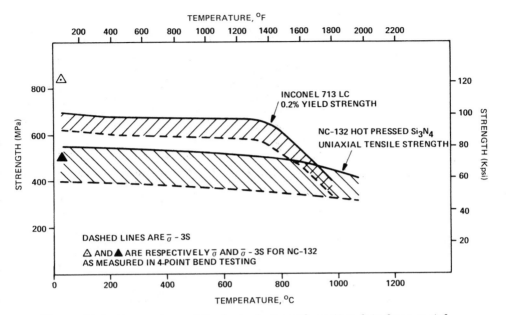

Figure 11.1 Comparison of the typical strength scatter data for a metal and a ceramic. (Data compiled from Garrett Turbine Engine Company, Phoenix, Ariz.)

specimens with a 0.32×0.64 cm (0.125×0.250 in.) cross section machined in the longitudinal direction with a 320-grit diamond grinding wheel and tested with an inner span of 1.9 cm (0.75 in.) and an outer span of 3.8 cm (1.5 in.). The Si_3N_4 tensile tests were conducted with specimens machined circumferentially (transverse direction) with a 320-grit diamond grinding wheel and then oxidized at 980°C (1800°F) for 50 hr to remove much of the grinding damage [3]. The specimens had a gauge length and diameter of 3.18 cm (1.25 in.) and 0.48 cm [0.188 in.), respectively, and were tested at Southern Research Institute using a tensile test apparatus with a sophisticated gas bearing load train to minimize parasitic bend stresses.

The information in Fig. 11.1 points out some of the problems in designing with ceramics and in particular in the use of the deterministic approach. First, a different strength is measured for ceramics depending on the size of the specimen, the volume and area under stress, and the nature of the surface finish. The deterministic approach has no way of effectively taking these factors into account. Second, the ceramics have a much larger scatter in strength data than metals do, which makes the selection of safety factors very difficult.

11.3 PROBABILISTIC DESIGN

Empirical and deterministic design approaches may be adequate for most ceramic applications, but are limited in cases where high stresses or complex stress distributions are present. In such cases, a probabilistic approach which takes into account the flaw distribution and the stress distribution in the material may be required.

Weibull Statistics

Currently, the most popular means of characterizing the flaw distribution is by the Weibull [4] approach. It is based on the weakest link theory, which assumes that a given volume of ceramic under a uniform stress will fail at the most severe flaw. It thus presents the data in a format of probability of failure F versus applied stress σ, where F is a function of the stress and the volume V or area S under stress

$$F = f(\sigma, \ V, \ S) \tag{11.3}$$

Weibull proposed the following relationship for ceramics:

$$f(\sigma) = \left(\frac{\sigma - \sigma_\mu}{\sigma_0}\right)^m \tag{11.4}$$

where σ is the applied stress, σ_μ the threshold stress (i.e., the stress below which the probability of failure is zero), σ_0 is a normalizing parameter (often selected as the characteristic stress, at which the probability of failure is 0.632), and m the Weibull modulus, which describes the flaw size distribution (and thus the data scatter). The probability of failure as a function of volume is

$$F = 1 - \exp\left[-\int_V \left(\frac{\sigma - \sigma_\mu}{\sigma_0}\right)^m dV\right] \tag{11.5}$$

This results in the shape of curve shown in Fig. 11.2. Such a curve can easily be plotted from experimental data by estimating F by n/(N + 1), where n is the ranking of the sample and N the total number of samples. This is plotted versus the measured strength value for each value of n as shown in Table 11.1 and Fig. 11.3 for a hypothetical set of data.

The curve in Fig. 11.3 provides only an approximation of the probability of failure and does not yield the m value. Plotting $\ln \ln [1/(1 - F)]$, calculated

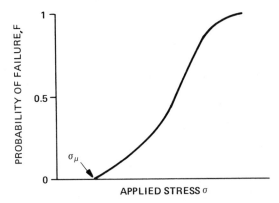

Figure 11.2 Typical Weibull distribution.

using equation (11.5), versus ln σ results in a straight line of slope m. This form of the Weibull curve is used extensively in depicting the reliability or predicted reliability of materials or components. Figure 11.4 shows the data distribution resulting from bend strength testing of 30 reaction-bonded Si_3N_4 specimens plotted in this fashion. The m value

Table 11.1 Organization of Experimental Data to
Plot Weibull Curve

Number of ordered data	Measured strength σ (MPa)	Estimated probability of failure, $F \sim \dfrac{n}{N+1}$
1	178	0.1
2	210	0.2
3	235	0.3
4	248	0.4
5	262	0.5
6	276	0.6
7	296	0.7
8	318	0.8
9	345	0.9

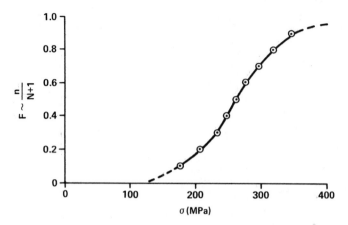

Figure 11.3 Weibull curve plotted from experimental data in Table 11.1.

determined from the slope is 5.5. This is a relatively low value and indicates substantial scatter.

Equations (11.4) and (11.5) represent three-parameter Weibull functions where σ_μ, σ_0, and m are the three parameters. Usually, a two-parameter form is used for ceramics, where the threshold stress σ_μ is set equal to zero. Thus the equation becomes

$$F = 1 - \exp\left[-\int_V \left(\frac{\sigma}{\sigma_0}\right)^m dV\right] \qquad (11.6)$$

Cracks initiate and propagate in ceramics under tensile loading rather than compressive loading, so that only the volume or area of material under tension is of concern in the Weibull equation. Therefore, if the full volume is under uniform uniaxial tension, the two-parameter equation becomes

$$F = 1 - \exp\left[-V\left(\frac{\sigma}{\sigma_0}\right)^m\right] \qquad (11.7)$$

If the loading is 3-point or 4-point bending, the effective volume under tensile stress is substantially lower. For 3-point bending, the effective volume is equal to $V/2(m + 1)^2$ and for 4-point bending is $V(m + 2)/4(m + 1)^2$. For an m of 10, the effective volume for 3-point and 4-point loading are, respectively, only 0.004 and 0.025 of the beam volume under load.

Use of the Weibull Distribution in Design

Plotting the Weibull curve from experimental data provides useful information about the probability of failure versus applied stress for the material,
but does not provide an assessment of failure probability of the component.
To do this, the Weibull distribution for the material must be integrated with
the stress distribution for the component. This can be done conveniently
with the use of finite element analysis [6, 7]. The material strength data in
the form of a Weibull probability curve is compared with each finite element
to determine the probability of failure of that specific element. The

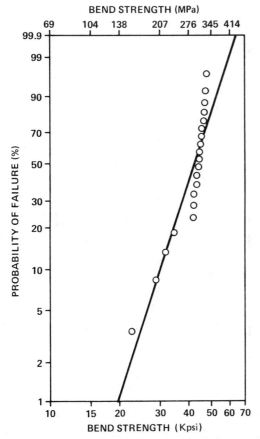

Figure 11.4 Example of Weibull curve generated from strength test data
for reaction-bonded Si$_3$N$_4$. (Data from Ref. 5.)

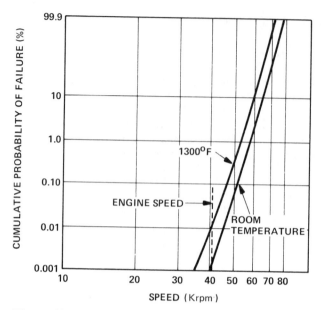

Figure 11.5 Weibull curve predicting reliability of gas turbine rotor blades, resulting from a comparison of the finite element analysis of the design stresses and the Weibull distribution of the material bend strength data. (From Ref. 19.)

probabilities, both volume and surface, for all the elements are then summed to determine the probability of failure for the component. This can also be plotted on a Weibull curve as a function of some component operation parameter. For instance, Fig. 11.5 shows a Weibull curve that predicts the reliability of hot-pressed Si_3N_4 rotor blades in an advanced gas turbine engine application [8] as a function of rotation speed.

Advantages of Probabilistic Design

The primary advantage of a probabilistic approach is that it allows designing closer to the properties of the material. For instance, in a component with high but localized stress, a material can be successfully used with a low probability of failure, even though its measured property scatter band overlaps the peak design stress. The probability is low that one of the more severe flaws in the material will be in the region of peak stress. Deterministic design in this case would probably conclude that the ceramic material was not acceptable.

Use of probabilistic design allows a trade-off in material selection between high strength and low scatter. For instance, an application requiring 0.99 reliability could be satisfied by a material with a characteristic strength of 120 MPa and an m of 8 or a material with a strength of 94 MPa and an m of 10 or a material with a strength of only 61 MPa and an m of 16. The advantage of probabilistic design is that these trade-offs are considered and can be integrated into the design analysis. This visibility and flexibility is not available in empirical and deterministic approaches.

Limitations of Probabilistic Design

Probabilistic design is limited primarily by inadequacy in defining peak stresses and stress distributions in the component and in defining the true strength-flaw size distribution in the candidate material. Stresses in the component arise from thermal and mechanical loading. The accuracy of predicting the magnitude and distribution of these stresses is restricted by the accuracy of defining the boundary conditions. For instance, the local heat transfer conditions of both the component and the environment plus all the effects of geometry (e.g., heat sinks) and thermal conductivity plus material anisotropy plus boundary layer effects must all be accurately defined or assumed before the thermal stress can be accurately calculated. Similarly, precise loads and load application angles plus contact areas and friction coefficients plus the effects of geometry and tolerances must all be accurately defined or assumed before the mechanical design stresses can be accurately calculated. Substantial inaccuracies in boundary conditions for metals can be tolerated because both thermal and mechanical stresses can redistribute slightly due to the ductility. Such is not the case with ceramics. Stresses are concentrated and cannot be adequately predicted unless the boundary conditions are accurately defined.

There are also pitfalls in defining the true strength-flaw size distribution in ceramics. Because of the large size range of flaws in a ceramic part and the resulting wide scatter in strength values, a large number of specimens must be strength-tested to determine adequately the Weibull parameters. For instance, calculation of the Weibull modulus based upon 10 bend strength tests is only accurate to approximately ±40%. About 300 bend strength tests are necessary to obtain an accuracy in the m-value calculation of about ±10%. An uncertainty of 40% in the Weibull modulus for the material strength results in a substantial uncertainty in the probability of failure for a component.

A further problem in accurately defining the m value is the effect of bimodal or multimodal flaw size distributions. The Weibull relationship in equation (11.5) assumes a uniform, random, unimodal flaw size distribution. As long as such a uniform unimodal distribution is present in the material,

the calculated m value makes sense and can be used by routine procedures
for probabilistic design. However, if the distribution is multimodal, but
the data are used as a single distribution in calculating the Weibull modulus,
the resulting m value will not be as suitable for design. This is especially
true when using the three-parameter Weibull approach.

Figure 11.4 shows a good example of treating a bimodal strength distribution with a unimodal Weibull curve. The Weibull modulus was only 5.5,
whereas the goal of the program was to achieve a material with an m value
of 10 or greater. Based only on the measured m value, the engineer would
probably have rejected this material. However, analysis of the fracture
surfaces of each test bar showed that the bottom four data points resulted
from a different flaw population than the rest of the data points. Specifically, these four lower strengths were due to large laminar porous flaws
that had occurred prior to the nitriding process step of the reaction-bonded
Si_3N_4 fabrication process and were of a size and type that could either be
detected and rejected by NDI or eliminated by further process development.
The latter was accomplished, resulting in a unimodal distribution and an
increase in the Weibull modulus of the material to greater than 10 [5]. The
major point to be made here is that the engineer should look for multimodal
distributions and then use techniques such as fractography to understand
the data rather than blindly using the calculated Weibull data. The importance and techniques of fractography are discussed in Chap. 12.

11.4 LINEAR ELASTIC FRACTURE
MECHANICS APPROACH

Linear elastic fracture mechanics is a useful approach to the design of
ceramics and other brittle materials. It takes into account that the material
(and thus the component) contains flaws and it treats fracture in terms of
fracture toughness, stress intensity factors, and flaw size rather than ultimate strength or yield strength.

The intent of this section is not to describe the theory and techniques
of fracture mechanics, but rather to alert the engineer to its existence and
provide some references. The technology is still evolving and the uses
expanding. Early work on the fundamentals of fracture mechanics was conducted by Griffith [9], Irwin [10], and Williams [11]. More recent work
relating to ceramics has been reported in Refs. 12 through 14. General
discussion and reviews are available in Refs. 15 through 18.

11.5 COMBINED APPROACHES

It is apparent from the prior discussions in this chapter that a variety of
design approaches are available. The approach selected will depend on the

severity of the application, the timing, the available budget, and the existing data base. Usually, the final approach will be a combination of empirical, deterministic, or probabilistic and fracture mechanics. As an example, empirical studies may be conducted to screen candidate materials in parallel with deterministic or probabilistic analysis. Once a configuration has been selected and is in the test phase, fracture mechanics and probabilistic analyses may be used to predict life and reliability.

REFERENCES

1. A. P. Boresi, O. M. Sidebottom, F. B. Seely, and J. O. Smith, Advanced Mechanics of Materials, 3rd ed., John Wiley & Sons, Inc., New York, 1978.
2. C. Lipson and N. J. Sheth, Statistical Design and Analysis of Engineering Experiments, McGraw-Hill Book Company, New York, 1973.
3. AiResearch Report No. 76-212188(9), Ceramic Gas Turbine Engine Demonstration Program, Interim Report Number 9, prepared under NASC Contract N00024-76-C-5352, May 1978, pp. 4-1 to 4-11.
4. W. Weibull, A statistical distribution function of wide applicability, J. Appl. Mech., $18[3]$, 293-297 (September 1951).
5. K. M. Johansen, D. W. Richerson, and J. J. Schuldies, Ceramic Components for Turbine Engines, AiResearch Mfg. Co. Report 21-2794(08), prepared under contract F33615-77-C-5171, 1980, Appendix A, p. 67.
6. G. J. DeSalvo, Theory and Structural Design Applications of Weibull Statistics, WANL-TME-2688, 1970.
7. D. J. Tree and H. L. Kington, Ceramic component design objectives, goals and methods, in Ceramic Gas Turbine Demonstration Engine Program Review (J. W. Fairbanks and R. W. Rice, eds.), MCIC Report MCIC-78-36, 1978, pp. 41-75.
8. F. B. Wallace, A. J. Stone, and N. R. Nelson, Ceramic component design, ARPA/Navy ceramic engine program, in Ceramics for High Performance Applications, II (J. J. Burke, E. N. Lenoe, and R. N. Katz, eds.), Brook Hill Publishing Co., Chestnut Hill, Mass., 1978, pp. 593-624. (Available from MCIC, Battelle Columbus Labs., Columbus, Ohio.)
9. A. A. Griffith, Philos. Trans. R. Soc. Lond. Ser. A $221(4)$, 163-198 (1920-1921).
10. G. R. Irwin, Analysis of stresses and strains near the end of a crack traversing a plate, J. Appl. Mech. $24(3)$, 361-364 (1957).
11. M. L. Williams, On the stress distribution at the base of a stationary crack, J. Appl. Mech. $24(1)$, 109 (1957).
12. R. W. Davidge, J. R. McLaren, and G. Tappin, Strength-probability-time (SPT) relationships in ceramics, J. Mater. Sci. $8(12)$, 1699-1705 (1973).

13. S. M. Wiederhorn, A. G. Evans, E. R. Fuller, and H. Johnson, Application of fracture mechanics to space-shuttle windows, J. Am. Ceram. Soc. 57(7), 319-323 (1974).

14. R. C. Bradt, D. P. H. Hasselman, and F. F. Lange, eds., Fracture Mechanics of Ceramics, Vols. 1 and 2 (1973), Vols. 3 and 4 (1978), Plenum Publishing Corp., New York.

15. J. B. Wachtman, Jr., Highlights of progress in the science of fracture of ceramics and glass, J. Am. Ceram. Soc. 57(12), 509-519 (1974).

16. J. F. Knott, Fundamentals of Fracture Mechanics, Butterworth & Company (Publishers) Ltd., Kent, England, 1973.

17. A. S. Tetelman and A. J. McEvily, Fracture of Structural Materials, John Wiley & Sons, Inc., New York, 1967.

18. W. H. Dukes, Handbook of Brittle Material Design Technology, AGARDograph No. 152, Dec. 1970.

19. D. W. Richerson and T. M. Yonushonis, Environmental effects on the strength of silicon nitride materials, in DARPA/NAVSEA Ceramic Gas Turbine Demonstration Engine Program Review, MCIC Report MCIC-78-36, 1978, pp. 247-271.

Failure Analysis

Failure analysis is extremely important in engineering, especially with
ceramics, because it is the only means of isolating the failure-causing
problem. In particular, failure analysis helps determine whether failure
or damage occurred due to a design deficiency or a material deficiency.
Until this has been determined, effort cannot be efficiently directed toward
finding a solution. The result is usually a "shotgun" approach that includes
a little design analysis, a little empirical testing, and a little material
evaluation and often ends up only in a repeat of the test or operating conditions that initially caused failure.

Much of the shotgun approach can often be avoided by fracture analysis.
Fracture analysis or fractography is the examination of the fractured or
damaged hardware in an effort to reconstruct the sequence and cause of
fracture. The path a crack follows as it propagates through a component
provides substantial information about the stress distribution at the time of
failure. Features on the fracture surfaces provide further information,
especially the position at which the fracture initiated (fracture origin), the
cause of fracture initiation (impact, tensile overload, thermal shock,
material flaw, etc.) and even the approximate local stress that caused
fracture. The primary objective of this chapter is to acquaint the reader
with these fracture surface features and the techniques to interpret the
cause of fracture in ceramic components.

12.1 FRACTOGRAPHY

The first step is to determine where the fracture initiated. Often, simply
reconstructing the pieces will pinpoint the fracture origin and may even give
useful information about the cause of fracture. After assembling the pieces,
look for places where a group of cracks come together or where a single
crack branches. Preston [1] has shown that the angle of forking is an indicator of the stress distribution causing fracture. Examples are shown
schematically in Fig. 12.1.

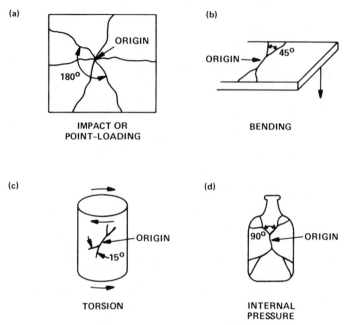

Figure 12.1 Information available by examining crack direction and crack branching. (Adapted from Ref. 7.)

The frequency of crack branching provides qualitative information about the amount of energy available during fracture. To branch, a crack must reach a critical speed. For glass, the critical speed is typically a little greater than half the speed of sound in the specific glass. At the instant of crack initiation, the crack velocity is zero, but quickly accelerates. The rate of acceleration is a function of the energy available either due to the stress applied or to energy stored in the part (such as residual stresses or prestresses, as in tempered glass). The more energy, the more rapidly the crack will reach its critical branching velocity and the more branching that will occur. A baseball striking a window will cause much more branching than a BB, due to the larger applied energy. Tempered glass will break into many fragments due to release of the high stored energy. On the other hand, a thermal shock fracture may not branch at all, especially if it initiates from a localized heat source and propagates into a relatively unstressed or compressively stressed region of the component. In this case, the fracture will tend to follow a temperature or stress contour and will have a characteristic wavy or curved appearance, as shown in Fig. 12.2 for a thermally fractured ceramic setter plate for a furnace.

Location of the Fracture Origin

The pattern of branching will often lead the engineer to the vicinity of the fracture origin. The engineer will then have to examine the fracture surfaces in this region, often under a low-power optical binocular microscope, to locate the precise point at which fracture initiated. This point of origin can be a flaw such as a pore or inclusion in the material, a cone-shaped Hertzian surface crack resulting from impact, a crack in a surface glaze, an oxidation pit, intergranular corrosion, a position of localized high stress, or a combination. Location and examination of the fracture origin will help determine which of these factors is dominant and provide specific guidance in solving the fracture problem.

As mentioned before, a fracture begins at zero velocity at the fracture origin and then accelerates as it travels through the part. As it does, it interacts with the microstructure, the stress field, and even acoustic vibrations and leaves distinct features on the fracture surface that can be used for locating the fracture origin [2]. The most important features include hackle, the fracture mirror, and Wallner lines.

The Fracture Mirror and Hackle   When a crack initiates at an internal flaw, it travels radially in a single plane as it accelerates. The surface formed is flat and smooth and is called the fracture mirror. When the crack reaches a critical speed, intersects an inclusion or encounters a shift in the direction of principal tensile stress, it begins to deviate slightly from the original plane, forming small radial ridges on the fracture surface.

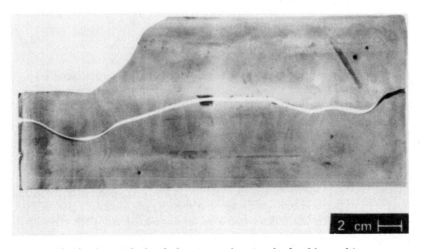

Figure 12.2 Thermal shock fracture showing lack of branching.

The first of these are very faint and are referred to as <u>mist</u>. Mist is usually visible on the fracture surfaces of glass, but may not be on crystalline ceramics. The mist transitions into larger ridges called <u>hackle</u>. Hackle is also referred to as <u>river patterns</u> because the appearance is similar to the branching of a river into tributaries and the formation of deltas. The hackle region transitions into macroscopic crack branching such that the remaining portion of the fracture surface is often on a perceptibly different plane than the mirror and hackle. Sometimes, this gives the appearance that the fracture origin is either on a step or a pedestal.

Figure 12.3a shows schematically the fracture mirror, mist, and hackle for a fracture that initiated in the interior of a part. The mirror is roughly circular and the fracture origin is at its center. Note that lines drawn parallel to the hackle will intersect at or very near the fracture origin. Similarly, Fig. 12.3b shows the fracture features for a crack that started at the surface of a part.

The hackle lines surrounding the mirror result from velocity effects and are sometimes called <u>velocity hackle</u>. Another form of hackle, called <u>twist hackle</u>, usually forms away from the mirror and results from an abrupt change in the tensile stress field, such as going from tension to compression. Twist hackle points in the new direction of crack movement and appears more as parallel cracks than as ridges and does not have to point to the fracture origin. Twist hackle is an important feature for deducing the stress distribution in the ceramic at the time of fracture.

The size of the fracture mirror is dependent on the material characteristics and the localized stress at the fracture origin at the time of fracture. Studies by Terao [3], Levengood [4], and Shand [5] on glass suggest that the fracture stress σ_f times the square root of the mirror radius r_m equals a constant A for a given material:

$$\sigma_f r_m^{1/2} = A \tag{12.1}$$

Kirchner and Gruver [6] determined that this relationship also provides a good approximation for polycrystalline ceramics, as long as the mirror is clearly visible and can be measured accurately. They obtained values of A ranging from 2.3 for a glass to 9.1 for a sintered Al_2O_3 to 14.3 for a hot-pressed Si_3N_4.

It is therefore possible to determine the stress causing failure of a ceramic component by comparing the mirror size with a graph of $r_m^{1/2}$ versus σ_f for the material. The graph can be compiled from bend strength or tensile strength data using scanning electron microscopy (SEM) of the fracture surfaces of the test bars to determine the mirror radius.

The stress causing the failure can also be estimated using the Griffith equation (described in Chap. 3) by measuring the flaw size on the fracture surface. Because inclusions and pores and other flaws are not symmetrical

(a) INTERNAL INITIATION

(b) SURFACE INITIATION

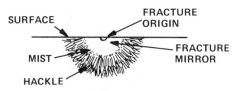

Figure 12.3 Schematic showing the typical fracture features that surround the fracture origin.

and their boundaries are often not well defined, stress estimates based on flaw size are only approximate. If a knowledge of the local stress at the fracture origin is needed, perhaps it should be calculated both from the mirror size and the flaw size and then the most appropriate value selected by good engineering judgment.

Figure 12.4 shows examples of fracture mirrors and flaws on the fracture surfaces of strength test specimens. The photomicrographs were taken by a scanning electron microscope. Since the ceramic specimens were not electrical conductors, a thin layer of gold was applied to the surface by sputtering to avoid charge buildup, which would result in poor resolution.

The mirror size cannot always be measured. If the fracture-causing stress is low and the specimen size is small, the mirror may cover the whole fracture surface. If the material has very coarse grain structure or a bimodal grain structure, the mirror and other fracture features may not be visible or distinct enough for measurement. Figure 12.5 shows examples of fracture surfaces with indistinct fracture features.

Wallner Lines Sonic waves are produced in a material during fracture. As each succeeding wave front overtakes the primary fracture crack, the principal stress is momentarily perturbed. This results in a series of

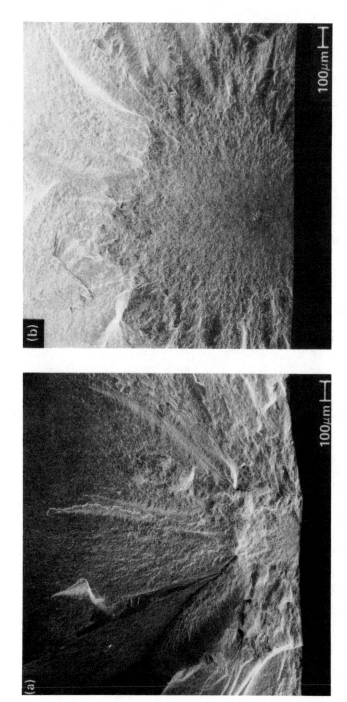

Figure 12.4 Examples of typical fracture mirrors for high-strength polycrystalline ceramics. (a) Initiation at a surface flaw in hot-pressed silicon nitride. (b) Initiation at an internal flaw in reaction-sintered silicon nitride.

faint arc-shaped surface lines which are termed <u>Wallner lines</u>. The curvature of each line shows the approximate shape of the crack front at the
time it was intersected by the sonic wave and provides information about
the direction of crack propagation and the stress distribution. The direction is from the concave to the convex side of the Wallner lines. The stress
distribution is inferred from the distance of each portion of a single line
from the origin. If the stress distribution were of uniform tension, each
portion of a line would be about equidistant from the origin. If a stress
gradient were present, the distance of various portions of the Wallner line
from the origin would vary, being farthest where the tensile stress was
highest. These effects are shown schematically in Fig. 12.6.

Wallner lines are not always present. For high-energy fractures,
where the fracture velocity is high and the surface is rough, Wallner lines
often cannot be distinguished. In very slow crack velocities, such as occur
in subcritical crack growth, Wallner lines are not present because the
sonic waves are damped and gone before the crack has propagated appreciably.

<u>Other Features</u>   Other fracture features besides the mirror, hackle, and
Wallner lines are useful in interpretation of a fracture. These include
arrest lines, gull wings, and cantilever curl.

An <u>arrest</u> line occurs when the crack front temporarily stops. The
reason for crack arrest is usually a momentary decrease in stress or a
change in stress distribution. When the crack starts moving again, its
direction invariably has changed slightly, leaving a discontinuity. This
line of discontinuity looks a little like a Wallner line, but is usually more
out of plane and more distinct. It is also called a <u>rib mark</u>. Arrest lines
or rib marks provide essentially the same information as Wallner lines,
i.e., the direction of crack movement and the stress distribution. Twist
hackle frequently is present after an arrest line.

The gull wing is a feature that occurs due to the crack intersecting a
pore or inclusion. As the crack travels around the inclusion, two crack
fronts result. These do not always meet on the same plane on the opposite
side of the inclusion or pore, resulting in a ridge where the two link up and
again become a single crack front. In some cases, the ridge is immediately in the wake of the inclusion or pore and looks like a tadpole. In other
cases, two ridges resembling a gull wing form in the wake.

Another very useful feature is referred to as the <u>cantilever curl</u> or
<u>compression lip</u>. It occurs when the material is loaded in bending. The
fracture initiates on the tensile side perpendicular to the surface and exits
on the compressive side not perpendicular to the surface. This is illustrated in Fig. 12.7. If the part were fractured under pure tension, the crack
would be straight through the thickness and would thus exit at 90°. This
information can be valuable in diagnosing the cause of fracture. For

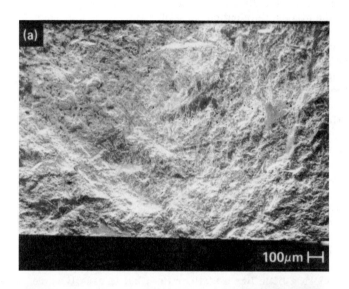

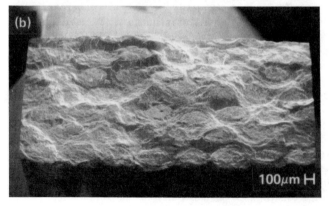

Figure 12.5 Examples of fracture surfaces with indistinct fracture features. (a) Sintered silicon carbide. (b) Silicon carbide-carbon-silicon composite. (c) Porous lithium aluminum silicate. (d) Bimodal grain distribution reaction-sintered silicon carbide.

instance, thermal fractures of plate-shaped parts typically approach a
stress state of pure tension near the point of origin and will not result in a
compression lip.   These same parts fractured mechanically will normally have some bend loading and will thus have a compression lip.
Another example is a part containing prestressing or residual internal
stresses.   The crack will not pass straight through the thickness, but
instead will follow contours consistent with the stress fields it encounters.   A third example is strength testing.   A problem with testing
a ceramic in uniaxial tension is avoiding parasitic bend stresses.

(a) UNIFORM TENSION

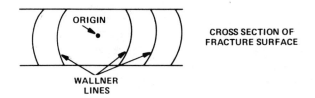

(b) NONUNIFORM TENSION

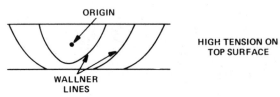

(c) TENSION-COMPRESSION

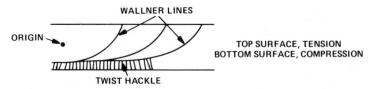

Figure 12.6   Relationship of Wallner lines on a fracture surface to the
stress distribution at the time of fracture.   (From Ref. 7.)

(a) BEND (FLEXURE) LOADING

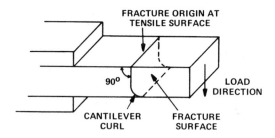

(b) PURE TENSILE LOADING

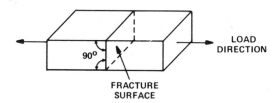

Figure 12.7 Difference in the fracture contour through the specimen thickness for bend loading versus pure tensile loading.

Examination of the fracture surface for signs of cantilever curl after tensile testing will help determine if pure tension was achieved or not.

Figure 12.8 shows the cross sections of typical specimens tested in 4-point bending. Note the variations in the shape of the compression lip.

Techniques of Fractography

The techniques of fractography are relatively simple and the amount of sophisticated equipment minimal. Often, the information required to explain the cause of fracture of a component can be obtained with only a microscope and a light source. In fact, sometimes an experienced individual can explain the fracture just by examining the fracture surfaces visually. At other times, a variety of steps and techniques including sophisticated approaches such as SEM, electron microprobe, and Auger analysis are required [8-10]. Where extensive fractography is necessary, the steps and procedures are as shown schematically in Fig. 12.9.

Step 1 involves visual examination of the fractured pieces and re- view of data regarding the test or service conditions under which the hardware failed. These data usually provide some hypotheses to guide the evalation. Reconstructing the broken pieces and sketching probable fracture origins and paths is also helpful.

A primary objective of visual examination is to locate the point of fracture origin. This can be done with the use of Wallner lines, hackle, and the fracture mirror, as described previously.

Visual examination determines the extent of additional evaluation that will likely be required. It also determines if cleaning procedures

Figure 12.8 Examples showing cantilever curl in 4-point bend specimens. Specimens 0.32 cm (0.125 in.) thick.

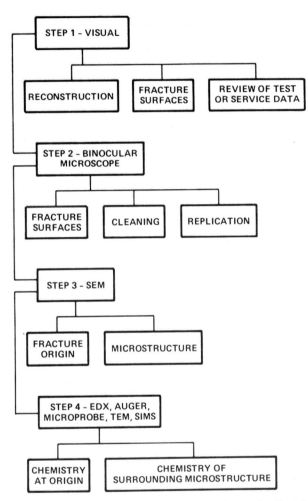

Figure 12.9 Major sequential steps in conducting fracture analysis.

are necessary prior to microscopy. Generally, one must be careful not to handle or damage the fracture surface. The origin and features can be fragile and key information explaining the fracture lost by improper handling. For instance, fingerprints can be mistaken for Wallner lines. Debris on the surface can obscure the true fracture surface.

Cleaning can sometimes be accomplished with compressed air, but the source must be considered. Some compressors mix small amounts of oil with the air, which could produce a thin surface film that would later result in interpretation difficulties during SEM analysis. Compressed air should not be used unless it is known to be clean.

Ultrasonic cleaning in a clean solvent such as acetone or methyl alcohol is frequently used. Caution and judgment must be exercised, however, because ultrasonic cleaning is quite vigorous due to the cavitation action at the fluid-specimen interface and can damage the fracture surface. For instance, if fracture initiated at a low-density region or a soft inclusion, this material might be removed during ultrasonic cleaning and either prevent interpretation or lead to an erroneous interpretation.

It is apparent that cleaning and handling should be avoided unless absolutely necessary. Before trying other cleaning approaches, try a soft camel's-hair brush.

Step 2 involves examination of the fracture surfaces under a low-power microscope. Usually, a binocular microscope with magnification up to 40× is adequate. Sometimes higher power and special lighting or contrast features are required. Stereo photography may also be useful, since it accentuates the fracture surface features.

Preparation of replicas can also be useful; a replica often provides better resolution of the fracture features than does examination of the original part. Several methods are available for preparing replicas. A room-temperature method makes use of cellulose acetate, acetone, and polyvinyl chloride (PVC). A thin sheet of cellulose acetate is placed on a piece of PVC and submerged in acetone for about 15 sec. It is then pressed against the fracture surface while being flooded with acetone and held for about 5 min. The acetone is then allowed to dry and the cellulose acetate replica peeled off.

Replicas can also be prepared with PVC alone. The specimen is heated and PVC pressed on with a Teflon rod. After cooling, the PVC replica is peeled off. This technique is quick, but should not be used if there is a chance that heating of the specimen will alter the fracture surface.

The fracture origin can usually be located by low-power optical microscopy using either the original specimen or a replica and an assessment made as to whether the fracture resulted from a material flaw or some other factor. However, details of the fracture origin, such as the nature of the flaw and interaction between the flaw and the microstructure require higher magnification. Optical microscopes do not have adequate depth of focus at high magnification, so the scanning electron microscope must be used.

Examination of the fracture surface, in particular the fracture origin, by SEM is the third step in fractography. The scanning electron microscope provides extremely large depth of focus (compared to the optical microscope) and a range in magnification from around 10× to well over 10,000×. Most fractography is conducted between 25× and 5000×.

Scanning electron microscopy shows the differences between the fracture origin and the surrounding material and helps the engineer to develop a hypothesis of the cause of failure. One can easily detect if the fracture initiated at a machining groove or at a material pore or inclusion. SEM shows if the surface region is different from the interior and provides visual evidence of the nature or cause of the differences.

As with other techniques, SEM requires interpretation and must be used with caution. Because of the large depth of focus, it is difficult sometimes to differentiate between a ridge and a depression or to determine the angle of intersection between two surfaces. This can be better appreciated by comparing a single SEM photomicrograph with a stereo pair of the same surface. Only after doing this does one understand how easy it is to misinterpret a feature on an SEM photomicrograph.

Difficulty in interpretation is inevitable but can be minimized if the engineer is present during the SEM analysis. This is especially true with respect to artifacts. An <u>artifact</u> is defined as extraneous material on the surface of the specimen. It can be a particle of dust or lint, a chunk of debris resulting from the fracture or handling, or a smeared coating resulting from oil contamination. If the engineer suspects that a feature is an artifact, he or she can instruct the SEM operator to look at it from different view angles and to examine surrounding areas in an effort to be sure.

As noted, SEM can usually locate the fracture origin and provide a photograph with a calibrated scale that allows accurate measurement of the size and shape of the flaw and the size of the fracture mirror. The engineer can then use equations (12.1) and (3.12) to estimate the magnitude of the tensile stress that caused failure.

Once the fracture origin has been located, evaluation of the localized chemistry is often desirable. This leads to step 4 of fractography, the use of sophisticated instruments to conduct microchemical analysis. Most SEM units have an energy-dispersive X-ray (EDX) attachment that permits chemical analysis of the x-rays that are emitted when the SEM electron beam excites the electrons within the material being examined. Each chemical element gives off x-rays under this stimulation that are characteristic of that element alone. These are detected by the EDX equipment and displayed as peaks by peripheral equipment. Comparison of the peak height of each element provides a semiquantitative chemical analysis of the microstructural feature being viewed.

Electron microprobe works on the same principle as EDX, except an alternate x-ray detection mechanism is used which provides better resolution and detects a wider range of elements. EDX cannot detect the lower-atomic-number elements.

Auger analysis provides additional features for chemical analysis of microstructural features. It can remove material by sputtering while it is conducting chemical analysis and can thus determine if the chemistry changes as we progress inward from the fracture surface. Auger analysis is especially useful in cases where oxidation, corrosion, slow crack growth, or other intergranular effects are suspected. In some cases, fracture of specimens can be conducted in the Auger apparatus under an inert atmosphere prior to conducting the chemical analysis. This allows examination of a fresh fracture surface that has not had a chance to pick up contamination from the atmosphere and handling.

In addition to analysis of the fracture surface, other material tests can help determine the cause of failure. These include surface and bulk x-ray diffraction analysis, reflected-light microscopy of polished specimens and, in occasional cases, transmission electron microscopy (TEM).

There is no guarantee that the four steps of fractography will explain the cause of a failure and suggest a solution. However, it is still the most effective technology available and should be used routinely.

Determining Failure Cause

As mentioned before, determining the cause of failure is critical. It is obviously important in liability suits, where responsibility for failure must be established, but it is also important for other reasons:

To determine if failure is resulting from design or material limitations
To aid in material selection or modification
To guide design modifications
To identify unanticipated service problems such as oxidation or corrosion
To identify material or material processing limitations and suggest direction for improvement
To define specification requirements for materials and operating conditions

Figure 12.10 shows schematically how fractography interacts with other sources of information to determine the cause of failure and lead to action in the right direction to achieve a solution. It is imperative that all sources of data be considered and that the source of fracture be isolated to determine as quickly as possible whether it is design or materials oriented.

The following paragraphs review some of the common causes of fracture and describe the typical fracture surfaces that result.

Material Flaws   As discussed in Chap. 3, flaws in a ceramic material concentrate the stress. When this concentrated stress at an individual flaw reaches a critical value that is high enough to initiate and extend a crack, fracture occurs. Therefore, the first thing to look for on the fracture surface is whether there is a material flaw at the fracture origin. If a flaw such as a pore or inclusion is present, the engineer can do the following:

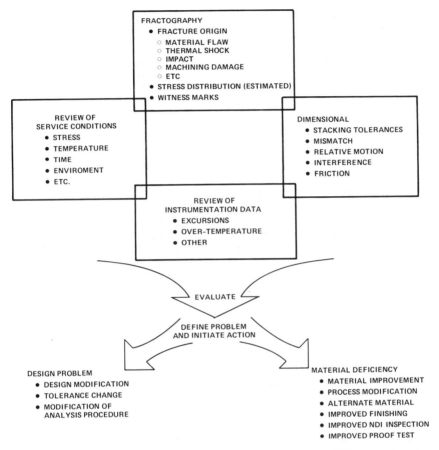

Figure 12.10 Failure analysis interaction to determine if the problem is design or materials oriented and to define a plan of action to solve the problem.

Compare the nature of the flaw with prior certification and service experience to determine if the flaw is intrinsic to the normal baseline material
or if it is abnormal. If the latter, a problem in the material fabrication
process probably exists, and the help of the component manufacturer
should be solicited.

Measure the flaw size and/or mirror size and estimate the fracture stress
σ_f using equations (3.12) and (12.1). Compare this with the stresses estimated by design analysis and with fracture stress distribution projected

for the material from prior strength certification testing. If the calculated fracture stress is within the normal limits specified for the material, a design problem should be suspected. If the fracture stress is
below the normal limits specified for the material, a material problem
is indicated.
If the flaw is at or near the surface, assess whether the flaw is intrinsic
   to the material or resulted from an extrinsic source such as machining,
   impact, or environmental exposure. Comparison of the chemical and
   physical nature of the flaw with the surrounding baseline microstructure
   will help in making this assessment.

   Figures 12.11 through 12.14 illustrate the types of intrinsic material
flaws that can cause fracture and compares whether they are normal or
abnormal for the material. Although the types of flaws vary depending on
the material, the fabrication process and the specific step in the process
in which they formed, it is relatively easy to distinguish between normal
and abnormal flaws and thus to determine if errors in processing contributed
to the failure [11].

Machining Damage   Surface flaws resulting from machining are a second
common source of failure in ceramic components, especially in applications
where high bend loads or thermal loads are applied in service. The flaws
resulting from machining were discussed in Chap. 8, the most important
being the median crack and the radial cracks.
   The median crack is elongated in the direction of grinding and is like
a notch, in that fracture usually initiates over a broad front. A broad mirror usually results, but no flaw is readily visible because of the shallow
initial flaw depth and because the median crack is perpendicular to the surface. The principal tensile stress is usually distributed such that the crack
will extend in a plane perpendicular to the surface. If the flaw is perpendicular to the surface to start with, it will be in the same plane as the fracture and will be difficult to differentiate from the rest of the fracture mirror.
   Figure 12.15 shows examples of specimens that fractured at transverse
grind marks such that the median crack was likely the strength-determining
flaw. Note the length of the surface involved in the fracture origin and the
lack of a distinguishable flaw. Note also that the fracture origin is at a
grinding groove and is elongated parallel to the direction of grinding.
   Radial cracks produced by machining are roughly perpendicular to the
direction of machining, are usually shallower than the median cracks, and
are usually semicircular rather than elongated. The resulting fracture
mirror is similar to one produced by a small surface pore or inclusion,
but a well-defined flaw is not visible. However, by examining the intersection of the fracture surface with the original specimen surface at the center
of the semicircular mirror under high magnification with the scanning
electron microscope, the source can usually be seen and interpreted as

machining damage. There are two features to look for. First, check to
see if a grinding groove, especially an unusually deep one, intersects the
fracture surface at the origin. Second, look for a small slightly out-of-
plane region at the origin. This could indicate machining damage, but
could also have other interpretations, such as contact damage or simply a
tensile overload.

Machining damage often limits the strength of fine-grained ceramics
and determines the measured strength distribution of certification specimens.
Rice et al. [12] reported that hot-pressed Si_3N_4 specimens machined in the
transverse direction fractured at machining flaws 98% of the time and that
bars machined in the longitudinal direction failed at machining flaws greater
than 50% of the time. This is similar to results of the author and his co-
workers for both Si_3N_4 and SiC [13-15].

Residual Stresses   Many ceramics have residual stresses, which usually
result from the surface cooling down faster than the interior after sintering
or are due to chemical differences between the surface and interior. Often,
the interior is under residual tension and the exterior is under compression.
This can provide a strengthening effect if the material is used in service in
the as-fired, prestressed condition. However, most components require
some finish machining. Once the compressive surface zone has been penetrated, the component is substantially weakened, often to the point of spontaneous crack initiation during machining.

Frechette [16] recounts a case where simple blanks of boron carbide
cracked during machining. At the time, the cause of cracking was unknown,
but improper machining was the primary suspect. Fracture analysis showed
that the crack inititiated at the point of machining and entered the material
roughly perpendicular to the surface. However, the crack then quickly
changed direction and propagated through the interior of the material parallel
to the surface, and finally changed direction again and exited through the
bottom surface. The fracture path and fracture surface features (Wallner
lines and hackle) indicated that the surfaces were in compression and the
interior was in tension and suggested that the problem was not linked to
machining practice or material defects. Review of the processing history
showed that the boron carbide was allowed to freely cool after hot pressing
from 1950°C to 1800°C, followed by slow, controlled cooling thereafter.
The free cooling was within the creep temperature range of the material,
resulting in an effect comparable to tempering of glass, i.e., formation of
surface compression and internal tension. Based on this hypothesis, derived from fracture surface analysis alone, the hot-pressing operation was
modified to permit slow cooling from 1950°C ro 1750°C. This eliminated
the residual stress condition and permitted machining without cracking.

Thermal Shock   Little work has been reported on identification of thermal
shock fractures, and much additional work is needed. Therefore, the

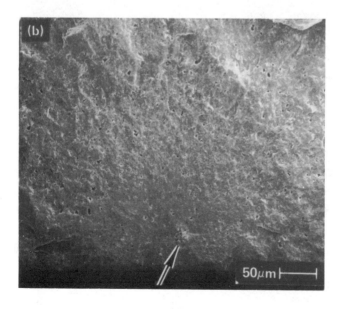

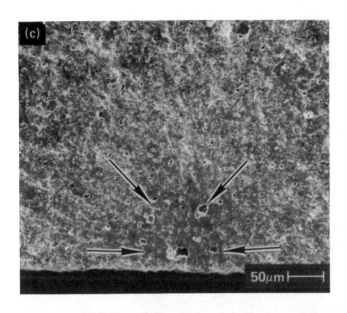

Figure 12.11 SEM photomicrographs of fracture-initiating material flaws in reaction-bonded Si_3N_4 (RBSN) which are typical and consistent with the normal microstructure and strength. Arrows point to the fracture origins.

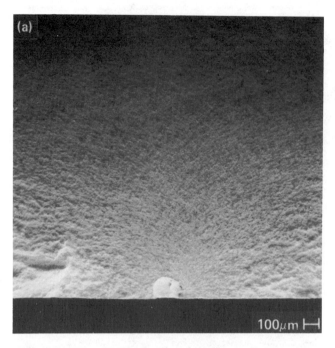

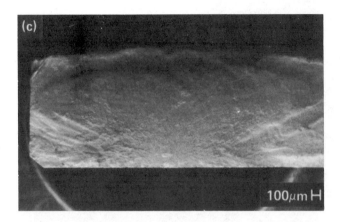

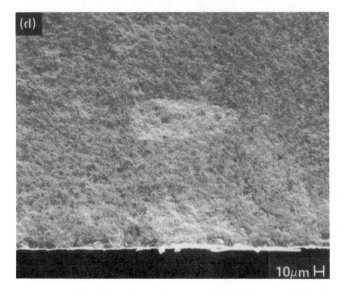

Figure 12.12 SEM photomicrographs of abnormal fracture–initiating material flaws in RBSN traceable to improper processing prior to nitriding.
(a) Large pore in slip–cast RBSN resulting from inadequate de airing.
(b) Crack in greenware prior to nitriding. (c) and (d) Low–density regions in slip–cast RBSN resulting from agglomerates in the slip.

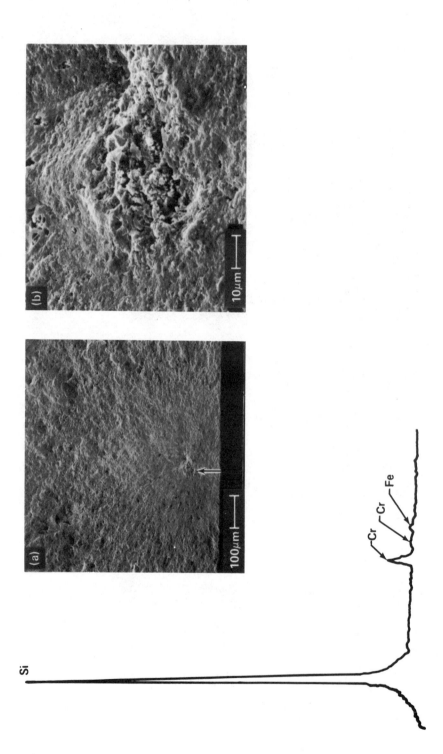

348

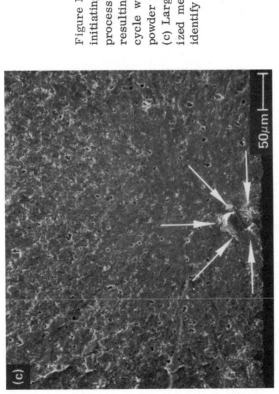

Figure 12.13 SEM photomicrographs of abnormal fracture-initiating material flaws in RBSN traceable to the nitriding process. (a) and (b) Porous aggregate rich in Cr and Fe, resulting from reaction of the silicon during the nitriding cycle with stainless steel contamination picked up during powder processing; energy dispersive x-ray analysis shown. (c) Large aggregate of unreacted silicon resulting from local-ized melting due to local exothermic overheating. Arrows identify the fracture origins.

349

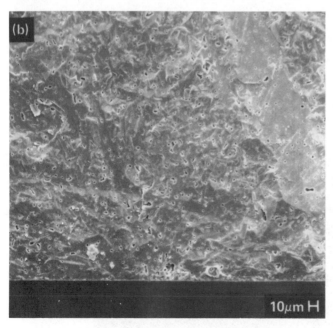

Figure 12.14 SEM photomicrographs comparing normal and abnormal material flaws in sintered SiC. (a) and (b) Typical microstructure of high-strength material. (c) Large pore resulting from powder agglomeration during powder preparation and shape forming. (d) Large grains resulting from improper control of temperature during sintering.

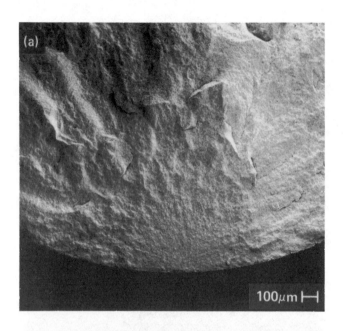

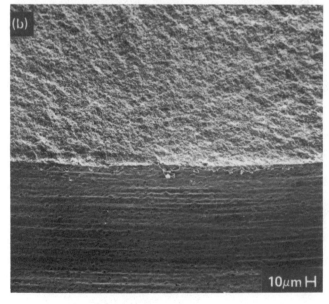

Figure 12.15 SEM photomicrographs showing fractures initiating at transverse machining damage. (a) The fracture surface of a tensile specimen of hot-pressed silicon nitride which had been machined circumferentially. (b) The intersection of this fracture surface with the machined surface, illustrating that the fracture origin is parallel to the grinding grooves. (c) and (d) The same situation for reaction-sintered silicon nitride.

comments here will be fairly general, and the reader should be aware that the distinguishing features described may not always occur and may not be the best ones in a specific case. As in all other interpretive studies, the best approach is to apply known principles, thoroughly analyze all available data and options, and then make a judgment decision, rather than depending on a cookbook procedure.

Thermal shock fractures tend to follow a wavy path with minimal branching and produce rather featureless fracture surfaces. This appears to be especially true for weak or moderate strength materials, including both glass and polycrystalline ceramics. It is common for thermal shock cracks not to propagate all the way through the part. These characteristics suggest that a nonuniform stress field is present during thermal fracture; i.e., the fracture initiates where the tensile stress is highest, follows stress or temperature contours (thus the wavy path), and stops when the stress drops below the level required for further extension or when a compressive zone is intersected. These conditions would also suggest a low crack velocity, which would account for lack of branching and lack of fracture surface features.

In line with the foregoing general considerations, several actual cases of fracture or material damage due to severe thermal transients are now analyzed.

The first case is the water quench thermal shock test described in Chap. 4 for comparing the relative temperature gradient required to cause damage in simple rectangular test bars of different ceramic materials. The test bars are heated in a furnace to a predetermined temperature and then dropped into a controlled-temperature water bath. The bars generally do not break due to the quench. Instead, a large number of microcracks are produced. These become critical flaws that lead to fracture during subsequent bend testing; the size of the flaws is dependent on the material properties and the ΔT, and the residual strength is dependent on the size of these cracks. Further discussion of this test procedure and the results on specific materials is available in Refs. 17 and 18.

Ammann et al. [19] conducted cyclic fluidized bed thermal shock tests on wedge-shaped specimens of hot-pressed Si_3N_4. They reported that multiple surface cracking occurred and that the cracks grew in depth as a function of the total number of cycles.

Oxyacetylene torch thermal shock testing and gas turbine rig testing of reaction-bonded Si_3N_4 stator vanes by the author and his co-workers [20] has shown that thermal shock fractures of actual components under service conditions can be quite varied. Figure 12.16a shows a typical featureless fracture surface. The origin is shown by the arrow and the crescent-shaped ink mark and occurred in a region predicted by three-dimensional finite element analysis to have the peak thermal stress during rapid heat-up. No material flaw or machining damage is visible, even at high magnification, as shown in Fig. 12.16b. The crack apparently initiated at the material

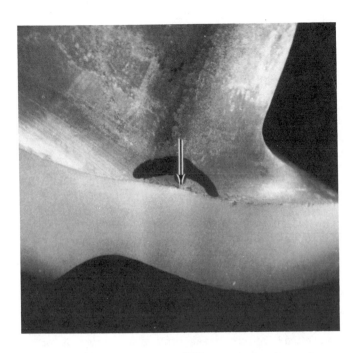

Figure 12.16 SEM photomicrograph showing a typical featureless thermal shock fracture surface. (a) Overall surface at low magnification. (b) Fracture origin at higher magnification. (Courtesy of Garrett Turbine Engine Company, Phoenix, Ariz., Division of The Garrett Corporation.)

surface where the stress was maximum and propagated at moderate velocity without branching or making any abrupt changes in direction.

Figure 12.17a shows a thermal shock failure that initiated at the trailing edge of a stator vane airfoil. Finite element analysis also determined this position to be under high thermal stress during heat-up and steady-state service conditions. However, in this case the thermal shock fracture initiated at a preexisting material flaw, as shown more clearly in Fig. 12.17b at higher magnification. The flaw was a penny-shaped crack nearly normal to the surface of the airfoil, but out of plane to the principal tensile stress. The fracture initiated at this crack due to the thermal stress and then quickly changed direction to follow the plane of maximum tensile stress. Hackle marks can be seen in the upper part of Fig. 12.17a which point to the vicinity of the origin.

The fracture surface shown in Fig. 12.17 is not what one would expect for thermal shock conditions. It has relatively well defined fracture features and could just as easily have been interpreted as a mechanical overload. Similarly, the fracture surface in Fig. 12.16 looks very much like fractures resulting from contact loading (discussed later). How does the engineer make the distinction? At our current state of knowledge, he or she does not, at least not based solely on fracture surface examination. The engineer needs other inputs such as stress analysis, controlled testing (such as the calibrated oxyacetylene torch thermal shock tests), and a thorough knowledge of the service conditions. The engineer then needs to evaluate all the data concurrently and use his or her best judgment.

Impact   Impact can cause damage or fracture in two ways: (1) localized damage at the point of impact and (2) fracture away from the point of contact due to cantilevered loading. The former will have distinctive features and can usually be identified as caused by impact. The latter will appear like a typical bend overload (with a compressive lip on the exit end of the fracture) and can only be linked to impact by supporting data such as location of the local damage at the point of impact.

The damage at the point of impact may be so little as to resemble a scuff mark on the ceramic or a smear of the impacting material (both cases referred to as a witness mark) or it may be as severe as complete shattering [21]. The degree depends on the relative velocity, mass, strength and hardness of the impacting bodies. A baseball striking a plate glass window causes shattering; a BB striking a window only causes a series of concentric cone-shaped (Hertzian) cracks intersected by radial cracks. However, if the window shattered by the baseball could be reconstructed, it would also have conical and radial cracks.

The conical cracks are typical of impact and are thus a strong diagnostic fracture feature [22]. Figure 12.18a shows a fracture surface of a polycrystalline ceramic where impact damage occurred first and was followed by fracture due to a bend load. The origin is indicated by an arrow

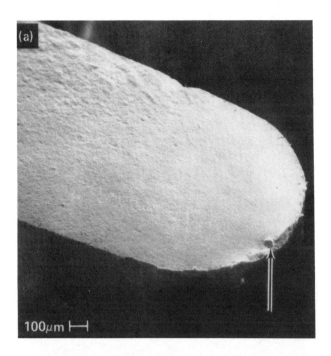

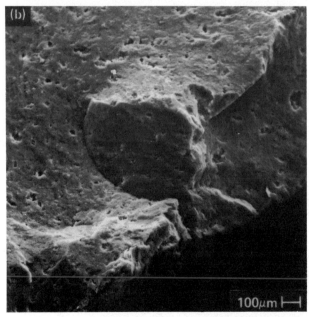

Figure 12.17 SEM photomicrograph of a thermal shock fracture initiating at a material flaw. (a) Overall surface at low magnification. (b) Preexisting crack at fracture origin. (Courtesy of Garrett Turbine Engine Company, Phoenix, Ariz., Division of The Garrett Corporation.)

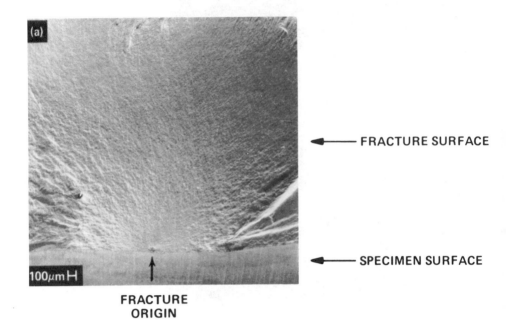

FRACTURE
ORIGIN

Figure 12.18 Typical Hertzian cone crack resulting from impact and act-
ing as the flaw which resulted in fracture under subsequent bend load. Shown
at increasing magnification from (a) to (c). (Courtesy of Garrett Turbine
Engine Company, Phoenix, Ariz., Division of The Garrett Corporation.)

and is easily located by observing the hackle lines and fracture mirror.
Figure 12.18b and c show the origin at higher magnification. A distinct
cone shape is present; the apex is at the surface where the impact occurred,
and the fracture flared out as the crack penetrated the material.

Figure 12.19 shows another example of fracture due to impact. In this
case, a ceramic rotor blade rotating at 41,000 rpm struck a foreign object
and was fractured by the impact. Note the Hertzian crack extending from
the fracture origin. Also note that additional damage is present at the
origin, possibly a series of concentric cracks, providing evidence that the
degree of impact was severe.

Biaxial Contact   Biaxial contact refers to a situation where normal and
tangential forces are being applied simultaneously at a ceramic surface.
Examples where this might occur are numerous and include the following:

Surface grinding
Sliding contact (such as in bearings, seals, and very many other applications)
Applications involving a shrink fit
Interfaces where the materials have different thermal expansion coefficients
       and operate under varying temperatures
Any high-friction interface

Tensile stress concentration at a biaxially loaded interface was discussed briefly in Chap. 4 and is discussed in more detail by Finger [23] and
Richerson et al. [24]. The fracture surface can be quite varied. If the
contact is concentrated at a point, a Hertzian cone crack can form and the
resulting fracture surface will be similar to the one in Fig. 12.18. However, if the contact is more spread out, as in high-friction sliding contact,
the damage will be spread over a larger surface area and will result in a
relatively featureless fracture surface similar to the one shown in Fig.
12.16 for a thermal shock fracture.

Distinguishing features that differentiate a contact failure from a thermal shock failure are just beginning to be defined [23] and appear to include
the following:

Contact cracks tend to enter the material surface at an angle other than 90°.
       An extreme case is shown in Fig. 12.20a for a ceramic specimen with a
       high-interference shrink fit axially loaded in tension.
Multiple parallel cracks often occur during a contact failure and can either
       be seen on the surface adjacent to the fracture origin or show up as chips
       obscuring the fracture origin (Fig. 12.20b).
Surface witness marks are often present at the fracture origin in cases of
       contact-initiated failure (Fig. 12.21a).
One or more clamshell shapes, faint Hertzian cones, or pinch marks occur
       at the origin of some contact-initiated fracture surfaces (Fig. 12.21b).

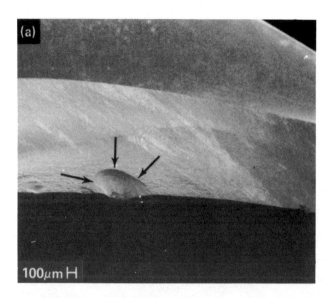

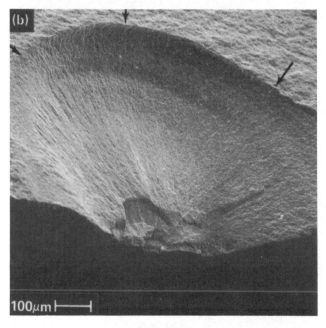

Figure 12.19 Impact fracture of a ceramic rotor blade showing Hertzian cone crack. (Courtesy of the Garrett Turbine Engine Company, Phoenix, Ariz., Division of The Garrett Corporation.)

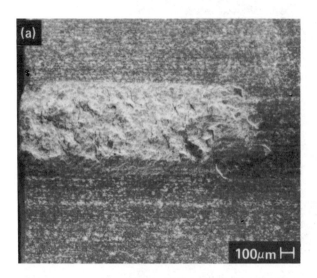

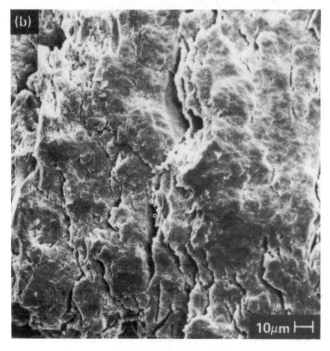

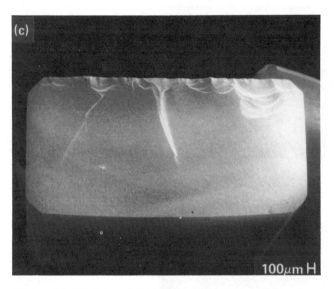

Figure 12.20 (a) and (b) Surface cracks resulting from relative movement between two contact surfaces under a high normal load and with a high co-efficient of friction. (c) Typical multiple chipping resulting from contact loading and visible on a fracture surface.

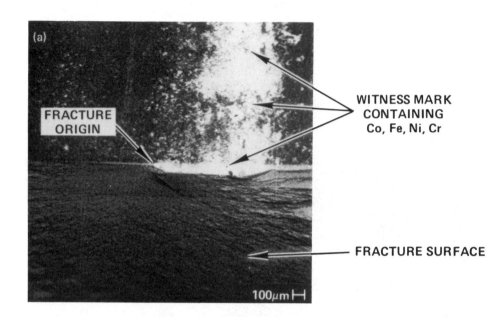

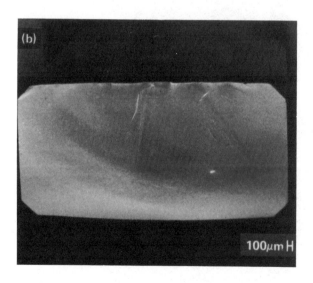

Figure 12.21 (a) Witness mark on the surface of the ceramic adjacent to the fracture origin, suggesting fracture due to contact loading. (b) and (c) Multiple cone features resulting from a contact fracture.

<u>Oxidation-Corrosion</u>   Some types of oxidation or corrosion are easy to detect because they leave substantial surface damage that is clearly visible to the naked eye. In this case, the objective is to identify the mechanism of attack and find a solution. In other cases, especially where the oxidation or corrosion is isolated along grain boundaries, the presence and source of degradation may be more difficult to detect. In this case, the degree of attack may only be determined by strength testing, and the cause may be ascertained by controlled environment exposures and/or sophisticated instruments such as Auger spectroscopy, which can detect slight chemical variations on a microstructural level.

Let us first examine some examples of oxidation and corrosion where visible surface changes have occurred. Figure 12.22 shows the surface and fracture surface of NC-132 hot-pressed Si_3N_4[*] after exposure in a SiC resistance-heated, oxide-refractory-lined furnace for 24 hr at 1100°C (2012°F) [25]. Figure 12.22a shows the complete cross section of the test bar. The fracture origin is at the surface on the left side of the photograph and is easily located by the hackle marks and the fracture mirror (the dark spots on the fracture surface are artifacts that accidentally contaminated the surface in preparing the sample for SEM). The specimen surface appears at low magnification to have many small spots that were not present prior to the oxidation exposure. At higher magnification (Fig. 12.22b), these spots appear to be blisters or popped bubbles and one is precisely at the fracture origin. Still higher magnification (Fig. 12.22c) reveals that a glass-filled pit is at the base on the center of the blister. It also reveals that a surface layer less than 5 μm thick covers the specimen and that this layer appears to be partially crystallized.

By simply examining the specimen surface, especially the intersection of the oxidized surface and the fracture surface, we have obtained much insight into both the nature and sequence of oxidation. What else can we do to obtain further information? We can compare the strength of the oxidized specimen with that of unoxidized material. In this specific case, the oxidation exposure resulted in a reduction in strength from 669 MPa (97,000 psi) to 497 MPa (72,000 psi). We can also compare x-ray diffraction and chemical analyses for the original surface, the oxidized surface, and the bulk material. In this case, the oxidized surface contained much more Mg and Ca than the original surface or the bulk material. Energy dispersive x-ray (EDX) analysis verified that the glassy material in the pit also had high concentrations of Mg and Ca. X-ray diffraction revealed crystallized cristobalite (SiO_2) plus magnesium silicate and calcium magnesium silicate phases in the oxide layer. No sign of Mg or Ca contamination was detected in the furnace.

---

[*]Manufactured by the Norton Company, Worcester, Mass.

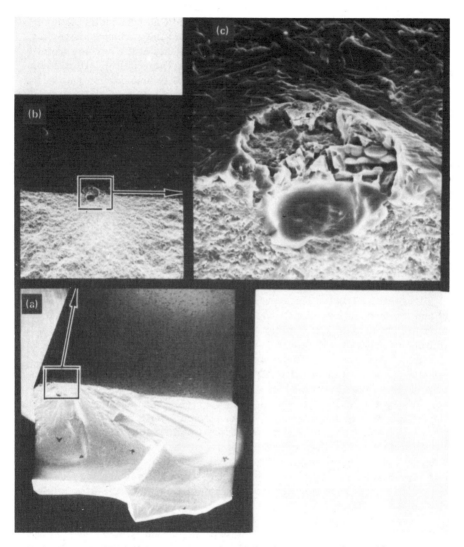

Figure 12.22 SEM photomicrographs of the fracture surface of hot-pressed Si_3N_4 exposed to static oxidation for 24 hr at 1100°C (2012°F). (a) Overall fracture surface showing hackle marks and fracture mirror (the irregular dark spots on the fracture surface are artifacts). (b) Higher magnification showing the fracture mirror with an oxidation/corrosion pit at the origin. (c) Higher magnification showing the nature of the pit and the surface oxidation layer. Specimen size 0.64 × 0.32 cm. (From Ref. 14.)

   Simultaneous evaluation of all the data led to a plausible hypothesis of
the mechanism of oxidation degradation. Mg and Ca, present as oxide or
silicate impurities in the Si_3N_4, were diffusing to the surface, where they
reacted with SiO_2 which was forming simultaneously at the surface from
reaction of the Si_3N_4 with oxygen from the air. The resulting silicate compositions apparently locally increased the solubility or oxidation rate of the
Si_3N_4. The reason for the formation of isolated pits was not determined,
but could have resulted from impurity segregation or other factors and
would have required additional studies to determine.

   A similar example of static oxidation for reaction-bonded Si_3N_4[*] is
illustrated in Fig. 12.23. In this case, the exposure was for 2 hr at 1350°C
(2462°F) plus 50 hr at 900°C (1652°F) [11]. Only isolated pits were present
on the surface and these appeared to occur where small particles of the furnace lining had contacted the specimen during exposure. The EDX analysis
included in Fig. 12.23 was taken in the glassy region at the base of the pit,
showing that Al, Si, K, Ca, and Fe were the primary elements present and
again indicating a propensity for Si_3N_4 to be corroded by alkali silicate
compositions. However, it should be noted that the size of the pit is much
smaller than in the prior example and resulted in only a small strength
decrease.

   Figures 12.24 and 12.25 show examples of more dramatic corrosion
of hot-pressed and reaction-bonded Si_3N_4 [25], resulting from exposure to
the exhaust gases of a combustor burning jet fuel and containing a 5-ppm
addition of sea salt. Exposure consisted of 25 cycles of 899°C (1650°F) for
1.5 hr, 1121°F (2050°F) for 0.5 hr, and a 5-min air quench. At 899°C
(1650°F), Na_2SO_4 is present in liquid form and deposits along with other
impurities on the ceramic surface. The EDX analyses taken in the glassy
surface layer near its intersection with the Si_3N_4 document the presence of
impurities such as Na, Mg, and K from the sea salt, S from the fuel, and
Fe, Co, and Ni from the nozzle and combustor liner of the test rig. An
EDX analysis for the Si_3N_4 on the fracture surface about 20 μm beneath the
surface layer is also shown in Fig. 12.25. Only Si is detected (nitrogen
and oxygen are outside the range of detection by EDX), indicating that the
corrosion in this case resulted from the impurities in the gas stream plus
the surface oxidation.

   The strength of the hot-pressed Si_3N_4 exposed to the dynamic oxidation
with sea salt additions decreased to an average of 490 MPa (71,000 psi from
a baseline of 669 MPa (97,000 psi). The reaction-bonded material decreased
to 117 MPa (17,000 psi) from a baseline of 248 MPa (36,000 psi). Repeat-
ing the cycle with fresh specimens and no sea salt resulted in an increase

---

[*] RBN-104 reaction-bonded Si_3N_4 from the AiResearch Casting Co., Torrance, Calif.

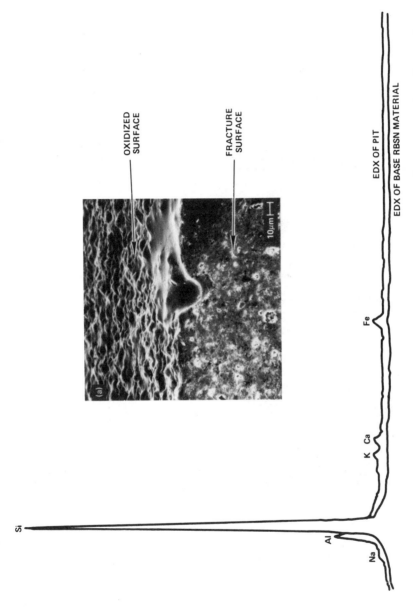

Figure 12.23 SEM photomicrograph of the fracture-initiating oxidation/corrosion pit on the surface of reaction-bonded Si$_3$N$_4$. The EDX graph shows the relative concentration of chemical elements in the glassy region at the base of the pit. (Courtesy of the Garrett Turbine Engine Company, Phoenix, Ariz., Division of The Garrett Corporation.)

369

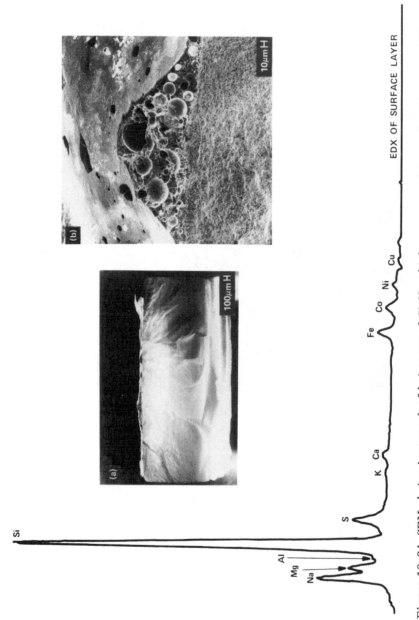

Figure 12.24 SEM photomicrograph of hot-pressed Si_3N_4 which was exposed to combustion gases with sea salt additions, showing that fracture initiated at the base of the glassy surface buildup. EDX analysis shows the chemical elements detected in the glassy material adjacent to the Si_3N_4. (Courtesy of the Garrett Turbine Engine Company, Phoenix, Ariz., Division of The Garrett Corporation.)

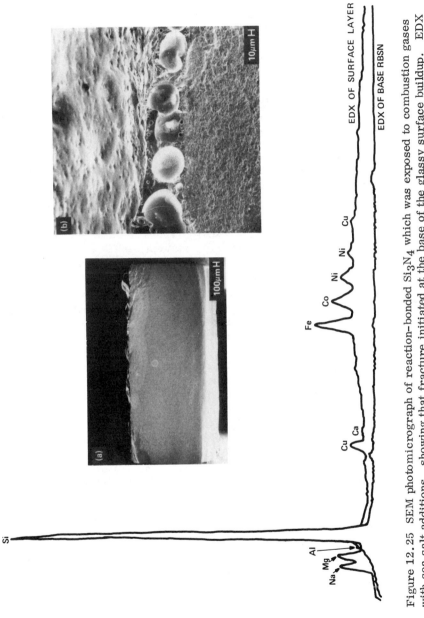

Figure 12.25 SEM photomicrograph of reaction-bonded Si_3N_4 which was exposed to combustion gases with sea salt additions, showing that fracture initiated at the base of the glassy surface buildup. EDX analysis shows the chemical elements detected in the glassy material adjacent to the Si_3N_4. (Courtesy of the Garrett Turbine Engine Company, Phoenix, Ariz., Division of The Garrett Corporation.)

to 690 PMa (100,000 psi) for the hot-pressed Si_3N_4 and only a decrease to 207 MPa (30,000 psi) for the reaction-bonded Si_3N_4.

The examples presented so far for oxidation and corrosion have had distinct features that help distinguish the cause of fracture from other mechanisms, such as impact or machining damage. Some corrosion-initiated fractures are more subtle. The corrosion or oxidation may only follow the grain boundaries and be so thin that it is not visible on the fracture surface. Its effects may not even show up in room-temperature strength testing since its degradation mechanism may only be active at high temperature. How do we recognize this type of corrosion? The following suggestions may be helpful:

Prepare a polished section of the cross section and try various etchants; this may enhance the regions near the surface where intergranular corrosion is present.

Conduct EDX, microprobe, or Auger analysis scans from the surface inward to determine if a composition gradient is present.

Use high-magnification SEM of the fracture surface to look for differences between the microstructure near the surface and in the interior; if the fracture surface near the specimen surface is intergranular and near the interior is intragranular, grain boundary corrosion is a possibility.

Conduct controlled exposures under exaggerated conditions in an effort to verify if the material is sensitive to attack.

Slow Crack Growth   Slow or subcritical crack growth was discussed in Chap. 4 and a typical fracture surface was shown in Fig. 4.8. The region of subcritical growth appears rough and intergranular and is very distinctive.

Slow crack growth can occur under sustained loading, as in the example in Chap. 4, and also under relatively fast loading, depending on the nature of the material, the temperature, the atmosphere, and the load. Figure 12.26 shows the fracture surface of a sintered Si_3N_4 material[*] (which was developed for low- to moderate-temperature applications) after 4-point bend testing at 982°C (1800°F) at a load rate of 0.05 cm/min (0.02 in./min). In spite of the rapid loading, substantial slow crack growth occurred. Examination of the fracture surface quickly tells the engineer that this material is not suitable for high-temperature application under a tensile load.

Sometimes, examination of the fracture surface by EDX or other surface chemical analysis technique can help identify the cause of slow crack growth. Specifically, the roughened region is analyzed separately from the rest of the fracture surface and bulk material. Chemical elements present in greater concentration in the slow-crack-growth region are probably

---

[*] Kyocera International, San Diego, Calif.

Figure 12.26 SEM photomicrograph of the fracture surface of a low-purity Si_3N_4 material sintered with MgO and showing slow crack growth. Region of slow crack growth identified by arrows.

associated with the cause. The tensile stress at fracture can also be approximated by assuming that the flaw size is equivalent to the slow-crack-growth region and using the Griffith equation (3.11). However, it should be realized that this is only an approximation and that the reported elastic modulus and fracture energy values for the material, when used in the Griffith equation, may not be good approximations for the material under slow-crack-growth conditions.

There are other limitations to the information available from the fracture surface. The size of the slow-crack-growth region provides no information about the time to failure, the rate of loading, or the mode of loading (cyclic versus static).

12.2 SUMMARY

Fractography is a powerful tool to the engineer in helping to determine the cause of a component or system failure. Well-defined features usually present on the fracture surface of a ceramic provide the engineer with useful information regarding the place where fracture initiated, the cause of fracture, the tensile stress at the point of failure, and the nature of the surrounding stress distribution. This information helps the engineer to

determine if the failure was design initiated or material initiated and provides direction in finding a solution. It can also help in achieving process
or product improvement. Finally, it can help determine legal liability
for personal or property damage.

REFERENCES

1. F. W. Preston, Angle of forking of glass cracks as an indicator of the
   stress system, J. Am. Ceram. Soc. 18, 175 (1935).
2. J. J. Mecholsky, S. W. Freiman, and R. W. Rice, in Fractography
   and Failure Analysis (D. M. Strauss and W. H. Cullen, Jr., eds.),
   ASTM STP 645, American Society of Testing, Philadelphia, 1978,
   pp. 363-379.
3. N. Terao, J. Phys. Soc. Jap. 8, 545 (1953).
4. W. C. Levengood, J. Appl. Phys. 29, 820 (1958).
5. E. B. Shand, Breaking stress of glass determined from dimensions of
   fracture mirrors, J. Am. Ceram. Soc. 42, 474 (1959).
6. H. P. Kirchner and R. M. Gruver, Fracture mirrors in polycrystalline ceramics and glass, in Fracture Mechanics of Ceramics, Vol. I
   (R. C. Bradt, D. P. H. Hasselman, and F. F. Lange, eds.), Plenum
   Publishing Corp., New York, 1974, pp. 309-321.
7. V. D. Frechette, Fracture of Heliostat Facets, presented at the ERDA
   Solar Thermal Projects Semiannual Review, Seattle, Wash., Aug. 23-
   24, 1977.
8. R. W. Rice, Fractographic identification of strength-controlling flaws
   and microstructure, in Fracture Mechanics of Ceramics, Vol. I (R. C.
   Bradt, D. P. H. Hasselman, and F. F. Lange, eds.), Plenum Publishing Corp., New York, 1974, pp. 323-345.
9. H. L. Marcus, J. M. Harris, and F. J. Szalkowsky, Auger spectroscopy of fracture surfaces of ceramics, in Fracture Mechanics of
   Ceramics, Vol. I (R. C. Bradt, D. P. H. Hasselman, and F. F.
   Lange, eds.), Plenum Publishing Corp., New York, 1974, pp. 387-
   398.
10. O. Johari and N. M. Parikh, in Fracture Mechanics of Ceramics, Vol.
    I (R. C. Bradt, D. P. H. Hasselman, and F. F. Lange, eds.), Plenum
    Publishing Corp., New York, 1974, pp. 399-420.
11. K. M. Johansen, D. W. Richerson, and J. J. Schuldies, Ceramic
    Components for Turbine Engines, Phase II Final Report, AiResearch
    Manufacturing Co. Report No. 21-2794(08), Feb. 29, 1980, prepared
    under Air Force contract F33615-77-C-5171.
12. R. W. Rice, S. W. Freiman, J. J. Mecholsky, and R. Ruh, Fracture
    sources in Si_3N_4 and SiC, in Ceramic Gas Turbine Demonstration
    Engine Program Review (J. W. Fairbanks and R. W. Rice, eds.),
    MCIC Report MCIC-78-36, 1978, pp. 665-688.

13. D. W. Richerson, T. M. Yonushonis, and G. Q. Weaver, Properties of silicon nitride rotor materials, in Ceramic Gas Turbine Demonstration Engine Program Review (J. W. Fairbanks and R. W. Rice, eds.), MCIC Report MCIC-78-36, 1978, pp. 193-217.

14. T. M. Yonushonis and D. W. Richerson, Strength of reaction-bonded silicon nitride, in Ceramic Gas Turbine Demonstration Engine Program Review (J. W. Fairbanks and R. W. Rice, eds.), MCIC Report MCIC-78-36, 1978, pp. 219-234.

15. W. D. Carruthers, D. W. Richerson, and K. Benn, 3500 Hour Durability Testing of Commercial Ceramic Materials, Interim Report, July 1980, NASA CR-159785.

16. V. D. Frechette, Fractography and quality assurance of glass and ceramics, in Quality Assurance in Ceramic Industries (V. D. Frechette, L. D. Pye, and D. E. Rase, eds.), Plenum Publishing Corp., New York, 1979, pp. 227-236.

17. R. W. Davidge and G. Tappin, Trans. Br. Ceram. Soc. 66, 8 (1967).

18. G. Q. Weaver, H. R. Baumgartner, and M. L. Torti, Thermal shock behavior of sintered silicon carbide and reaction-bonded silicon nitride, in Special Ceramics 6 (P. Popper, ed.), British Ceramic Research Association, Stoke-on-Trent, England, 1975, pp. 261-281.

19. C. L. Ammann, J. E. Doherty, and C. G. Nessler, Mater. Sci. Eng. 22, 15-22 (1976).

20. K. M. Johansen, L. J. Lindberg, and P. M. Ardans, Ceramic Components for Turbine Engines, 8th Interim Report, AiResearch Manufacturing Co. Report No. 21-2794, 10, June 5, 1980, prepared under Air Force contract F33615-77-C-5171.

21. Ceramic Gas Turbine Engine Demonstration Program Interim Report No. 11, Nov. 1978, AiResearch Report 76-212188(11), prepared under Contract N00024-76-C-5352, pp. 3-47 to 3-64.

22. J. M. Wimmer and I. Bransky, Impact resistance of structural ceramics, Ceram. Bull. 56(6), 552-555 (1977).

23. D. G. Finger, Contact Stress Analysis of Ceramic-to-Metal Interface, Final Report, Contract N00014-78-C-0547, Sept. 1979.

24. D. W. Richerson, W. D. Carruthers, and L. J. Lindberg, Contact stress and coefficient of friction effects on ceramic interfaces, in Surfaces and Interfaces in Ceramic and Ceramic-Metal Systems (J. A. Pask and A. G. Evans, eds.), Plenum Publishing Corp., New York, 1981.

25. D. W. Richerson and T. M. Yonushonis, Environmental effects on the strength of silicon nitride materials, in Ceramic Gas Turbine Demonstration Engine Program Review (J. W. Fairbanks and R. W. Rice, eds.), MCIC Report MCIC-78-36, 1978, pp. 247-271.

13

Applications: Material Selection

The objective of this final chapter is to provide a practical review of the prior 12 chapters by selecting ceramic materials for a variety of applications. The applications are the same ones that appeared in the exercise in the introduction to this book. Just for fun, redo that exercise and compare your answers with those you gave before reading the book. Then read the following more detailed sections on each application.

13.1 SANDBLAST NOZZLE

Sandblasting is used extensively in industry for cleaning metal surfaces prior to finishing, for removing the ceramic molds from metal castings, for cleaning ceramic parts after sintering, for applying inscriptions to memorials, and for many other applications. Particles of SiO_2, Al_2O_3, or other abrasives are carried by room-temperature high-pressure air through a nozzle. The primary requirements of the nozzle are high wear resistance and relatively low cost.

Sandblast nozzles have been made out of a variety of ceramic-based materials. Cobalt-bonded WC cermet nozzles are manufactured by either a sintering or a high-temperature casting technique. This material is very tough and hard and has excellent wear resistance. Its main disadvantage is cost. Sintered Al_2O_3 nozzles cost less, but have shorter life and thus result in more labor because they must be replaced more frequently. Hot-pressed B_4C has a higher cost of fabrication, but a much longer life than either the Al_2O_3 or cobalt-bonded WC because of its extreme hardness.

Therefore, the best answer to the question of which ceramic material would be optimum for a sandblast nozzle would be Al_2O_3 or B_4C (most people do not think of Co-bonded WC as a ceramic), and the special property that makes either of these materials a good choice would be high hardness. These materials would then become prime candidates to the engineer for a sandblast nozzle application. The final selection would require further information, such as initial cost and life in the specific application. Another important evaluation factor might be contamination.

13.2 INSULATING REFRACTORY FURNACE LINING

The linings of high-temperature furnaces are referred to as <u>refractories</u>. A wide variety of refractories are available, and the selection for a specific application depends on temperatures, environment, thermal cycle, cost, and other factors. Perhaps the best way to select a refractory system for an application is first to define the operational conditions and then survey several refractories manufacturers. They have much experience, including many case histories, and can usually design the most cost-effective and energy-effective system.

Providing thermal protection for a high-temperature furnace, combustor, or incinerator usually involves several layers of refractory. The inner lining is exposed to the highest temperature and is in direct contact with the furnace contents, which can consist of molten metal, slag, corrosive or high-velocity gases, and fluidized particles. Therefore, this inner lining must be chemically resistant and erosion resistant and is thus relatively high in bulk density. The basic oxygen furnace (BOF) discussed in Chap. 4 provides a good example of the severe conditions to which an inner lining can be exposed and the nature of the refractory. Other examples would be glass-melting-tank refractories, coal gasifier refractories, and MHD refractories.

The inner lining is selected for corrosion and erosion resistance, but because of the relatively high bulk density, does not provide adequate thermal insulation. Therefore, most furnaces have an outer layer of insulating refractory. It is protected from the furnace environment by the inner lining and thus does not have to be erosion resistant or chemically resistant. Usually, the furnace is designed so that this insulating layer does not have to carry high structural loads, so the material does not have to be strong.

We can now answer the question: Which ceramic material or materials would be optimum as an insulating refractory for a furnace lining, and which properties are most critical? First, the critical properties are low thermal conductivity and high melting or decomposition temperature. These properties can be achieved with porous aggregates or bricks made up of oxides, silicates, or combinations of oxides and silicates. Insulation can also be achieved with ceramic fibers, which are available as aggregates, blankets, and fiberboards. Fibers are especially good for lining small laboratory furnaces where rapid heating and cooldown are desired. Much air space is present between the fibers, so that the lining has extremely low thermal conductivity, which keeps heat in the furnace, but also low thermal mass, so that it does not store up heat as a heat sink. A small fiber-lined furnace can be taken up to 1500°C in less than 15 min and cooled down in about 1 hr. Brick-lined furnaces require much longer heat-up and cooldown time. They also consume more energy.

13.3 SEALS

Ceramic seals are currently being manufactured by the millions in a variety of sizes, types, and shapes, yet most people (even engineers) would have difficulty naming a specific example. The properties of ceramics that make them suitable for seals are hardness (resulting in dimensional stability and abrasion resistance), low friction when machined to a fine surface finish, high resistance to corrosion, and higher temperature capability than materials such as rubber, nylon, and Teflon.

Carbon graphite is one of the best seal materials. It can run in face seals against itself, metals, or ceramics without galling or seizing. It is dimensionally stable over a wide temperature range and has excellent corrosion resistance. It has a high thermal conductivity and helps to dissipate heat generated at the rub face of the seal. Its low thermal expansion, together with its high thermal conductivity, provide excellent thermal shock resistance [1].

Face seals essentially provide a seal at a rotating interface which prevents passage of liquids or gases on one side of the seal to the other. For instance, the compressor seal in an automotive air conditioner seals halogenated hydrocarbons and oil at pressures up to 250 psig and surface speeds of 1800 ft/min. Graphite against Al_2O_3 provides a low-cost, reliable seal for this application. Graphite against graphite has also been used.

A more severe application is the main rotor bearing seal in jet engines. It must seal the oil lubricating system from 120-psig hot air at temperatures up to 1100°F and surface speeds up to 20,000 ft/min. Graphite impregnated with other materials to increase the strength and oxidation resistance is required for this application [2].

Another severe application is in recovery of crude oil by the saltwater pressure system. Salt water is pumped into the ground at about 2500 psi to force crude oil out of the rock formations so that it can be recovered in adjacent wells. The face seal in the pump must survive the 2500-psi pressure plus temperatures up to 600°F plus surface rub speeds of 5000 ft/min.

Face-type seals are used in many applications, including sand slurry pumps (which pump approximately 35% solids), chemical processing and handling, fuel pumps, torque converters, washing machines, dishwashers, and garbage disposals.

Face seals require two compatible surfaces. One of the various grades of graphite is frequently used for one surface. The mating surface can also be graphite or can be a metal, ceramic, or plastic, depending on the temperature and other conditions of application. Stainless steels, hard chromeplated steels, nickel-bonded tungsten carbide, and ceramics such as Al_2O_3

and SiC are all in common use. The metals and SiC are especially good because they have high thermal conductivity and help dissipate heat. If a hard surface is required on the graphite, it can be achieved by reacting the graphite at high temperature with silicon monoxide. This converts the outer layer to SiC.

Surface finish and surface flatness are extremely important for seal applications. For noncritical applications, a flatness to within 0.002 in. (17 helium light bands) is usually acceptable. For higher pressure and more critical applications, a flatness to within six or three light bands is usually required.

Graphite has been referred to as self-lubricating. Some people assume that this is due to the sheet structure and the weak van der Waals bonds between these sheets. This may be a factor in some cases, but generally the lubricating results from formation of a hydrodynamic or transfer film between the graphite and the mating surface. This film appears dependent on the presence of polar liquids, oxygen, or water vapor. Graphite operating in vacuum, dry gases, cryogenic fluids, or at high temperatures does not form a suitable interfacial film. In these cases, special grades of graphite containing impregnants are used. An engineer requiring a seal material should consult with experienced seal manufacturers to make sure that both the material and design are suitable for the application.

In summary, graphite, SiC, and Al_2O_3 have all been successfully used as face seals and other types of seals. Although the key property requirements will vary with the application, wear resistance, chemical resistance, and ability to be produced with close surface flatness tolerances are especially critical.

13.4 POTTERY

Perhaps the earliest ceramic articles fabricated by human beings were pottery, using natural clay minerals. The essential properties of the clay were its plasticity or workability when water was added and its ability to become hard and impermeable when fired.

Pottery is still fabricated and used today in large quantity. The basic essentials are still the same. The major differences are that the raw materials are usually processed to a higher degree, the compositions are more varied and sophisticated, firing is done in electric or gas furnaces, and the surfaces are usually coated with a glass glaze. The primary factor is cost. The clay minerals, feldspar, and silica sand which make up most pottery compositions are "dirt cheap" and require relatively little processing. These pottery compositions are the basic ingredients of art pottery, such as vases, mugs, and knickknacks, and of flowerpots, dishes, bricks, sewer pipe, and a variety of other articles.

13.5 HIGH-TEMPERATURE HEAT EXCHANGER

With current and projected energy conservation requirements, the use of heat exchangers to recover waste heat is growing in importance. In many cases, such as metal and glass melting furnaces, incinerators, and coal-burning furnaces, the discharge gases are either too high in temperature or too corrosive for metal heat exchangers. Thus two important design requirements of the material are high-temperature capability and corrosion resistance. Another is thermal shock resistance. Industrial heat exchangers are massive and involve substantial thermal stresses.

Another desirable property of a heat exchanger material is high thermal conductivity to maximize the rate of heat transfer and optimize the efficiency of the unit. However, this property is usually of less importance than thermal shock resistance, temperature capability, and corrosion resistance.

The major material being used and evaluated for industrial heat exchangers is SiC, primarily in tubular form [3]. SiC has high thermal conductivity, excellent high-temperature and corrosion resistance, and moderately good thermal shock resistance. Although the thermal shock resistance is substantially better than most oxide ceramics, it is marginal for some heat exchanger applications and requires very careful design to minimize thermally induced strains.

A second important limitation of SiC (and other candidate ceramic materials) is fabricability in the required size. Scale-up is required to produce the size components needed for industrial heat exchangers. As always, cost is an important factor.

Smaller heat exchangers are being developed to recover heat in vehicular engines and in small power generators. The most advanced technology is for a rotary regenerator for a truck gas turbine engine. Over 10,000 hr of life has been demonstrated for regenerator cores of an aluminum silicate composition at Ford Motor Company [4], and acceptable durability in a vehicle has been demonstrated by Detroit Diesel Allison Division of General Motors [5].

The rotary regenerator consists of a thin-walled honeycomb-type configuration (see Fig. 6.15) that continually rotates through the exhaust and inlet gases of the gas turbine. The exhaust gases heat the regenerator material to over 982°C (1800°F). This then rotates into the inlet region, where the heat exchanger gives up its heat to preheat the air prior to combustion. Preheating the combustion air provides a substantial improvement in efficiency and reduction in fuel consumption.

A major requirement of the rotary regenerator material is thermal shock resistance. Large temperature gradients exist through the thickness of the heat exchanger and across the diameter. The initial material selected for development was lithium aluminum silicate (LAS) because of its near-zero coefficient of thermal expansion and the resulting extremely good thermal shock resistance. Thermally, LAS worked fine, but a corrosion problem was identified during engine testing. The corrosion consisted of ion exchange of Na^+ and H^+ ions for Li^+ ions at different regions of the heat exchanger. This resulted in localized changes in properties which led to distortion and cracking. Work at Corning Glass, the manufacturers of the LAS core material, resulted in a procedure for acid leaching the LAS honeycomb to remove the Li prior to application. The resulting aluminum silicate (AS) material had acceptable thermal expansion properties and did not distort or crack in the engine.

Rotary regenerators have also been manufactured out of magnesium aluminum silicate (MAS, cordierite). It has a higher-temperature capability than LAS or AS and no corrosion problem. However, it has a higher thermal expansion and requires design modifications to reduce the thermal stresses in the regenerator core.

Additional materials with potential for various types of heat exchangers include Si_3N_4 and mullite. Like SiC, these materials have higher thermal expansion than LAS and MAS and require more careful design to solve thermal shock problems.

In summary, the key properties for the selection of a heat exchanger material include thermal shock resistance, high-temperature capability, corrosion resistance, and in some cases high thermal conductivity. Materials with a good combination of properties include LAS, MAS, SiC, Si_3N_4, and mullite.

13.6 ARMOR

Ceramic armor was developed during the Viet Nam war for helicopter armor and personnel armor. The requirements were light weight and capability to defeat small arms armor-piercing projectiles (projectiles containing a tungsten carbide core). The ceramic that was successfully developed for this application was boron carbide (B_4C).

The armor consisted of a composite system. B_4C was bonded to a fiberglass or Kevlar backing and covered with a fabric spall shield. When the projectile struck the B_4C, it was shattered by the high hardness of the B_4C. The energy of this impact was absorbed by localized fracture of the B_4C. The momentum of the debris was then absorbed by the fiberglass, in much the way that a baseball glove deforms and absorbs the momentum of a baseball. The spall shield prevents chips and particles from rebounding and causing secondary damage.

A plate of B_4C about 0.64 cm (0.25 in.) thick with a similar backing of fiberglass can stop a .30 caliber armor-piercing projectile. A much heavier layer of steel would be required to defeat the same armor-piercing projectiles.

Hot-pressed B_4C has a hardness of about 3500 kg/mm^2 and a density of only 2.4 to 2.5 g/cm^3. The low density and moderate thickness requirements make B_4C armor feasible for infantry armor and helicopter armor. Other ceramics such as Al_2O_3 have also been used for armor, but are not as light and do not have equivalent protection capability.

We can now answer the question of which ceramic material would be suitable for armor and which properties would be most important. Near-theoretical density B_4C would be the optimum candidate because of its high hardness and low density.

13.7 PERMANENT MAGNET

Ceramic materials with magnetic properties are important in many applications, as described in Chap. 2. The hexagonal ferrites (especially barium, strontium, and lead hexaferrites) are frequently used for permanent magnets because of their high magnetization, compact size, and low cost.

Each application requires different magnetic properties, and a composition and crystal structure having the optimum properties must be selected or developed. For instance, the magnesium-zinc and nickel-zinc spinel ferrites have high permeability and low loss and are well suited for transformer and inductor applications. Some spinel and garnet ferrites have a square loop hysteresis curve and are suitable for switching devices.

13.8 CERAMIC QUENCH BLOCK

A quench block is used in brazing and other applications where localized heating needs to be quickly dissipated. Copper is frequently used because of its high thermal conductivity. In some cases copper or other metals are not suitable because of temperature, dimensional stability, corrosion resistance, or other factors and a ceramic is required. The primary criterion is still high thermal conductivity.

Diamond has extremely high thermal conductivity, but is not available in suitable configurations or cost. Perhaps the next best candidate is BeO, followed more distantly by SiC. For some applications, graphite may be the best candidate. It has the advantage of being easily machined or drilled into complex configurations and is excellent if a large number of braze junctions must be made simultaneously, such as in some electrical devices. Sometimes, graphite with the surface converted to SiC works best for this application.

13.9 REENTRY-VEHICLE THERMAL PROTECTION

When a vehicle such as the space shuttle reenters the atmosphere, surface temperatures up to about 1650°C ($\sim$3000°F) can result at the surface due to

friction. Two design approaches have been used to protect critical reentry
surfaces. One approach makes use of ablative materials which dissipate
the heat, but are slowly consumed and must be replaced after each mission.
The second uses materials that can withstand the temperatures generated
by reentry. This second approach was selected for protection of the space
shuttle orbiter [6, 7].

Figure 13.1 shows the approximate temperature contours that the orbiter must withstand during ascent and reentry and be reusable for 100 mis-
sions. The highest temperatures are in the range 1425 to 1650°C ($\sim$ 2600
to 3000°F) and are on the nose and the leading edge of the wings. A reinforced carbon-carbon composite with a surface SiC-based coating to prevent
oxidation has been developed by Vought Corporation for these areas. The

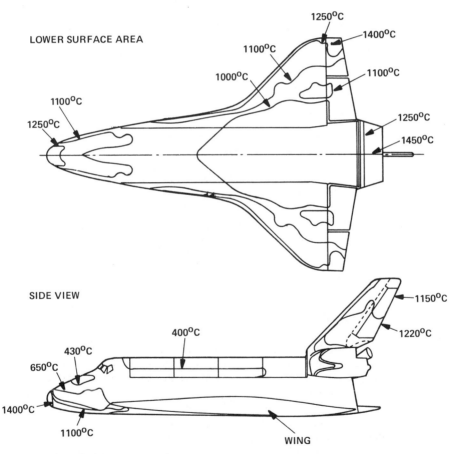

Figure 13.1   Approximate temperature distribution that the space shuttle
orbiter must withstand during ascent and reentry. (Source: Ref. 6.)

nose cap of this material is about 1.4 m (4.5 ft) in diameter. Each wing tip contains an additional 36 m^2 (400 ft^2) of the carbon-carbon panels. The carbon-carbon material is backed by insulation blankets which limit the temperature of the adjacent metal to 177°C (350°F).

For surfaces that will be exposed to temperatures below 1260°C (2300°F), a lightweight, porous, fused silica material has been developed by Lockheed Missiles and Space Co. The composition provides the required temperature resistance, the highly porous structure provides the light weight (~ 9 lb/ft^3) and low thermal conductivity, and a combination of the two provides thermal shock resistance and strain tolerance. A coating is applied to prevent moisture absorption and to provide erosion resistance, since the porous material has relatively low strength and poor erosion resistance. The coating is about 93% SiO_2, 5% B_2O_3, and 2% boron silicide.

Tiles of the thermal protection system are cemented with an elastomer to the surface of the shuttle. Tile sizes are 15.2 × 15.2 cm (6 × 6 in.) and range in thickness from 0.5 to 11.4 cm (0.2 to 4.5 in.). Very high reliability is required of the complete system. Loss of a single tile could result in disaster.

13.10 ISOTROPIC TRANSPARENT MATERIAL

Isotropic means that the material has uniform properties in all directions. Crystals having the cubic crystal structure are optically isotropic. Polycrystalline solids that do not have preferred orientation of the grains behave in bulk as if they were isotropic, even though effects of anisotropy will be present on a microstructure level. Both single-crystal and polycrystalline ceramics can be transparent and have applications which depend on the combination of transparency and isotropism. However, by far the most important material with this combination of properties is glass.

13.11 MATERIAL WITH ANISOTROPIC
THERMAL EXPANSION

All noncubic crystals will have different thermal expansion properties in the various crystallographic directions. Single-crystal or oriented polycrystalline graphite has one of the widest variations in thermal expansion. Parallel to the sheet structure the expansion is very low (1.0×10^{-6} per °C) due to the strong covalent bonding. Perpendicular to the sheets the expansion is high (27×10^{-6} per °C) because of the weak van der Waals bonding between sheets. However, a block of synthetic graphite will have very different thermal expansion properties from single-crystal graphite. If the block was manufactured by pressure compaction, the individual graphite crystals will be in somewhat random orientation, so that the bulk thermal expansion will not vary greatly as a function of direction. If the block was extruded, the individual crystals will tend to align and result in substantial anisotropy.

As discussed in Chap. 2, other single-crystal ceramics with large anisotropy include aluminum titanate, calcium carbonate (calcite), β-spodumene, and eucryptite.

13.12 CERAMIC NOT CHANGING DIMENSIONS DURING DENSIFICATION

Reaction-bonded Si_3N_4 and reaction-sintered SiC are two ceramics that undergo very little dimensional change during densification. The fabrication steps for both materials were discussed in Chap. 7. Densification of the Si_3N_4 is achieved by reaction of nitrogen gas with a compact of silicon particles whereby the Si_3N_4 crystals that form initially produce a skeletal structure which prevents shrinkage and then subsequent crystals grow into the pores to achieve an increase in density. This material has the advantage of net shape fabrication, but the disadvantage that open pores are inherently present. The open pores adversely affect the strength and oxidation resistance of the Si_3N_4.

Densification of SiC by reaction sintering is achieved by infiltration of a SiC-C compacted mixture with molten or vapor-phase silicon. The silicon reacts with the carbon to form new SiC, thus bonding the original SiC particles together. Any pores are filled with silicon, resulting in a strong, nonporous material. Dimensional change during this process is a little greater than for reaction-bonded Si_3N_4, but still below 1%. Shrinkage of this level is easier to control than the typical linear shrinkage of 10 to 20% encountered during conventional sintering.

13.13 GRINDING MEDIA FOR A BALL MILL

As previously noted, grinding media are the balls or cylinders that are tumbled in a ball mill to achieve particle size reduction of the powder being milled. Size reduction is achieved as the particles are pinched between adjacent balls and against the mill wall. Grinding action is enhanced by increase in specific gravity of the media and contamination is minimized by increase in hardness. The selection of media depends on a compromise between grinding time and efficiency and allowable contamination. Contamination can be minimized by using media of the same composition as the powder. Rapid grinding can be achieved by selecting high-specific-gravity materials such as WC or iron. In the latter case, much contamination results and acid leaching is usually necessary.

Porcelain media are used widely for compositions made up of clay, silica, and feldspar. Al_2O_3 media are frequently used for alumina-containing compositions and have also been used for Si_3N_4 and SiAlON compositions. Al_2O_3 provides a good compromise because it has reasonable hardness and intermediate density.

13.14 LOW-COST FIBER-ORGANIC COMPOSITE

The best candidate for this category is standard fiberglass. It consists of laminates of woven fiberglass cloth bonded together by organic resin. The glass fibers provide high strength and the plastic provides protection to the fibers from surface damage and impact as well as imparting strain tolerances to the system.

13.15 THERMAL BARRIER COATING

Thermal barrier coatings are gaining importance in gas turbine applications as a means of increasing the system operating temperature where cooled metal components are used or allowing substitutions of lower-service-temperature metals containing less strategic materials such as Co, Ni, Cb, Mo, and Hf. The coating must have low thermal conductivity and low emissivity and adhere to the metal even under severe thermal shock conditions.

Best results for gas turbine components have been achieved with stabilized ZrO_2 applied by plasma spray. ZrO_2 has very low thermal conductivity, low emissivity, and a high enough coefficient of thermal expansion to avoid a large mismatch with the metal substrate. Additional improvement in thermal expansion match and adherence is achieved with the use of an intermediate bond layer, such as nickel aluminide or "CoCrAlY." Alternately, deposition of a graded coating may be used. The graded coating consists of a sequence of surface layers grading from all metal to all ceramic.

An increase in service temperature capability of about 204°C (400°F) has been demonstrated, resulting in an important decrease in fuel consumption. Initial development was restricted to stationary components, but has now been extended successfully to rotor blades.

Thermal barrier coatings applied to metals are effective only in high-heat-flux situations, such as that which occurs when cooling is applied to the back surface (for a liner-type component) or interior (for a rotor blade or stator vane) of a metal. If the heat is not being drawn out, the temperature of the metal will approach that of the ceramic coating and no benefit will be gained.

Ceramic coatings have also been applied to ceramic components to alter the thermal response. For instance, zircon and mullite have been plasma spray deposited on SiC in cases where a low emissivity was required. Zircon, mullite, and SiC all have similar coefficients of thermal expansion.

Development of thermal barrier coatings has been relatively recent. Molten particle deposition of ceramic coatings for lower-temperature applications such as wear resistance has been known much longer and is widely practiced.

13.16 LOW-COST SLIP-CASTING MOLD

As discussed in Chap. 6, plaster is the ideal low-cost mold material for slip casting. Mold plaster powder is readily available, inexpensive, and reproducible and requires only the addition of water and stirring to produce the required mix for mold fabrication. The plaster mold sets quickly and accurately duplicates intricate details and surface finish of the pattern.

13.17 THERMOCOUPLE WIRE PROTECTION

Ceramics are used to separate thermocouple wires and to protect them from direct exposure to the high-temperature environment. Extruded sintered tubes of Al_2O_3, MgO, and other compositions are often used. These tubes are solid except for the holes required for the wires in the thermocouple system. The thermocouple bead is formed first and then the opposite end of the wires are slipped through the ceramic tube.

Integral shielded thermocouples and other instrumentation are also widely used. These are fabricated by drawing or swaging. In this case, the thermocouple manufacturer starts with oversize wire, loosely bonded MgO powder preforms with the required number of holes, and oversized metal tubing of the shield (sheath) material. The MgO preforms are slipped over the wires and inserted into the sheath tubing. This assembly is then heated and passed through a drawing or swaging tool, which reduces the diameter of the assembly by about 15%. The wire and sheath elongate in a ductile fashion and the porous, lightly bonded MgO preform compacts. The drawing operation is repeated until the specified diameter has been achieved.

13.18 THERMAL-SHOCK-RESISTANT MATERIAL

Thermal shock damage results in a material by buildup of thermal stresses, usually during rapid heating or rapid cooling. As discussed in Chap. 4, the major materials factors influencing thermal shock resistance are coefficient

of thermal expansion, elastic modulus, thermal conductivity, strength, and
fracture toughness. Thermal shock resistance is increased by decreasing
the coefficient of thermal expansion or the elastic modulus and by increasing the thermal conductivity, strength, or fracture toughness.

Fused silica and β-spodumene (lithium aluminum silicate, LAS) have
extremely good thermal shock resistance because of their low expansion
coefficients, even though they have relatively low strength, low thermal
conductivity, and low fracture toughness. Low thermal expansion is especially effective at reducing thermal shock damage because it limits the amount
of dimensional change and thus limits the amount of strain.

Fused silica is used in applications ranging from impervious liners in
high-temperature vacuum furnaces to the low-density thermal protection
tiles on the space shuttle. LAS is used in applications ranging from Corningware to high-temperature heat exchangers.

Both Si_3N_4 and SiC have moderately good thermal shock resistance.
The Si_3N_4 has moderately low thermal expansion and high strength together
with moderately high thermal conductivity and elastic modulus. The SiC
has a higher coefficient of thermal expansion and elastic modulus but com-
pensates with higher thermal conductivity.

Many materials used as refractories and kiln furniture have good thermal shock resistance, but for reasons other than those mentioned so far.
These materials have low strength, low thermal conductivity, often high
thermal expansion of the constituent materials, and contain much porosity
and microcracks. They are not resistant to crack initiation, but are very
resistant to crack propagation, which provides the degree of strain tolerance necessary for good thermal shock resistance.

Another approach to achieving good thermal shock resistance is to increase the fracture toughness of the material. High fracture toughness has
been achieved in dense sintered ZrO_2 by mixing stabilized and unstabilized
powders prior to sintering. The unstabilized ZrO_2 undergoes phase transformation during cooldown from the sintering temperature and causes either
very fine microcracks or internal stress. This results in a multifold increase in fracture toughness.

13.19 SUBSTRATE FOR ELECTRICAL DEVICE

A variety of ceramic materials are used as substrates or mounting fixtures
for electrical circuits and devices. Although selection depends on the requirements of the application, high electrical resistance is a common requirement. Other important properties are the dielectric constant, the
dielectric strength, the thermal conductivity, and the surface finish achievable in the as-sintered condition. Porcelain, steatite, and various grades
of Al_2O_3 are commonly used. High-purity Al_2O_3 is used for the most demanding applications.

13.20 KILN FURNITURE FOR DIODE MANUFACTURE

Kiln furniture refers to the setter plates and support structures inside a
furnace on which the material being sintered or heat treated sets. Diodes
are heat-treated in a controlled atmosphere at high temperature to diffuse
a dopant into the surface of the diode to achieve a predetermined level of
semiconductor properties. Parts per million impurities can poison the
operation. Nonporous high-purity kiln furniture that does not degas at high
temperature or under a subatmospheric pressure is required. Pyrolytic
graphite, glassy carbon, CVD SiC-coated graphite, and high-purity SiC
have been used successfully for this and related applications.

13.21 HIGH CHARGE-STORAGE CAPABILITY

The primary property required for high charge-storage capability in a capacitor is a high dielectric constant. $BaTiO_3$ has a dielectric constant of
about 1600, $BaTiO_3$-10% $CaZrO_3$-1% $MgZrO_3$ has a dielectric constant of
about 5000 and $BaTiO_3$-10% $CaZrO_3$-10% $SrTiO_3$ has a dielectric constant
of about 9500. This can be compared to organic materials, mica, MgO,
and Al_2O_3, all with dielectric constants under 10, and TiO_2, with a diele-
tric constant around 100. Development of the $BaTiO_3$ dielectrics has been
an important factor in the miniaturization of electronic devices.

13.22 LOW-DENSITY FURNACE INSULATION

In the introduction to the book, the requirement for "a very low density material used for insulation in high-temperature furnace construction" was
considered. The best answer is ceramic fiber insulation. Recently developed
oxide and silicate fiber compositions have high purity and high temperature
stability and are available in fiber bundles, woven cloth or blankets, and
hardboard. The fiber bundles, in particular, contain much dead air space
between the fibers and have very low density and thermal conductivity. A
furnace lined with this low-thermal-conductivity insulation has low thermal
inertia and can therefore be heated and cooled rapidly and requires much
less power input than does a refractory-brick-lined furnace.

13.23 RADOMES

A radome is essentially a protective covering and window for electronic
guidance and detection equipment on missiles, aircraft, and spacecraft.

The radome must be transparent to the wavelengths of electromagnetic radiation used by the equipment. MgO, Al_2O_3, and fused SiO_2 are transparent to ultraviolet wavelengths and a portion of the infrared and radar wavelengths. MgF_2, ZnS, ZnSe, and CdTe are transparent to infrared and radar wavelengths.

A radome must also be resistant to high-velocity impact by rain and other possible atmospheric particulates. For some applications, it must also be resistant to high temperature and thermal shock.

13.24 GAS TURBINE STATOR

The stator in a gas turbine engine is a nonmoving airfoil-shaped component that directs the hot-gas flow from the combustor at an optimum angle to the rotor to achieve peak aerodynamic performance. The first-stage stator is therefore exposed to the peak temperature from the combustor, including hot streaks, and must have high temperature stability. The stator is also exposed to severe temperature gradients circumferentially, radially, and axially and must have extremely good thermal shock resistance. Finally, the stator material must be stable in oxidizing, corrosive, high-velocity, high-temperature gases.

Many ceramic materials have been evaluated for gas turbine stators, but most have been rejected due to inadequate thermal shock resistance. The most promising have been various forms of Si_3N_4 and SiC. These materials have a combination of high strength, moderate-to-low thermal expansion, relatively high thermal conductivity, and good oxidation/corrosion resistance. Successful tests have been conducted in development engines, but further development is required before durability in commercial applications is demonstrated.

13.25 HIGH-TEMPERATURE CEMENT

High-temperature ceramic cements were discussed briefly in Chap. 7. The most important properties are adhesion and high temperature capability. The ability to apply the cement on-site either as a repair or as a reconstruction is often another important consideration.

Many ceramic cements are available commercially and others can be compounded easily by an individual to meet special requirements. The selection depends on the specific application. Some commonly used high-temperature cements are calcium aluminate, sodium silicate, and mono-aluminum phosphate.

13.26 ABRASIVE FOR CUTOFF AND GRINDING WHEELS

The primary property requirement of the abrasive is high hardness, or at least higher hardness than the material being machined. Cutoff and grinding wheels used for machining of densified ceramics usually require diamond or cubic BN abrasive. Bonded Al_2O_3 or SiC are normally used for metals.

The hardness of the abrasive is not the only criterion for a cutoff or grinding wheel to function efficiently. Controlled breakdown of either the abrasive or the bond holding the abrasive particles together is also necessary to provide fresh, sharp cutting surfaces. Otherwise, the exposed tips

Table 13.1 Some of the Variations Available in Cutoff and Grinding Wheels

---

Type of abrasive material

  Diamond
  Al_2O_3
  SiC
  SiO_2
  Cubic BN
  Others

Size of abrasive material

Concentration of abrasive particles

Type of bond material

  Metal
  Glass
  Resin
  Rubber

Properties of bond material

  Hardness
  Elasticity
  Porosity

Configuration of wheel

  Solid with center bore
  Formed-surface
  Abrasive only as coating
  Spindle-mounted

---

of the abrasive particles would wear smooth and cutting efficiency would decrease. Controlled breakdown can be aided by the operator by "dressing"
the grinding wheel with porous SiC.

Cutoff or grinding wheels usually must be optimized or carefully selected for each specific material and machining operation. A wide variety are
available, as indicated by Table 13.1.

Ceramics are also used extensively as inserts for cutting tools used
for milling, lathing, and other machining operations of metals. The use of
WC-Co cermets is well known, but the further improvement that can be
achieved in many operations with Al_2O_3, Al_2O_3-TiC, or SiAlON cutting
inserts is not as well known. Ceramic physical or vapor deposition coatings
on the surface of metal or ceramic cutting tools also provide substantial
improvements in life, cutting efficiency and other parameters for some
machining operations.

REFERENCES

1. Pure Carbon Technical Information Pamphlet PC-5393-5M (1979).
2. R. R. Paxton, Electrochem. Technol. 5(5-6), 174-182 (1967).
3. A. Pietsch and K. Styhr, Ceramic heat exchanger applications and de-
    velopments, in Ceramics for High Performance Applications-II (J. J.
    Burke, E. N. Lenoe, and R. N. Katz, eds.), Brook Hill Publishing Co.,
    Chestnut Hill, Mass., 1978, pp. 385-395. (Available from MCIC,
    Battelle Columbus Labs., Columbus, Ohio.)
4. C. A. Fucinari and V. D. N. Rao, Ceramic Regenerator Systems
    Development Program, NASA Contract DEN3-8, NASA CR-159707,
    Oct. 1979.
5. H. E. Helms and F. A. Rockwood, Heavy Duty Gas Turbine Engine
    Program, Progress Report for period July 1, 1976 - Jan. 1978, Report
    No. DDA EDR 9346, prepared under NASA Contract NAS3-20064, Feb.
    1978.
6. J. J. Svec, Orbiter has ceramic skin, Ceram. Ind. 107(4), 20-24
    (1976).
7. L. J. Korb, C. A. Morant, R. M. Calland, and C. S. Thatcher, The
    shuttle orbiter thermal protection system, Bull. Amer. Ceram. Soc.
    Vol. 60, No. 11 (1981), pp. 1188-1193.

Abrasive , 260, 391
Absorption, 58, 282
Acheson process, 152
Acoustic emissions, 300
Acoustic holography, 297
Acoustic waves, 287
Adhesive, 256
Agglomerates, 175, 191, 221
Air classification, 158
Alginates, 173
Al_2O_3
  applications of, 51, 53, 59, 61,
    116, 123, 137, 151, 154, 209,
    241, 261, 376-379, 387-392
  creep of, 103, 106, 111
  elastic modulus of, 72
  hardness of, 85
  machining of, 269
  melting temperature of, 37
  source of, 151, 166
  strength of, 92, 271
  structure of, 10
Anisotropy, 265, 384
Annealing, 270
Armor, 381
Arrest line, 331
Artifact, 339
Attenuation, 294
Attrition milling, 163

Ball milling, 160, 385
$BaTiO_3$
  applications of, 54, 250, 389
  electrical properties of, 54, 55
  polymorphs of, 55
  structure of, 10, 55
Basic oxygen steelmaking process,
  124
Bauxite, 151
Bayer process, 151, 166
Bearings, 270
Bend strength, 86
BeO
  applications of, 382
  hardness of, 85
  structure of, 10
  thermal conductivity of, 39,
    40
Biaxial stress, 90, 264, 360
Binder removal, 238
Binders, 173, 178
Bonding, atomic
  covalent, 13, 15
  ionic, 7, 12, 13
  metallic, 6, 7
  van der Waals, 17
B_4C, 376, 381
Boundary conditions, 321
Brittle fracture, 70